Circuit-Technology Co-Optimization of SRAM Design in Advanced CMOS Nodes

Hsiao-Hsuan Liu • Francky Catthoor

Circuit-Technology Co-Optimization of SRAM Design in Advanced CMOS Nodes

Springer

Hsiao-Hsuan Liu
IMEC
Leuven, Belgium

Francky Catthoor
IMEC and KU Leuven
Leuven, Belgium

ISBN 978-3-031-76108-9 ISBN 978-3-031-76109-6 (eBook)
https://doi.org/10.1007/978-3-031-76109-6

This Springer imprint is published by the registered company Springer Nature Switzerland AG
The registered company address is: Gewerbestrasse 11, 6330 Cham, Switzerland

To my family—Lisa, Simon, and Eric.

Preface

Modern computing engines, such as central processing units (CPUs), graphics processing units (GPUs), and neural processing units (NPUs), play a pivotal role in advancing general-purpose computing, artificial intelligence (AI), machine learning (ML), and deep learning (DL) applications. These silicon-based microprocessors necessitate substantial amounts of SRAM for level-2 (L2) and level-3 (L3) caches. For instance, the AMD CPU Ryzen™ 7 5800X3D features 96-MB L3 cache, while Nvidia's GPU GeForce RTX 4090 carries 72-MB L2 cache. The increasing demand for L2 and last-level cache (LLC) size underscores the urgency for improved on-die SRAM density, thereby exerting significant pressure to enhance speed and energy consumption metrics. Simultaneously, level-1 (L1) cache is essential for each core to store frequently accessed data and instructions, operating at the same speed as the CPU. The increasing number of on-chip cores (e.g., from 4 to 8 cores from AMD Zen 1 to Zen 4) necessitates an increase in the number of L1 caches. These L1 caches dictate CPU performance and consume a significant portion of power due to the high activity factor in modern multi-core system-on-chips (SoCs). Therefore, this book focuses on identifying the power-performance-area (PPA) booster in deeply scaled nodes, covering two primary areas: (1) the continuation of SRAM bitcell scaling and (2) the exploration of alternative SRAM subarray design. This book leverages both the impacts of bitcell scaling and alternative subarray design on achieving notable PPA enhancements with technology scaling.

The progression of SRAM bitcell scaling encompasses a systematic research cycle that involves identification of scaling bottlenecks, development of novel device architectures utilizing complementary field-effect transistor (CFET) technologies, assessment of associated challenges (e.g., process and routing complexities), and proposition of potential solutions from a bottom-up design-technology co-optimization (DTCO) standpoint. A variety of design alternatives have been scrutinized using a dedicated PPA simulation framework aimed at selecting the most promising candidate for future SRAM bitcell scaling strategies. The conventional write margin methodology has proven inadequate in resistance-dominated technology nodes, as it fails to account for the time-dependent effects of a complete set of parasitic bitline (BL) components. To address this, a novel write margin methodology has been developed to mitigate the risk of underestimating yield loss. The scaling of cell height in CFET SRAM bitcells is constrained by gate-cut (GC) occupancy, which occupies up to 44% of the total cell height in 5-Å-compatible technology (A5). By replacing the GC with dielectric isolation wall (DIW) and assuming self-aligned M0A/gate merges, up to 29% bitcell area scaling can be achieved when transitioning to 3-Å-compatible technology (A3). Moreover, challenges associated with insufficient back-end-of-line (BEOL) space for routing signals and power lines in functional SRAM bitcell designs have been effectively resolved through innovative approaches. To further enhance performance and power, hybrid CFET has been integrated into SRAM bit cell design, resulting in significant reductions of 41%, 25%, and 35% in wordline (WL) resistance, BL capacitance, and WL capacitance, respectively, compared to sequential (seq) and monolithic (mono) counterparts. A5 CFET SRAM macro demonstrates a 50~70% improvement in energy-delay-area product (EDAP) compared to its 14-Å-compatible technology (A14) counterpart, underscoring the importance of continuous bitcell-level scaling. However, macro-level EDAP improvement saturates beyond A5. A3 CFET SRAM macros offer only a marginal additional improvement of 10~20% over A5 counterparts at the expense of a more intricate process and heightened cost concerns. This finding highlights the need to explore alternative SRAM subarray designs as more potent avenues for optimizing PPA metrics in subsequent phases of this book.

In the second part of this book, the design strategy has shifted towards a top-down DTCO approach, driven by the diminishing returns of bitcell scaling beyond A5 at the macro level. As technology scales, resistance and capacitance (RC) increase. To mitigate this, designers often opt for smaller subarray sizes to reduce the effective RC and enhance PPA at the subarray level. However, this approach actually increases the overhead of inter-subarray interconnects, potentially offsetting the improvements at the macro level. In this book, we have examined the impact of various subarray sizes on macro-level PPA. We found out that the adoption of a larger subarray size is anticipated to yield a significant energy-delay-area product (EDAP) gain at the macro level due to reduced inter-subarray interconnect overhead. However, attaining this advancement in standard (Std.) subarray design is impossible owing to the write failures induced by the increased RC. Hence, there is a strong motivation to investigate the feasibility of enabling large-size subarray designs that can address potential write failure concerns. Initially, conventional (Conv.) divided BL and divided WL techniques are employed to decrease parasitic RC through hierarchical subarray design in A14. Optimization of 12 distinct subarray sizes and corresponding design parameters is conducted with a focus on ensuring write ability. This study demonstrates that the Conv. divided subarray design can effectively mitigate the write failure risk, allowing for the extension of the upper limit of subarray size from 256-columns×256-rows to 1024-columns×512-rows compared to the Std. subarray. However, the Conv. divided design incurs a substantial subarray area penalty of up to 30%, leading to macro-level PPA degradation despite subarray-level PP and write margin benefits. To address this challenge, the concept of active interconnect (AIC) is introduced, aimed at relocating additional logic gates from the front-end-of-line (FEOL) to the BEOL active layer. Integration of A14 carbon nanotube FET (CNFET) as BEOL-compatible devices facilitates remarkable macro-level EDAP improvements of up to 65% when transitioning from Std. subarrays (128-columns × 128-rows) to AIC divided subarrays (1024-columns × 512-rows). These compelling outcomes illuminate the future trajectory of SRAM subarray design roadmap in deeply scaled nodes.

Leuven, Belgium
August, 2024

Hsiao-Hsuan Liu
Francky Catthoor

Acknowledgements

First and foremost, we are deeply appreciative of the opportunity to work with the Logic Technology Enablement (LTE) team and to collaborate with esteemed research groups at Interuniversity Microelectronics Centre (imec). We extend our heartfelt thanks to Julien, Geert, Marie, Pieter, Shairfe, Rongmei, and Fernando for their exceptional leadership and insightful feedback throughout the completion of this book.

We would also like to acknowledge those who collaborated with us to bring this work to fruition. Special thanks to Zhenlin Pei and Prof. Chenyun Pan from the University of Texas at Austin for their advancements in the CACTI tool for macro-level simulation and their innovative interconnect design. Our gratitude also goes to Mahta Mayahinia and Prof. Mehdi Tahoori from the Karlsruhe Institute of Technology (KIT) for supporting the macro-level simulation framework with workload dependencies research. We are equally thankful to Carlo Gilardi and Prof. Subhasish Mitra from Stanford University for providing the simulation model and essential background knowledge on carbon nanotube transistors. Without their contributions, this book would not have been possible.

Lastly, we wish to convey our deepest gratitude to our families for their unwavering support and encouragement. We extend our heartfelt wishes to our families and friends for enduring health, happiness, and fulfilment in life.

Contents

Acronyms

6T	Six-Transistor
AIC	Active Interconnect
BEOL	Back-End-of-Line
BL	Bitline
BLB	Bitline-Bar
BPR	Buried Power Rail
BSM	Backside Metal
BSPDN	Backside Power Delivery Network
BTI	Bias Temperature Instability
BWL	Boosted Wordline
CFET	Complementary Field-Effect Transistor
CLK	Clock
CP	Centre Pin
CP-E	Centre Pin Access with E-tree inter-subarray interconnect
CP-H	Centre Pin Access with H-tree inter-subarray interconnect
CPP	Contacted Poly Pitch
Cu	Copper
DBL	Divided Bitline
DC	Direct Current
DRAM	Dynamic Random Access Memory
DRV	Data Retention Voltage
DTCO	Design-Technology Co-Optimization
DWL	Divided Wordline
EDAP	Energy-Delay-Area Product
EDP	Energy-Delay Product
FEOL	Front-End-of-Line
FS	Forksheet
FSFET	Forksheet Field-Effect Transistor
GBL	Global Bitline
GBLB	Global Bitline-Bar
GDS	Graphic Data Stream

GND	Ground
GWL	Global Wordline
HD	High Density
HP	High Performance
HSNM	Hold Static Noise Margin
HVT	High Threshold Voltage
L1	Level-1
L2	Level-2
L3	Level-3
LBCS	Local Block Column Select
LBL	Local Bitline
LBLB	Local Bitline-Bar
LBRS	Local Block Row Select
LLC	Last-level Cache
LP	Low Power
LVT	Low Threshold Voltage
LWL	Local Wordline
MC	Monte Carlo
MEOL	Middle-End-of-Line
Mono	Monolithic
MP	Metal Pitch
MRAM	Magnetic Random Access Memory
NBL	Negative Bitline
N-PG	NMOS Pass-Gate Device
NS	Nanosheet
NSFET	Nanosheet Field-Effect Transistor
nTSV	Nano-scale Through Silicon Via
PD	Pull-Down Device
PDN	Power Delivery Network
PG	Pass-Gate Device
PP	Power-Performance
PPA	Power-Performance-Area
P-PG	PMOS Pass-Gate Device
PU	Pull-Up Device
RD	Read Delay
RE	Read Energy
RSNM	Read Static Noise Margin
SBL	Suppressed Bitline
Seq	Sequential
SP	Side Pin
SP-E	Side Pin Access with E-tree inter-subarray interconnect
SP-H	Side Pin Access with H-tree inter-subarray interconnect
SRAM	Static Random Access Memory
STCO	System-Technology Co-Optimization
SVT	Standard Threshold Voltage

TSV	Through Silicon Via
VTC	Voltage Transfer Characteristics
WD	Write Delay
WE	Write Energy
WL	Wordline
WLUD	Wordline Underdrive
WM	Write Margin
R_{BL}	Bitline Resistance
C_{BL}	Bitline Capacitance
R_{WL}	Wordline Resistance
C_{WL}	Wordline Capacitance
$R_{\mathrm{BLPERCELL}}$	Bitline Resistance Per Cell
$C_{\mathrm{BLPERCELL}}$	Bitline Capacitance Per Cell
$R_{\mathrm{WLPERCELL}}$	Wordline Resistance Per Cell
$C_{\mathrm{WLPERCELL}}$	Wordline Capacitance Per Cell
R_{LBL}	Local Bitline Resistance
C_{LBL}	Local Bitline Capacitance
R_{LWL}	Local Wordline Resistance
C_{LWL}	Local Wordline Capacitance
R_{GBL}	Global Bitline Resistance
C_{GBL}	Global Bitline Capacitance
R_{GWL}	Global Wordline Resistance
C_{GWL}	Global Wordline Capacitance
C_{via}	Via Capacitance
C_{boost}	Boosting Capacitance
$C_{\mathrm{boost-th}}$	Threshold of Boosting Capacitance
R_{VDD}	Supply Voltage Resistance
R_{VSS}	Ground Voltage Resistance
R_{power}	Power Rail Resistance
V_{B}	Body Voltage
V_{D}	Drain Voltage
V_{G}	Gate Voltage
V_{S}	Source Voltage
V_{DS}	Drain-to-Source Voltage
V_{GD}	Gate-to-Drain Voltage
V_{GS}	Gate-to-Source Voltage
$V_{\mathrm{GS-max}}$	Maximum Operating Gate-to-Source Voltage
V_{DD}	Supply Voltage
V_{SS}	Ground Voltage
V_{WD}	Write Driver Voltage
V_{BL}	Bitline Voltage
V_{BLB}	Bitline-bar Voltage
V_{WL}	Wordline Voltage
V_{NBL}	Negative Bitline Write-Assist Voltage
V_{BWL}	Boosted Wordline Write-Assist Voltage

$V_{\mathrm{NBL-max}}$	Maximum Negative Bitline Write-Assist Voltage
$V_{\mathrm{BWL-max}}$	Maximum Boosted Wordline Write-Assist Voltage
I_{on}	On Current
I_{off}	Off Current
S_{N}	Stress of N-Type Channel
S_{P}	Stress of P-Type Channel
S_{BL}	Bitline Ramp-Down/Up Slope
t_{BL}	Bitline Ramp-Down/Up Time
A_{VT}	Pelgrom's Coefficient
V_{T}	Threshold Voltage
V_{min}	Minimum Operating Voltage
L_{g}	Gate Length
L_{ch}	Channel Length
L_{eff}	Effective Channel Length
W_{eff}	Effective Channel Width
W_{BL}	Bitline Width
L_{BL}	Bitline Length
W_{WL}	Wordline Width
L_{WL}	Wordline Length
W_{cell}	Cell Width
H_{cell}	Cell Height
T_{wire}	Wire Thickness
W_{wire}	Wire Width
L_{wire}	Wire Length
W_{NS}	Nanosheet Width
W_{fin}	Fin Width
H_{fin}	Fin Height
NNS	Number of Stacked Nanosheets
NNS_{CNT}	Number of Stacked Nanosheets in Carbon Nanotube Device
NNS_{Si}	Number of Stacked Nanosheets in Silicon Device
d_{CNT}	Channel Diameter in Carbon Nanotube Device
p_{CNT}	Channel Pitch in Carbon Nanotube Device
N_{row}	Number of Row
N_{col}	Number of Column
$N_{\mathrm{row-LB}}$	Number of Row Per Local Block
$N_{\mathrm{col-LB}}$	Number of Column Per Local Block
V_{SP}	Spacing between the active region and VBPR
$Via_{\mathrm{LBL-TG}}$	Via Connection from Local Bitline to Transmission Gate
$Via_{\mathrm{TG-GBL}}$	Via Connection from Transmission Gate to Global Bitline
$Via_{\mathrm{LWL-AND}}$	Via Connection from Local Wordline to AND Gate
$Via_{\mathrm{AND-GWL}}$	Via Connection from AND Gate to Global Wordline

Chapter 1
Introduction

In modern system-on-chip (SoC) and high-memory-demand designs, the static random-access memory (SRAM) area is crucial because it occupies a significant portion of the total chip area. For example, level 1 (L1), level 2 (L2), and level 3 (L3) caches occupy at least 50% of the entire die area, as depicted in the die photos of AMD Zen family microprocessors (Fig. 1.1) [6, 41, 59, 60]. Table 1.1 illustrates the increasing demand for cache size and core count from the first to the fourth generations. The growing need for expanded capacity in L2 and L3 caches underscores the critical importance of enhancing on-die SRAM area density, driving research into performance and power (PP) improvements in ultra-scaled nodes. Additionally, the rising trend in core count signifies an increase in the number of L1 and L2 caches. As L1 caches must match the CPU clock speed and exhibit high activity characteristics, they exert significant influence on the overall PP in multi-core SoCs. Furthermore, with SRAM scaling lagging behind logic scaling (Fig. 1.2), the relative area and PP occupied by SRAM (i.e., L1, L2, and L3 caches) are anticipated to expand in advanced nodes, thereby significantly impacting node-to-node power, performance, and area (PPA) enhancements. Consequently, this book primarily focuses on two main storylines: (1) scaling the 6-transistor (6T) SRAM bitcell for high-density (HD) applications, and (2) exploring alternative subarray designs for high-performance (HP) and low-power (LP) applications. The optimized designs are determined through comprehensive PPA evaluation at the bitcell, subarray, and macro levels. Notably, the utilized macro architecture in this study consists of surrounding cores and data arrays [43–45]. This data array incorporates a hierarchical design including bank, mat, subarray, and bitcell arrays while excluding the tag array. Section 1.1 introduces the memory hierarchy. Section 1.2 outlines the SRAM device scaling trend and major challenges in deeply scaled complementary metal-oxide-semiconductor (CMOS) technology. Notably, this roadmap and PPA results were proposed 2 years ago, with relatively aggressive assumptions compared to the recent trend of slowing down in scaling [9]. The conclusions remain valid, contingent upon the contact poly pitch (CPP) and metal

H.-H. Liu, F. Catthoor, *Circuit-Technology Co-Optimization of SRAM Design in Advanced CMOS Nodes*, https://doi.org/10.1007/978-3-031-76109-6_1

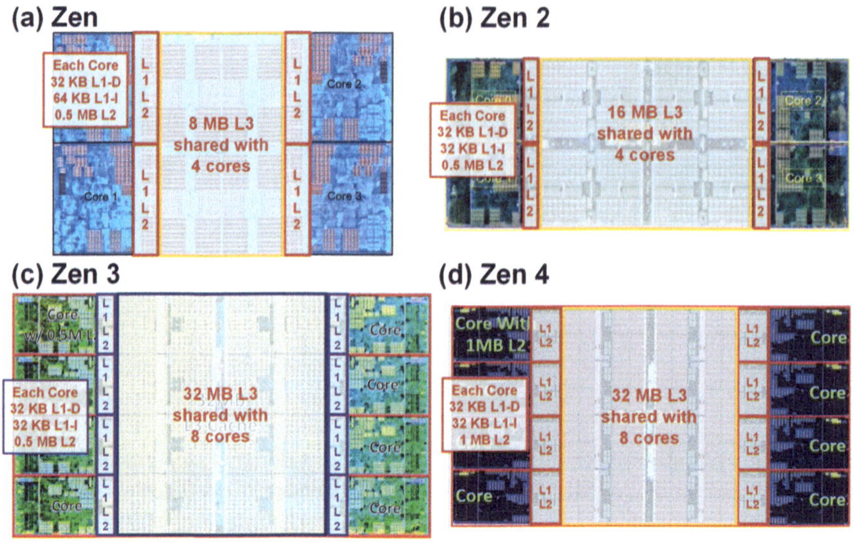

Fig. 1.1 Die photo of AMD (**a**) Zen, (**b**) Zen 2, (**c**) Zen 3, and (**d**) Zen 4 x86-64 microprocessor core [6, 41, 59, 60]

Table 1.1 Die information of AMD Zen family of microprocessors [6, 41, 59, 60]

	Zen	Zen 2	Zen 3	Zen 4
Technology	Global foundries 14-nm FinFET	TSMC 7-nm FinFET	TSMC 7-nm FinFET	TSMC 5-nm FinFET
Die area (mm^2)	44	31.32	68	55
Transistors counts (B)	1.4	1.9	4.08	6.5
Core counts	4	4	8	8
L1 data cache per core	32 KB	32 KB	32 KB	32 KB
L1 instruction cache per core	64 KB	32 KB	32 KB	32 KB
L2 cache per core	0.5 MB	0.5 MB	0.5 MB	1 MB
L3 cache (shared)	8 MB	16 MB	32 MB	32 MB

pitch (MP) developments over time. Sections 1.3 and 1.4 present the research aims, objectives, and an overview of the research structure of this book.

1.1 Memory Hierarchy

In computing system design, achieving simultaneous HD, HP, large capacity, and low cost remains a challenge. Consequently, the memory hierarchy is introduced to efficiently organize and optimize memory devices based on speed, cost, and capacity, as depicted in Fig. 1.3 [4, 25, 42, 58, 69, 74]. Generally, memory types positioned closer to the apex of this hierarchy exhibit higher speed, greater cost, proximity to the processor core, and lower capacity.

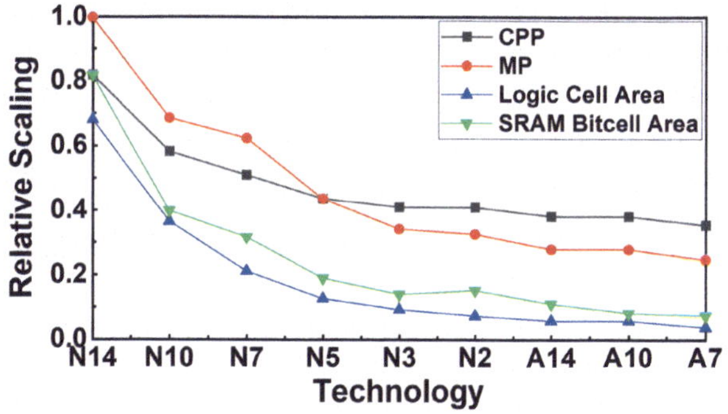

Fig. 1.2 Relative scaling from N14 to A7 for CPP, MP, logic cell area, and 6T SRAM bitcell area based on imec data [10, 36, 39]. (CPP: contacted poly pitch, MP: metal pitch)

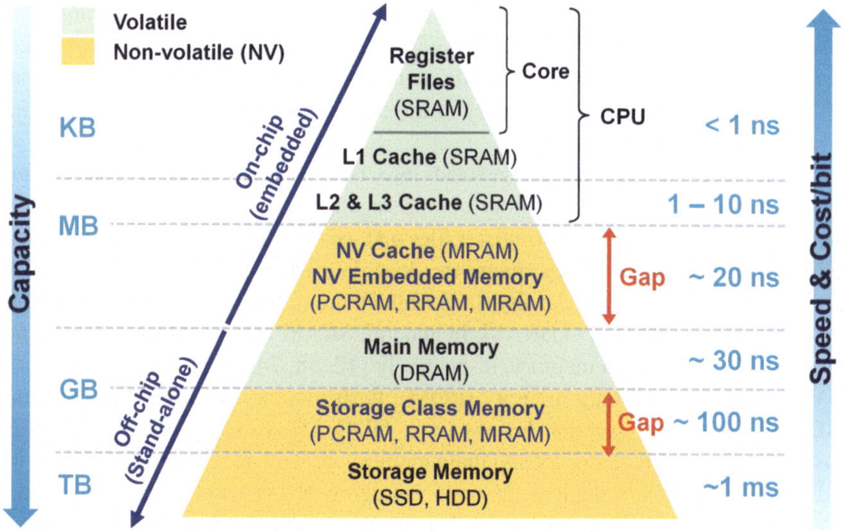

Fig. 1.3 Memory hierarchy of a modern computer system [4, 25]

At the pinnacle of this memory hierarchy pyramid are the register files, located closest to the processor core. These memories store frequently accessed data or instructions, necessitating the fastest access time to match the operation speed of the processor core. Register files, constructed using SRAM, incur the highest cost while offering the smallest storage capacity (16 to 64 bits) in the memory hierarchy [42, 58, 74].

On the second tier of the pyramid, the cache memory serves as a compact, fast, and high-cost storage unit facilitating temporary duplication between the CPU and the lower-tier main memory. It is positioned near the CPU to reduce access time for retrieving stored data or instructions currently in use. Cache memory, made of SRAM, is classified into three categories: L1, L2, and L3 caches based on their accessibility, access time, and capacity. In cache design, a tag memory is essential for identifying the address regions stored within the cache memory.

The main memory, situated at the third tier of the hierarchy, serves as a repository for data and instructions actively accessed by the processor. Dynamic random-access memory (DRAM) is mostly used for main memory, distinguished by its lower speed, lower price, and higher capacity compared to first and second-level memories. Program execution entails the transfer of data and instructions from a secondary storage device to the main memory. The capacity of the main memory directly influences the number of programs that can be concurrently executed and the volume of data accessible to a given program.

At the base of the pyramid lies the secondary storage memory, situated externally on another chip such as a hard disk drive (HDD), solid-state drive (SSD), optical disk, and backup tapes. Secondary storage memory is characterized by the longest access time, lower cost, and larger capacity in comparison to alternative memory devices. It fulfills long-term storage requirements, aiding the transfer of unneeded data and instructions from main memory. This process enhances system performance and frees up space in main memory. Typically, non-volatile memories (NVMs) are employed for secondary storage, ensuring data retention even when power is off.

Memory design can be categorized into two primary groups: volatile and non-volatile (NV). Volatile memory comprises cache memory and random-access memory (RAM), with SRAM (register files and cache) and DRAM (main memory) being notable examples. These memories lose data when power is cut off. SRAM exhibits faster speed, lower density (larger bitcell area), higher energy consumption, and higher cost compared to DRAM. In general, SRAM and DRAM are utilized for cache memory and main memory, respectively [42, 58, 74]. Non-volatile memories (NVMs) include read-only memory (ROM), flash memory (e.g., NAND flash and SSD), and computer storage devices (e.g., HDD). Historically, speed disparities existed between cache memory and main memory, as well as between main memory and secondary storage devices. Various emerging NVM technologies have been researched to bridge these gaps [53, 73], such as magnetic random-access memory (MRAM) [8, 18, 57], phase-change random-access memory (PCRAM) [70], resistive random-access memory (RRAM) [71], and ferroelectric random-access memory (FeRAM) [23], among others.

1.2 SRAM Technology Scaling Trend and Challenge

According to Moore's Law, the number of transistors per square inch on integrated circuits (IC) has doubled every 18 months [40]. Historically, as outlined in Dennard's Law, the voltage, current, and energy decrease as transistor dimensions scale down [12]. However, scaling is no longer a free lunch but now faces various bottlenecks, including the scaling wall (limited lithography capability), memory wall (limited memory bandwidth fails to align with processor performance), power

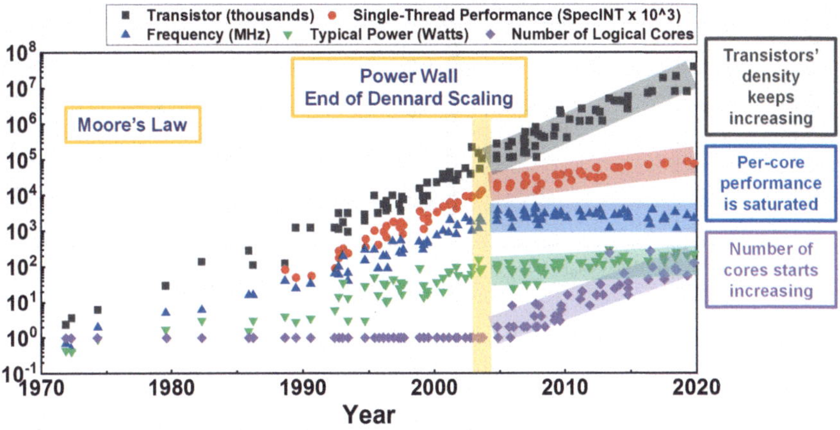

Fig. 1.4 50 years of microprocessor trend data [50]

wall (increased leakage and heat), sustainability wall (increased environmental footprint due to semiconductor manufacturing), and cost wall (increased process complexity and steps) [22]. As depicted in Fig. 1.4, the trends of frequency and power have reached saturation, while transistor density continues to rise based on 50 years of microprocessor trend data [50]. Initially, increasing the per-core frequency of CPUs was the primary method for enhancing processor performance. However, this approach results in heightened heat and energy overhead. To mitigate these issues, the escalation of per-core frequency must be halted. As a compromise, there is a shift towards increasing the number of cores to achieve improved CPU performance.

The pace of device scaling has ceased to deliver holistic PP benefits. Instead, it encounters challenges such as short channel effects (SCEs), increased metal resistance, leakage, and process variability at the device level [42, 58, 74]. These scaling penalties have ripple effects that significantly impact various aspects of SRAM technology, including bitcell, subarray, and macro-level margins and PPA in deeply scaled nodes [42, 58, 74]. Hence, this section presents an overview of the trends and challenges in SRAM technology, including CMOS architecture evolution, interconnect scaling, supply voltage (V_{DD}) scaling, and variability in continuous scaling technology nodes.

1.2.1 CMOS Transistor and SRAM Bitcell Technology Evolution

In the realm of very large-scale integrated (VLSI) circuit design, silicon metal-oxide-semiconductor field-effect transistors (MOSFETs) have historically held prominence. Achieving node-to-node PP enhancement in planar transistors has

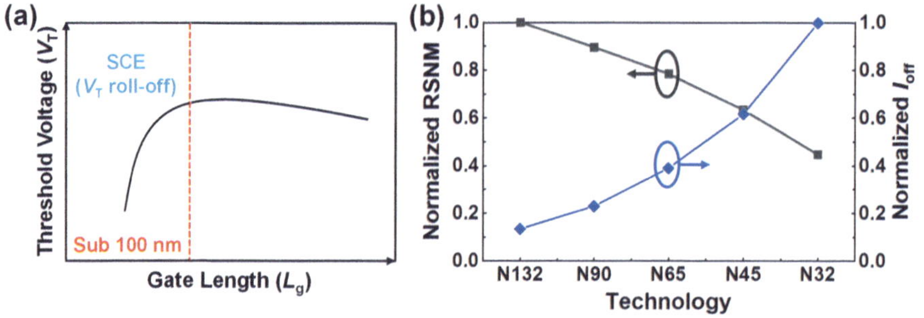

Fig. 1.5 (**a**) Impact of L_g scaling on V_T with strong SCEs for short channel devices. (**b**) Scaling trends of read static noise margin (RSNM) and leakage current (I_{off}) of 6T SRAM bitcell [58]

often involved reducing both the channel length (L_{ch}) and gate length (L_g) based on Moore's and Denard's laws. However, as L_g scales down below 1 μm, the gate begins to lose control over channel current when the L_{ch} becomes comparable to the depletion-layer width between the source and drain (S/D) [65]. This gives rise to a significant challenge known as short channel effects (SCEs), which profoundly impact various parameters including the threshold voltage (V_T) (Fig. 1.5a), drain voltage, velocity saturation, quantum confinement, and hot carrier degradation [65]. Furthermore, the escalation of sub-threshold slope (SS) in scaled bulk SRAM adversely affects device functionality due to inadequate on-off ratio (i.e. increased leakage and decreased on-current). Thus, this phenomenon causes degraded bitcell stability (Fig. 1.5b) [7, 16].

To address SCEs in planar devices, the FinFET with a tri-gate structure has been introduced to enhance electrostatic control [20, 52, 62], initially implemented in the market at the 22 nm node [3]. By reducing fin width and increasing fin height, FinFET maintains device scaling for a smaller footprint and higher performance at constant power down to 3 nm node (Fig. 1.5) [2]. However, as the L_g scales down to 15 nm, the degraded electrostatic control and the constrained effective channel width (stemming from the reduced number of fins but limited increased fin height) have begun impeding node-to-node PP enhancement [24].

In the post-FinFET era, the vertically stacked nanosheet (NS) FET with gate-all-around (GAA) structure (Fig. 1.6) has emerged as a potential solution for circuit design, offering a larger effective width and reduced parasitic capacitance per footprint compared to single-fin devices [24, 37]. As highlighted in [52], the NSFET demonstrates superior peak active power and performance compared to the single-fin FinFET at both 3 nm (N3) and 2 nm (N2) nodes, while maintaining similar process complexity. Consequently, the NSFET presents itself as a promising candidate for achieving enhanced power, performance, and area (PPA) benefits in future scaling beyond N3. In this book, we have intentionally excluded N2 NSFET SRAM bitcell designs from our PPA comparison. This exclusion stems from the observed reverse scaling phenomenon when transitioning from N3 FinFET to N2 NSFET. This reverse scaling is mainly due to a wider active width, resulting in

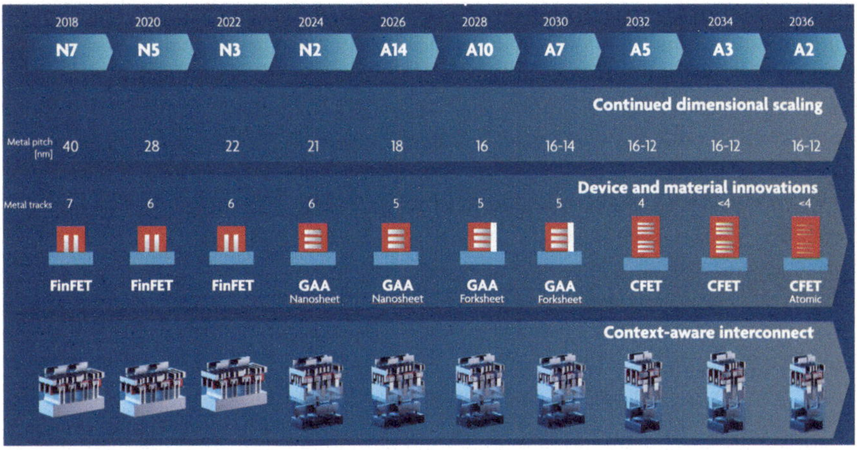

Fig. 1.6 Potential logic scaling roadmap extension [22]

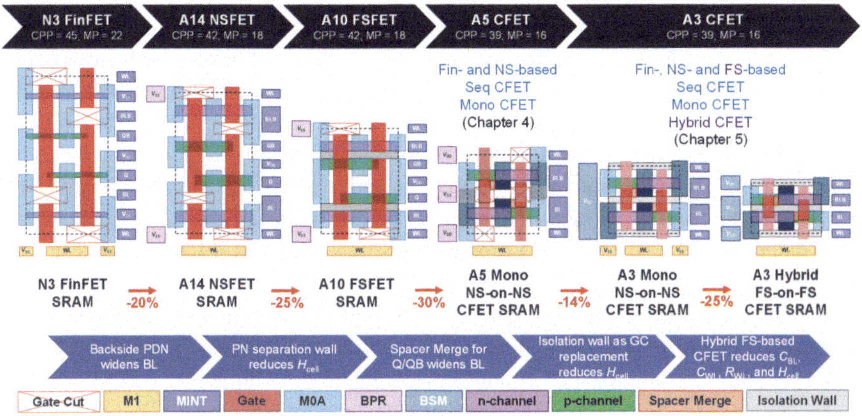

Fig. 1.7 Potential 6T SRAM scaling roadmap extension [34–36]

an enlarged bitcell area. Such enlargement could potentially lead to degradation in both subarray- and macro-level PPA due to the increased area and total RC of WL and BL. Therefore, A14 NSFET is selected to represent NS-based SRAM bitcell designs. This choice offers bitcell area scaling with a relatively larger potential for node-to-node gain from FinFET to NSFET SRAM. Figure 1.7 illustrates the 14-Å-compatible technology (A14) NSFET SRAM bitcell layout design (with an NS width of 11 nm), showcasing a 20% reduction in bitcell area compared to its N3 FinFET counterpart (with a fin width of 5 nm).

Although NS architecture offers potential for PP improvement, the scalability of cell height remains constrained by the required separation distance between p- and n-type channels (PN separation) [67]. Ensuring an adequate PN separation is imperative due to various factors such as lithography limitations (to avoid mask edge placement error and mask undercutting), drive-in anneal for ground plane doping implantation (to cut-off the parasitic planar channel underneath floating channels by increasing its V_T) [49], lateral etch control during replacement metal gate (RMG) formation (to properly form the metal work function in the available fill-metal space) [48], and the risk of merging S/D epitaxy between PMOS and NMOS during lateral growth [67].

To address these challenges and advance scalability towards 10-Å-compatible technology (A10), forksheet (FS) FET was proposed to reduce the PN separation by incorporating a dielectric isolation wall between the channels (PN separation wall) (Fig. 1.6) [67]. Several benefits are realized by integrating this natural barrier before gate stack formation, including self-aligned gate patterning, simplified RMG process with eliminated parasitic paths, and mitigated risk of S/D epitaxy merge. Hence, the PN separation can be significantly diminished, thereby decreasing the cell height of FS devices or enabling increased channel width by utilizing the saved area for enhanced performance compared to NS devices. Additionally, FS devices exhibit a 10% speed improvement at constant power and a 24% power reduction at constant speed compared to NS counterparts [62]. Figure 1.7 illustrates the A10 FSFET SRAM bitcell layout design, showcasing a 25% reduction in bitcell area compared to A14 NSFET counterparts with NS width of 11 nm for both cases [36]. It is worth noting that 7-Å-compatible technology (A7) FSFET SRAM bitcell designs have been omitted from our PPA comparison. Despite offering a 9% bitcell area scaling advantage, A7 technology comes with a narrower BL compared to A10 counterparts. This reduction in BL width may lead to an increase in R_{BL}, potentially resulting in a poorer write margin and higher risk of write failure. Consequently, we have opted to use A10 FSFET to represent FS-based SRAM bitcell designs in our analysis.

The journey of device scaling continues through the adoption of complementary FET (CFET) with a three-dimensional (3D) stacked architecture, aimed at achieving 5-Å-compatible technology (A5) and 3-Å-compatible technology (A3) as well as nodes beyond [32, 51, 55, 56, 64]. By vertically stacking NMOS and PMOS devices on top of each other, the need for PN separation is eliminated, leading to significant device scaling. Further device insights are elaborated in Chaps. 4 and 5, covering sequential (seq), monolithic (mono), and hybrid CFET configurations, as outlined in [10, 51, 55, 56, 64]. As depicted in Fig. 1.7, A5 and A3 mono stacked NS CFET SRAM bitcell layout designs demonstrate a 30% and 14% reduction in bitcell area scaling from one node to another, respectively [34, 35]. Furthermore, the incorporation of a hybrid stacked FS CFET SRAM bitcell layout design yields an additional 25% reduction in bitcell area scaling compared to its A3 mono stacked NS CFET counterpart [35].

1.2.2 Interconnect Scaling

The scaling of the middle-end-of-line (MEOL) and back-end-of-line (BEOL) interconnects correlates directly with the device area, resulting in increased resistance

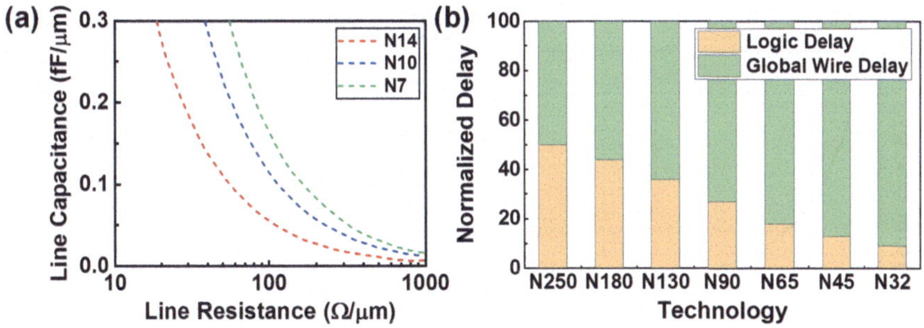

Fig. 1.8 (**a**) Scaling trend of BEOL RC based on 2013 ITRS roadmap projections. Dotted lines represent contours of constant RC [5]. (**b**) Scaling trends of logic and interconnect delay [28]

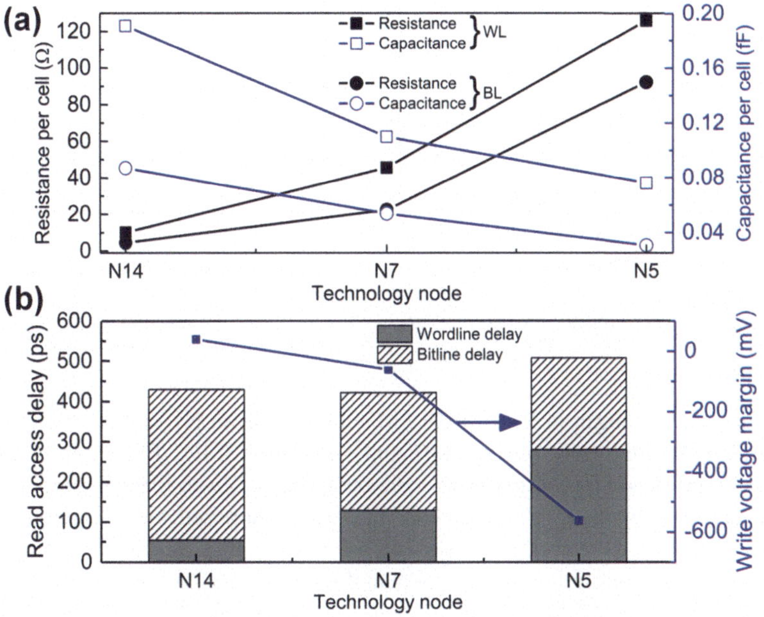

Fig. 1.9 (**a**) Scaling trends of parasitic RC for BL and WL in 6T FinFET SRAM bitcells. (**b**) Scaling trends of read delay and write margin for 256 rows and 136 columns 6T FinFET SRAM subarray [54]

(R) and decreased capacitance (C) in both global and local interconnects (Figs. 1.8a and 1.9a). While transistor RC delay typically improves with scaling, the dominance of global wire delay in overall system-level performance escalates as technology moves towards a more "interconnect-centric" paradigm due to significantly increased R (Fig. 1.8b) [5, 43–45]. This underscores the necessity for design-technology co-optimization (DTCO) analysis in Chaps. 8 and 9, aimed at optimizing SRAM subarray designs based on macro-level PPA metrics. In addition to global wire delay, increased bitlines (BL) and wordlines (WL) resistances degrade read

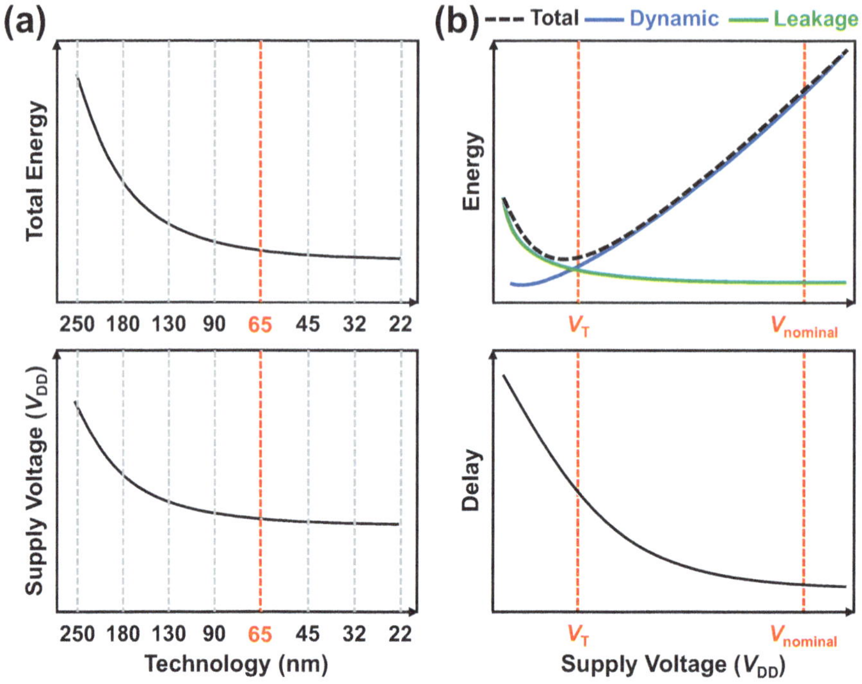

Fig. 1.10 (**a**) Scaling trends of energy and V_{DD} [13]. (**b**) Impact of V_{DD} scaling on energy and delay [13, 14]

delay and write margin in FinFET SRAM subarrays from N7 to N5, as illustrated in Fig. 1.9. Furthermore, novel transistors such as NSFET, FSFET, and CFET have been proposed to continuously scale device area, given the saturated scaling of contact poly pitch (CPP) and metal pitch (MP) imposed by lithography constraints (Fig. 1.2). However, the shrinking SRAM bitcell area poses challenges in accommodating all required metal tracks and interconnects at both the MEOL and frontside BEOL as technology advances towards A14 [36]. Thus, the adoption of backside delivery network (PDN) and buried power rail (BPR) strategies [17, 54] has been introduced, while the spacer merge technique [51] is implemented in CFET SRAM bitcell design (Fig. 1.7). Beyond A5, a comprehensive MEOL and backside interconnect solution, along with PPA evaluation for all the A3 CFET SRAM bitcell design options, are detailed in Chaps. 4, 5, 6, and 7.

1.2.3 Supply Voltage (V_{DD}) Scaling

The primary approach for reducing power consumption in SoC design is V_{DD} scaling at the cost of increased delay [13, 47]. Significant power reduction has been achieved by V_{DD} down to the 65 nm node (N65), as depicted in Fig. 1.10a [13].

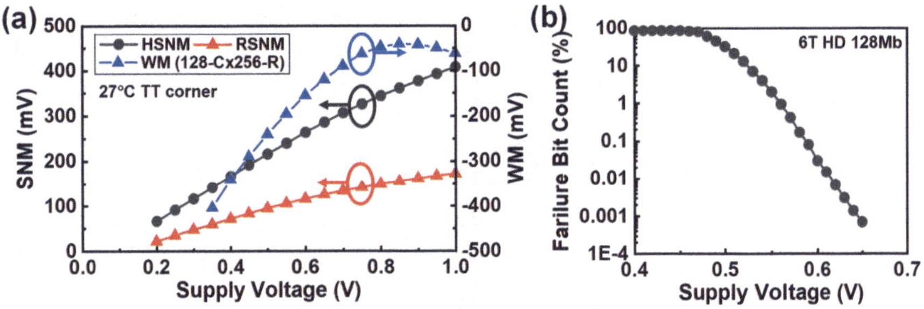

Fig. 1.11 (**a**) Impact of V_{DD} scaling on hold static noise margin (HSNM), read static noise margin (RSNM), and write margin (WM) of 6T HD A14 SRAM bitcell and subarray (128-column×256-row). (**b**) Impact of V_{DD} scaling on failure bit count of 6T 128Mb SRAM array [61]

However, beyond N65, V_{DD} scaling encounters challenges due to increased leakage current and V_T random variation with device scaling [14, 61, 68]. As illustrated in Fig. 1.11b, leakage energy begins to dominate total energy consumption when V_{DD} scales near or below V_T, eventually elevating total energy despite reduced dynamic energy. V_T scaling should be halted to prevent a rise in leakage current in nanometer technologies, which also poses a constraint on V_{DD} scaling. Figure 1.11 demonstrates that V_{DD} saturation is necessary to maintain SRAM cell stability, write ability, and failure bit count [15, 61, 75]. Various approaches have been proposed in the literature to address the issue of degraded stability in SRAM design, including the exploration of different cell topologies such as 8T and 10T bitcell configurations [11], as well as the utilization of read- and write-assist techniques [77]. These considerations serve as the motivation for Chap. 6 to study the limitations of V_{DD} scaling through a comprehensive analysis of hold stability, read stability, and write ability in ultra-scaled CFET SRAM bitcells and subarrays.

1.2.4 Variability

As L_{ch} and L_g shrink, the influence of process and device variabilities on SRAM device margin, performance, and yield becomes more pronounced, primarily due to increased V_T variations (ΔV_T) (Fig. 1.12) [1, 21, 72]. Key sources of variability include random dopant fluctuations (RDF), line edge roughness (LER), work function variation (WFV), channel dimension variation, random telegraph noise (RTN), time-dependent degradation, and aging [1, 31]. The substantial impact of ΔV_T on read stability and write ability, which increases as V_{DD} scales down, has been documented in previous studies [29, 38, 42]. Furthermore, the data destruction may be induced due to the increased device mismatch with dimensional scaling at nominal V_{DD} [58]. The standard deviation of V_T variation ($\sigma\Delta V_T$) between adjacent cells decreases inversely with gate area, expressed as $\sigma\Delta V_T = A_{VT}/\sqrt{L_{eff}W_{eff}}$, where

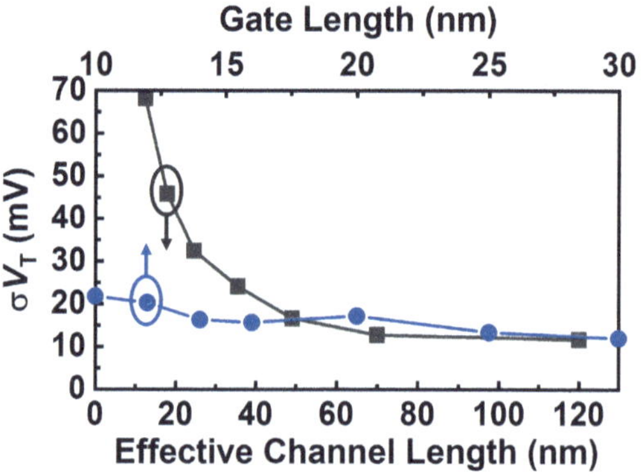

Fig. 1.12 Impact of L_{ch} and L_g scaling on $\sigma\Delta V_T$ and channel dopants [1, 21]

A_{VT} is a process-determined constant and the slope of the Pelgrom plot [31, 46]. It is documented that the worst-case (μ-6σ) SNM needs to exceed $0.04 \times V_{DD}$ to achieve a 90% yield on 1 MB SRAM [63]. Typically, this implies that the worst-case SNM should be at least 20% of the nominal SNM [42]. Moreover, the worst-case SNM should also exceed the thermal voltage [26]. Notably, transitioning to 3D device structure can alleviate RDF-induced V_T fluctuations owing to fully depleted and undoped channels as opposed to planar devices [21, 30, 76]. In the case of FinFET and GAAFET devices, it has been observed that WFV predominantly contributes to ΔV_T variations [76]. The adoption of GAAFET technology yields notable benefits, including an 18% reduction in ΔV_T, a 35% decrease in A_{VT}, and a potential reduction of up to 90 mV in minimum operating voltage (V_{min}) [21, 76]. This improvement can be attributed to enhanced confinement and carrier concentration within the GAAFET structure when compared to FinFET devices. To assess the impact of process and device variabilities on NS, FS, and CFET SRAM bitcell designs, 6-sigma Monte Carlo simulations with random V_T variation are conducted to guarantee the read and write functionalities in Chap. 6.

In addition to time-zero process variability, it is important to consider time-dependent variability and reliability, as they can also impact ΔV_T, V_{min} degradation, and SRAM design yield [19, 27, 66]. Notable failure mechanisms for transistors include bias temperature instability (BTI), time-dependent dielectric breakdown (TDDB), and hot carrier degradation (HCI). In sub-14 nm nodes, negative bias temperature instability (NBTI) is observed to prevail over positive bias temperature instability (PBTI) due to significant reduction of PBTI effects with high-k (HK) scaling [33]. In a conventional SRAM bitcell configuration, NBTI affects the two PMOS pull-up (PU) devices, while PBTI affects the two NMOS pull-down (PD) and two NMOS pass-gate (PG) devices. Prior research has demonstrated that writability is solely influenced by PG devices and remains unaffected by NBTI [19]. Thus, the subsequent discussion on the impact of these time-dependent factors primarily focuses on read stability and readability [27]. The probability of read stability failure increases due to: (1) heightened read disturbance, primarily stemming from V_T mismatch in transistors affected by both time-zero RDF and NBTI-induced

PMOS V_T increment; (2) reduced BL differential voltage caused by TDDB events in PG devices. Conversely, HCI can diminish read disturbance and alleviate the risk of read stability failure. The probability of readability failure rises due to: (1) diminished nodal voltage between internal nodes resulting from TDDB events in PU devices; (2) decreased saturation current in PG and PD devices due to HCI. Notably, readability failure remains unaffected by NBTI, as both PG and PD transistors on the read path are NMOS, not PMOS. Furthermore, the probability of readability failure can be mitigated by reducing the BL differential voltage during TDDB events in PG devices. A notable distinction could arise with CFET SRAM bitcell designs, as they utilize PMOS PG rather than the conventional NMOS type. This configuration introduces the possibility of NBTI affecting the PG devices, potentially influencing the write margin results. Further investigation into the comprehensive assessment of time-dependent variability and reliability's impact on margins and PP for A14 to A3 SRAM bitcells is recommended for future studies, extending beyond the scope of this book.

1.3 Research Aims

The increasing demand for L1, L2, and L3 cache utilization accompanying technology scaling underscores the significance of (1) continuous SRAM bitcell scaling and (2) alternative SRAM subarray design as primary strategies to enhance node-to-node PPA in SoC design. This book endeavours to extend SRAM bitcell scaling into CFET technology beyond 1 nm node by presenting various bitcell layout design alternatives, alongside an examination of key process challenges and potential interconnect routing solutions. Establishing a comprehensive DTCO PPA analysis framework encompassing bitcell, subarray, and macro levels is imperative to identify the most promising CFET SRAM bitcell design. Furthermore, the increased BL resistance (R_{BL}) observed in resistance-dominant technology nodes highlights the heightened risks of write failures, necessitating a thorough investigation into write ability extraction using an appropriate write margin methodology. On top of the conventional strategy for SRAM bitcell scaling, the optimization direction of subarray design should be reoriented toward an "interconnect-centric" perspective to achieve the best PPA outcomes for cache design. Enlarging the subarray size enhances area efficiency and reduces inter-subarray interconnects, thereby bolstering PPA metrics at the macro level. However, the increased R_{BL} poses a challenge to adjusting the subarray size within standard design architectures, constrained by the limited range for successful write operations. Consequently, exploring alternative subarray designs capable of accommodating a broader range of subarray sizes constitutes another focal point of this book.

1.4 Research Objectives and Overview of the Research Structure

The book evaluates the impact of various bitcell and subarray designs in deeply scaled CFET SRAM technologies, organized into four parts (Fig. 1.13): (1) SRAM basic principles and simulation methodologies; (2) SRAM bitcell scaling roadmap towards CFETs; (3) SRAM subarray- and macro-level DTCO PPA analysis; and (4) Beyond traditional SRAM bitcell-level scaling—future design direction for SRAM data array towards A14. We primarily examine how the proposed process technology affects the scaling of bitcell area in bitcell layout design, as well as how the design of subarray size influences macro-level outcomes. Our DTCO PPA analysis focuses on the innovation of bitcells and bitcell arrays, rather than the periphery circuit design. In this work, the macro-level PPA results account for contributions from the subarray, mat, bank, data array, and the surrounding core.

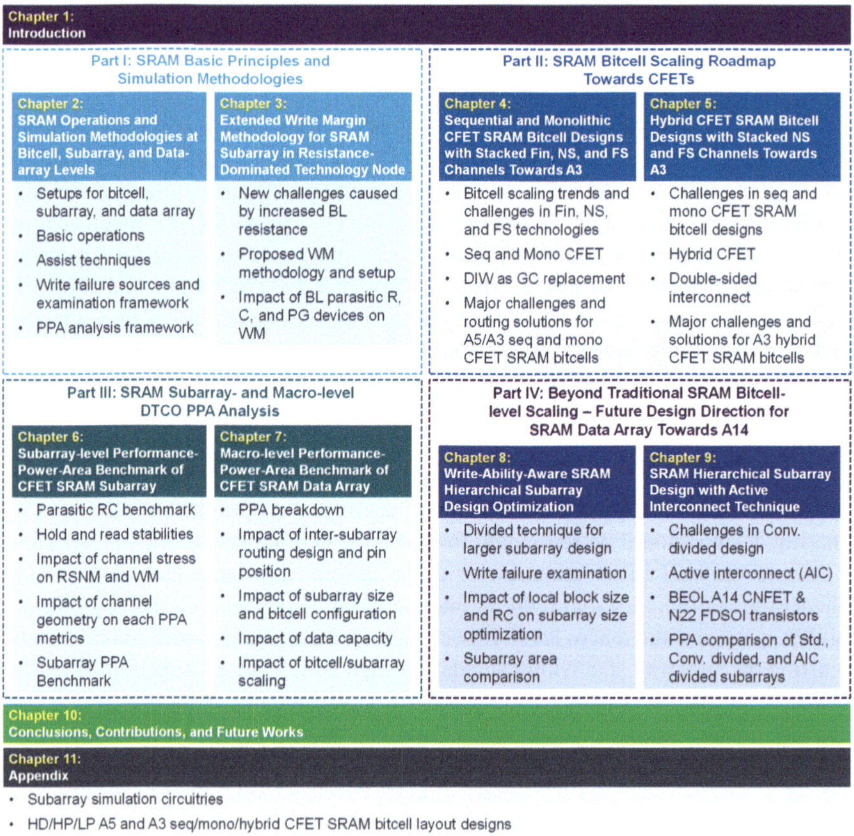

Fig. 1.13 Overview of the research structure

1.4.1 Part I: SRAM Basic Principles and Simulation Methodologies

- Chapter 2 introduces the simulation circuit setup, operational principles, and auxiliary techniques for hold, read, and write operations to facilitate comprehension of the SRAM PPA results. Moreover, a consistent simulation framework for DTCO PPA analysis at the subarray and macro levels is established to enable fair comparisons across diverse technologies and design alternatives. This chapter encompasses explanations of utilized compact models, Monte Carlo simulations, RC extraction for bitcells, definition and methodologies for subarray-level simulation terms, sources and examination setups of write failure, and macro-level simulation.
- In Chap. 3, the heightened risk of write failure accompanying technology scaling underscores the limitations of existing write margin (WM) methodologies, which fail to account for the influence of parasitic resistance (R), capacitance (C), and pass-gate (PG) devices on write ability. To mitigate the underestimation of WM degradation, an extended WM simulation methodology is proposed, primarily tailored for ultra-scaled SRAM subarray designs.

1.4.2 Part II: SRAM Bitcell Scaling Roadmap Towards CFETs

- CFET SRAM bitcell design is limited by the large gate-cut (GC) requirement due to the stacked architecture compared to the non-stacked SRAM devices. Therefore, the dielectric isolation wall (DIW) has been proposed to replace the GC for extending the bitcell scaling towards A3 in Chap. 4.
- In seq and mono CFET SRAM bitcell designs, the tall via connection from frontside BEOL signals to the bottom-tier PG devices causes high capacitance for both BL and WL (C_{BL} and C_{WL}). Moreover, the WL becomes narrower due to the insufficient backside routing space when bitcel scales from A5 to A3, resulting in increased WL resistance (R_{WL}). Thus, hybrid CFET has been introduced into the SRAM bitcell design to reduce the R_{WL}, C_{BL}, and C_{WL} by the double-sided interconnect routing design, as discussed in Chap. 5.

1.4.3 Part III: SRAM Subarray- and Macro-Level DTCO PPA Analysis

- In Chap. 6, the assessment of different bitcell design options takes place at the subarray level. This evaluation encompasses parasitic RC benchmarking, assessments of worst-case hold and read stabilities, examination of the impact of channel stress on RSNM and WM, analysis of channel geometric influences on each metric, and PPA benchmarking.

- Chapter 7 presents the identification of the most viable options pertaining to inter-subarray routing type, pin position, bitcell design, subarray configuration, and data capacity based on macro-level PPA evaluations. However, the advancement in node-to-node PPA enhancement has reached a saturation point after A5 at the macro level, accompanied by increased process costs and complexity. Hence, an alternative approach for PPA enhancement is explored in the latter part of this dissertation.

1.4.4 Part IV: Beyond Traditional SRAM Bitcell-Level Scaling—Future Design Direction for SRAM Data Array Towards A14

- The transition towards an "interconnect-centric" technology paradigm reduces the node-to-node macro-level PPA enhancement solely achieved through conventional bitcell scaling, as discussed in Chap. 7. These modest node-to-node PPA gains resulting from bitcell-level scaling necessitate a research focus on identifying new methods to boost PPA improvement. The proposed approach in this study entails enabling larger subarray size designs to mitigate inter-subarray interconnect overhead, thereby enhancing macro-level PPA. However, the adoption of large standard (Std.) subarray designs may lead to write failures, a challenge that can be addressed by employing conventional hierarchy (Conv. divided) subarray design. Consequently, Chap. 8 investigates the optimization of different subarray sizes utilizing both Std. and Conv. divided subarray designs, alongside an examination of write failure occurrences.
- Although the Conv. divided subarray design enhances subarray-level PP and enables larger subarray sizes, it leads to degraded macro-level PPA results due to significant subarray area overhead compared to the Std. counterpart. Therefore, Chap. 9 proposes a novel active interconnect (AIC) divided subarray design to mitigate subarray area overhead and enhance macro-level PPA compared to both Std. and Conv. divided counterparts.

References

1. Aburahma, M., & Anis, M. (2012). *Nanometer variation-tolerant SRAM: circuits and statistical design for yield* (pp. 1–170). ISBN: 978-1-4614-1748-4. https://doi.org/10.1007/978-1-4614-1749-1
2. Aoyagi, Y., et al. (2023). A 3-nm 27.6-Mbit/mm2 Self-timed SRAM enabling 0.48–1.2 V wide operating range with far-end pre-charge and weak-bit tracking. In *2023 IEEE Symposium on VLSI Technology and Circuits (VLSI Technology and Circuits)* (pp. 1–2). https://doi.org/10.23919/VLSITechnologyandCir57934.2023.10185429
3. Auth, C., et al. (2012). A 22 nm high performance and low-power CMOS technology featuring fully-depleted tri-gate transistors, self-aligned contacts and high density MIM capacitors. In

2012 Symposium on VLSI Technology (VLSIT) (pp. 131–132). https://doi.org/10.1109/VLSIT.2012.6242496
4. Barnum, J. (2020). *Semiconductor engineering —MRAM process development and production briefing*. https://semiengineering.com/mram-process-development-and-production-briefing/
5. Bonilla, G., et al., Interconnect scaling challenges, and opportunities to enable system-level performance beyond 30 nm pitch. In *2020 IEEE International Electron Devices Meeting (IEDM)* (pp. 20.4.1–20.4.4). https://doi.org/10.1109/IEDM13553.2020.9372093
6. Burd, T., et al. (2022). Zen3: The AMD 2nd-generation 7 nm x86-64 micro-processor core. In *2022 IEEE International Solid-State Circuits Conference (ISSCC)* (Vol. 65, pp. 1–3). https://doi.org/10.1109/ISSCC42614.2022.9731678
7. Burnett, D., et al. (2014). FinFET SRAM design challenges. In *2014 IEEE International Conference on IC Design Technology* (pp. 1–4). https://doi.org/10.1109/ICICDT.2014.6838606
8. Cargnini, L. V., et al. (2013). Embedded memory hierarchy exploration based on magnetic RAM. In *2013 IEEE Faible Tension Faible Consommation* (pp. 1–4). https://doi.org/10.1109/FTFC.2013.6577780
9. Chang, J., et al. (2023). A 3 nm 256Mb SRAM in FinFET technology with new array banking architecture and write-assist circuitry scheme for high-density and low-VMIN applications. In *2023 IEEE Symposium on VLSI Technology and Circuits (VLSI Technology and Circuits)* (pp. 1–2). https://doi.org/10.23919/VLSITechnologyandCir57934.2023.10185287
10. Chehab, B., et al. (2021). Design-technology co-optimization of sequential and monolithic CFET as enabler of technology node beyond 2 nm. In C.-M. Yuan & R.-H. Kim (Eds.), *Design-process-technology co-optimization XV* (Vol. 11614, p. 116140D). International Society for Optics and Photonics. SPIE. https://doi.org/10.1117/12.2583395
11. Chen, Y.-H., et al. (2016). A 16 nm dual-port SRAM with partial suppressed word-line, dummy read recovery and negative bit-line circuitries for low VMIN applications. In *2016 IEEE Symposium on VLSI Circuits (VLSI-Circuits)* (pp. 1–2). https://doi.org/10.1109/VLSIC.2016.7573459
12. Dennard, R. H., et al. (1947). Design of ion-implanted MOSFET's with very small physical dimensions. *IEEE Journal of Solid-State Circuits, 9*(5), 256–268. https://doi.org/10.1109/JSSC.1974.1050511
13. Dreslinski, R., et al. (2009). Near threshold computing: Overcoming performance degradation from aggressive voltage scaling". In: *Workshop on Energy-Efficient Design held in conjunction with ISCA 36* (pp. 44–49).
14. Dreslinski, R. G., et al. (2008). Reconfigurable energy efficient near thresh-old cache architectures. In *2008 41st IEEE/ACM International Symposium on Microarchitecture* (pp. 459–470). https://doi.org/10.1109/MICRO.2008.4771813
15. Grossar, E., et al. (2006). Read stability and write-ability analysis of SRAM cells for nanometer technologies. *IEEE Journal of Solid-State Circuits, 41*(11), 2577–2588. https://doi.org/10.1109/JSSC.2006.883344
16. Guo, Z., et al. (2005). FinFET-based SRAM design. In *ISLPED '05. Proceedings of the 2005 International Symposium on Low Power Electronics and Design, 2005* (pp. 2–7). https://doi.org/10.1109/LPE.2005.195476
17. Gupta, A., et al. (2021). Buried power rail metal exploration towards the 1 nm Node. In *2021 IEEE International Electron Devices Meeting (IEDM)* (pp. 22.5.1–22.5.4). https://doi.org/10.1109/IEDM19574.2021.9720684
18. Gupta, M., et al. (2023). Ultimate MRAM scaling: Design exploration of high-density, high-performance and energy-efficient VGSOT for last level cache. In *2023 International Electron Devices Meeting (IEDM)* (pp. 1–4). https://doi.org/10.1109/IEDM45741.2023.10413886
19. Gupta, M. K., et al. (2017). SRAM enablement beyond N7: A BTI study. In: *2017 IEEE International Reliability Physics Symposium (IRPS)* (pp. CR-4.1–CR-4.6). https://doi.org/10.1109/IRPS.2017.7936353
20. Hisamoto, D. et al. (2000). FinFET-a self-aligned double-gate MOSFET scalable to 20 nm. *IEEE Transactions on Electron Devices, 47*(12), 2320–2325. https://doi.org/10.1109/16.887014

21. Huynh-Bao, T., et al. (2017). SRAM designs for 5 nm node and beyond: Opportunities and challenges. In *2017 IEEE International Conference on IC Design and Technology (ICICDT)* (pp. 1–4). https://doi.org/10.1109/ICICDT.2017.7993502
22. Imec (2023). *Potential Roadmap Extension*. https://www.imec-int.com/en/articles/smaller-better-faster-imec-presents-chip-scaling-roadmap
23. Ishiwara, H. (2012). Ferroelectric random access memories. *Journal of Nanoscience and Nanotechnology, 12*, 7619–27. https://doi.org/10.1166/jnn.2012.6651
24. Jang, D., et al. (2017). Device exploration of nanosheet transistors for sub-7-nm technology node. *IEEE Transactions on Electron Devices, 64*(6), 2707–2713. https://doi.org/10.1109/TED.2017.2695455
25. Jovanovic, B., Brum, R., & Torres, L. (2015). MTJ-based hybrid storage cells for "normally-off and instant-on" computing. *Facta Universitatis—Series: Electronics and Energetics, 28*, 465–476. https://doi.org/10.2298/FUEE1503465J
26. Keyes, R. W. (1969). Physical problems and limits in computer logic. *IEEE Spectrum, 6*(5), 36–45. https://doi.org/10.1109/MSPEC.1969.5213809
27. Khan, S., & Hamdioui, S. (2010). Trends and challenges of SRAM reliability in the nano-scale era. In *5th International Conference on Design Technology of Integrated Systems in Nanoscale Era* (pp. 1–6). https://doi.org/10.1109/DTIS.2010.5487565
28. Kil, J., Gu, J., & Kim, C. H. (2008). A high-speed variation-tolerant interconnect technique for sub-threshold circuits using capacitive boosting. *IEEE Transactions on Very Large Scale Integration (VLSI) Systems, 16*(4), 456–465. https://doi.org/10.1109/TVLSI.2007.915455
29. Krishnan, A. T., et al. (2006). SRAM cell static noise margin and VMIN sensitivity to transistor degradation. In *2006 International Electron Devices Meeting* (pp. 1–4). https://doi.org/10.1109/IEDM.2006.346778
30. Kuhn, K. J. (2007). Reducing variation in advanced logic technologies: Approaches to process and design for manufacturability of nanoscale CMOS. In *2007 IEEE International Electron Devices Meeting* (pp. 471–474). https://doi.org/10.1109/IEDM.2007.4418976
31. Kuhn, K. J., et al. (2011). Process technology variation. *IEEE Transactions on Electron Devices, 58*(8), 2197–2208. https://doi.org/10.1109/TED.2011.2121913
32. Liao, S., et al. (2023). Complementary field-effect transistor (CFET) demonstration at 48nm gate pitch for future logic technology scaling. In *2023 International Electron Devices Meeting (IEDM)* (pp. 1–4). https://doi.org/10.1109/IEDM45741.2023.10413672
33. Liu, C., et al. (2015). Systematical study of 14nm FinFET reliability: From device level stress to product HTOL. In *2015 IEEE International Reliability Physics Symposium* (ppp. 2F.3.1–2F.3.5). https://doi.org/10.1109/IRPS.2015.7112693
34. Liu, H.-H., et al. (2023). CFET SRAM DTCO, interconnect guideline, and benchmark for CMOS scaling. *IEEE Transactions on Electron Devices, 70*(3), 883–890. https://doi.org/10.1109/TED.2023.3235701
35. Liu, H.-H., et al. (2023). CFET SRAM with double-sided interconnect design and DTCO benchmark. In *IEEE Transactions on Electron Devices* (pp. 1–8). https://doi.org/10.1109/TED.2023.3305322
36. Liu, H.-H., et al. (2023). DTCO of sequential and monolithic CFET SRAM. In: R.-H. Kim & N. V. Lafferty (Eds.), *DTCO and computational patterning II* (Vol. 12495, p. 124950Z). International Society for Optics and Photonics. SPIE. https://doi.org/10.1117/12.2657524
37. Loubet, N., et al. (2017). Stacked nanosheet gate-all-around transistor to enable scaling beyond FinFET. In *2017 Symposium on VLSI Technology* (pp. T230–T231). https://doi.org/10.23919/VLSIT.2017.7998183
38. Makino, H., et al. (2011). Reexamination of SRAM cell write margin definitions in view of predicting the distribution. *IEEE Transactions on Circuits and Systems II: Express Briefs, 58*(4), 230–234. https://doi.org/10.1109/TCSII.2011.2124531
39. Mirabelli, G., et al. (2023). Design-technology co-optimization overview of CFET architecture. In R.-H. Kim & N. V. Lafferty (Eds.), *DTCO and computational patterning II* (Vol. 12495, p. 124950M). International Society for Optics and Photonics. SPIE. https://doi.org/10.1117/12.2657726

40. Moore, G. E. (2006). Cramming more components onto integrated circuits, Reprinted from electronics, volume 38, number 8, April 19, 1965, pp.114 ff. *IEEE Solid-State Circuits Society Newsletter, 11*(3), 33–35. https://doi.org/10.1109/N-SSC.2006.4785860
41. Munger, B., et al. (2023). "Zen 4": The AMD 5nm 5.7GHz x86-64 microprocessor core. In *2023 IEEE International Solid-State Circuits Conference (ISSCC)* (pp. 38–39). https://doi.org/10.1109/ISSCC42615.2023.10067540
42. Pavlov, A., & Sachdev, M. (2008). *CMOS SRAM circuit design and parametric test in nano-scaled technologies: Process-aware SRAM design and test* (1st ed.). Springer. ISBN: 14-02083-62-9.
43. Pei, Z., et al. (2023). Emerging interconnect exploration for SRAM application using non-conventional H-tree and center-pin access. In *2023 24th International Symposium on Quality Electronic Design (ISQED)* (pp. 1–1). https://doi.org/10.1109/ISQED57927.2023.10129316
44. Pei, Z., et al. (2023). Graphene-based interconnect exploration for large SRAM caches for ultrascaled technology nodes. *IEEE Transactions on Electron Devices, 70*(1), 230–238. https://doi.org/10.1109/TED.2022.3225512
45. Pei, Z., et al. (2023). Technology/memory co-design and co-optimization using E-tree interconnect. In *Proceedings of the Great Lakes Symposium on VLSI 2023*. GLSVLSI '23 (pp. 159–162). Association for Computing Machinery. ISBN: 9798400701252. https://doi.org/10.1145/3583781.3590311
46. Pelgrom, M. J. M., Duinmaijer, A. C. J., & Welbers, A. P. G. (1989). Matching properties of MOS transistors. *IEEE Journal of Solid-State Circuits, 24*(5), 1433–1439. https://doi.org/10.1109/JSSC.1989.572629
47. Qazi, M., Sinangil, M., & Chandrakasan, A. (2011). Challenges and directions for low-voltage SRAM. *IEEE Design Test of Computers, 28*(1), 32–43. https://doi.org/10.1109/MDT.2010.115
48. Ragnarsson, L.-Å., et al. (2015). RMG nMOS 1st process enabling 10x lower gate re-sistivity in N7 bulk FinFETs. In *2015 Symposium on VLSI Technology (VLSI Technology)* (pp. T148–T149). https://doi.org/10.1109/VLSIT.2015.7223656
49. Ritzenthaler, R., et al. (2017). Isolation of nanowires made on bulk wafers by ground plane doping. In *2017 47th European Solid-State Device Research Con- ference (ESSDERC)* (pp. 300–303). https://doi.org/10.1109/ESSDERC.2017.8066651
50. Rupp, K., et al., *50 Years of Microprocessor Trend Data*. https://github.com/karlrupp/microprocessor-trend-data/tree/master/50yrs
51. Ryckaert, J., et al. (2018). The complementary FET (CFET) for CMOS scaling beyond N3. In *2018 IEEE Symposium on VLSI Technology* (pp. 141–142). https://doi.org/10.1109/VLSIT.2018.8510618
52. Ryckaert, J., et al. (2019). Enabling Sub-5nm CMOS technology scaling thinner and taller! In *2019 IEEE International Electron Devices Meeting (IEDM)* (pp. 29.4.1–29.4.4). https://doi.org/10.1109/IEDM19573.2019.8993631
53. Salahuddin, S., Ni, K., & Datta, S. (2018). The era of hyper-scaling in electronics. *Nature Electronics, 1*, 442–450. https://doi.org/10.1038/s41928-018-0117-x
54. Salahuddin, S. M., et al. (2019). SRAM with buried power distribution to improve write margin and performance in advanced technology nodes. *IEEE Electron Device Letters, 40*(8), 1261–1264. https://doi.org/10.1109/LED.2019.2921209
55. Schuddinck, P., et al. (2019). Device-, circuit-block-level evaluation of CFET in a 4 track library. In *2019 Symposium on VLSI Technology* (pp. T204–T205). https://doi.org/10.23919/VLSIT.2019.8776513
56. Schuddinck, P., et al. (2022). PPAC of sheet-based CFET configurations for 4 track design with 16nm metal pitch. In *2022 IEEE Symposium on VLSI Technology and Circuits (VLSI Technology and Circuits)* (pp. 365–366). https://doi.org/10.1109/VLSITechnologyandCir46769.2022.9830492
57. Senni, S., et al. (2014). Exploration of magnetic RAM based memory hierarchy for multicore architecture. In *2014 IEEE Computer Society Annual Symposium on VLSI* (pp. 248–251). https://doi.org/10.1109/ISVLSI.2014.29

58. Singh, J., Mohanty, S., & Pradhan, D. (2013). *Robust SRAM designs and analysis*. ISBN: 978-1-461408-17-8. https://doi.org/10.1007/978-1-4614-0818-5
59. Singh, T., et al. (2017). 3.2 Zen: A next-generation high-performance ×86 core. In *2017 IEEE International Solid-State Circuits Conference (ISSCC)* (pp. 52–53). https://doi.org/10.1109/ISSCC.2017.7870256
60. Singh, T., et al. (2020). 2.1 Zen 2: The AMD 7nm energy-efficient high-performance x86-64 microprocessor core. In *2020 IEEE International Solid-State Circuits Conference—(ISSCC)* (pp. 42–44). https://doi.org/10.1109/ISSCC19947.2020.9063113
61. Song, T., et al. (2017). A 10 nm FinFET 128 Mb SRAM with assist adjustment system for power, performance, and area optimization. *IEEE Journal of Solid-State Circuits, 52*(1), 240–249. https://doi.org/10.1109/JSSC.2016.2609386
62. Spessot, A., et al. (2020). Device scaling roadmap and its implications for logic and analog platform. In *2020 IEEE BiCMOS and Compound Semiconductor Integrated Circuits and Technology Symposium (BCICTS)* (pp. 1–8). https://doi.org/10.1109/BCICTS48439.2020.9392980
63. Stolk, P. A., et al. (2001). CMOS device optimization for mixed-signal technologies. In *International Electron Devices Meeting. Technical Digest (Cat. No.01CH37224)* (pp. 10.2.1–10.2.4). https://doi.org/10.1109/IEDM.2001.979469
64. Subramanian, S., et al. (2020). First monolithic integration of 3D complementary FET (CFET) on 300mm wafers. In *2020 IEEE Symposium on VLSI Technology* (pp. 1–2). https://doi.org/10.1109/VLSITechnology18217.2020.9265073
65. Taur, Y., & Ning, T. (2021). *Fundamentals of modern VLSI devices*. ISBN: 978-11-08480-02-4. https://doi.org/10.1017/9781108847087
66. Weckx, P., et al. (2015). Impact of time-dependent variability on the yield and performance of 6T SRAM cells in an advanced HK/MG technology. In *2015 International Conference on IC Design Technology (ICICDT)* (pp. 1–4). https://doi.org/10.1109/ICICDT.2015.7165896
67. Weckx, P., et al. (2019). Novel forksheet device architecture as ultimate logic scaling device towards 2nm. In *2019 IEEE International Electron Devices Meeting (IEDM)* (pp. 36.5.1–36.5.4). https://doi.org/10.1109/IEDM19573.2019.8993635
68. Weste, N., & Harris, D. (2010). *CMOS VLSI design: A circuits and systems perspective* (4th ed.). Addison-Wesley. ISBN: 0321547748.
69. Wong, H.-S., Salahuddin, S. (2015). Memory leads the way to better computing. *Nature Nanotechnology, 10*, 191–4. https://doi.org/10.1038/nnano.2015.29
70. Wong, H.-S. P., et al. (2010). Phase change memory. *Proceedings of the IEEE, 98*(12), 2201–2227. https://doi.org/10.1109/JPROC.2010.2070050
71. Wong, H.-S. P., et al. (2012). Metal–oxide RRAM. *Proceedings of the IEEE, 100*(6), 1951–1970. https://doi.org/10.1109/JPROC.2012.2190369
72. Yamauchi, H. (2010). A discussion on SRAM circuit design trend in deeper nanometer-scale technologies. *IEEE Transactions on Very Large Scale Integration (VLSI) Systems, 18*(5), 763–774. https://doi.org/10.1109/TVLSI.2009.2016205
73. Yu, S., & Chen, P.-Y. (2016). Emerging memory technologies: Recent trends and prospects. *IEEE Solid-State Circuits Magazine, 8*(2), 43–56. https://doi.org/10.1109/MSSC.2016.2546199
74. Zhang, K. (2009). *Embedded memories for nano-scale VLSIs*. ISBN: 978-0-387-88496-7. https://doi.org/10.1007/978-0-387-88497-4
75. Zhang, K., Karl, E., & Wang, Y. (2012). SRAM design in nano-scale CMOS technologies (Invited). In *2012 Symposium on VLSI Technology (VLSIT)* (pp. 85–86). https://doi.org/10.1109/VLSIT.2012.6242473
76. Zheng, P., et al. (2014). Variation-aware comparative study of 10-nm GAA versus FinFET 6-T SRAM performance and yield. *IEEE Transactions on Electron Devices, 61*(12), 3949–3954. https://doi.org/10.1109/TED.2014.2360351
77. Zimmer, B., et al. (2012). SRAM assist techniques for operation in a wide voltage range in 28-nm CMOS. *IEEE Transactions on Circuits and Systems II: Express Briefs, 59*(12), 853–857. https://doi.org/10.1109/TCSII.2012.2231015

Part I
SRAM Basic Principles and Simulation Methodologies

Chapter 2
SRAM Operations and Simulation Methodologies at Bitcell, Subarray, and Macro Levels

This book aims to propose 6-transistor (6T) novel static random-access memory (SRAM) bitcell designs, leveraging the complementary field-effect transistor (CFET) architecture within the 3-Å-compatible technology (A3). A comprehensive simulation framework encompassing margin, power, performance, and area (PPA) parameters tailored for deeply scaled nodes is indispensable for identifying the optimized design. Hence, this chapter elaborates the development of simulation setups, methodologies, fundamental operations, and figures of merit at the bitcell, subarray, and macro levels. Common figures of merit include critical parameters such as hold static noise margin (HSNM) [50], read static noise margin (RSNM) [50], write margin (WM) [32], read delay (RD) [33, 34], write delay (WD) [33, 34], read energy (RE) [33, 34], and write energy (WE) [33, 34]. It is important to highlight that the macro-level PPA analysis in this book integrates contributions from the data array and the surrounding cores while excluding the tag array in the initial phase. The data array is a hierarchy design including multiple banks, mats, subarrays, and bitcells.

Figure 2.1 illustrates the PPA simulation framework, along with the corresponding sections and chapters within this dissertation. Section 2.1.1 provides concise model information for 14-Å-compatible technology (A14), 10-Å-compatible technology (A10), 5-Å-compatible technology (A5), and 3-Å-compatible technology (A3) across various transistor architectures, including nanosheet (NS) FET, forksheet (FS) FET, and CFET architectures, as documented in [23, 49, 64]. Section 2.1.2 summarizes the simulation setups, circuit schematics, and procedures for parasitic extraction (PEX) of 6T SRAM bitcells. Section 2.1.3 introduces simulation setups, circuit schematics, and Monte Carlo (MC) simulation methodology of SRAM subarrays. The simulation techniques pertaining to bitcell- and subarray-level analysis are applied in Chaps. 6, 8, and 9. Section 2.1.4 provides a detailed discussion on the simulation setups, inter-subarray interconnects, access pin positions, and the CACTI simulation framework [42–44] used for analysing the SRAM macro in Chaps. 7, 8, and 9. Section 2.2 discusses data retention, leakage, and hold stability methodologies. In Sect. 2.3, read operation paths, read stability, RD, and RE

H.-H. Liu, F. Catthoor, *Circuit-Technology Co-Optimization of SRAM Design in Advanced CMOS Nodes*, https://doi.org/10.1007/978-3-031-76109-6_2

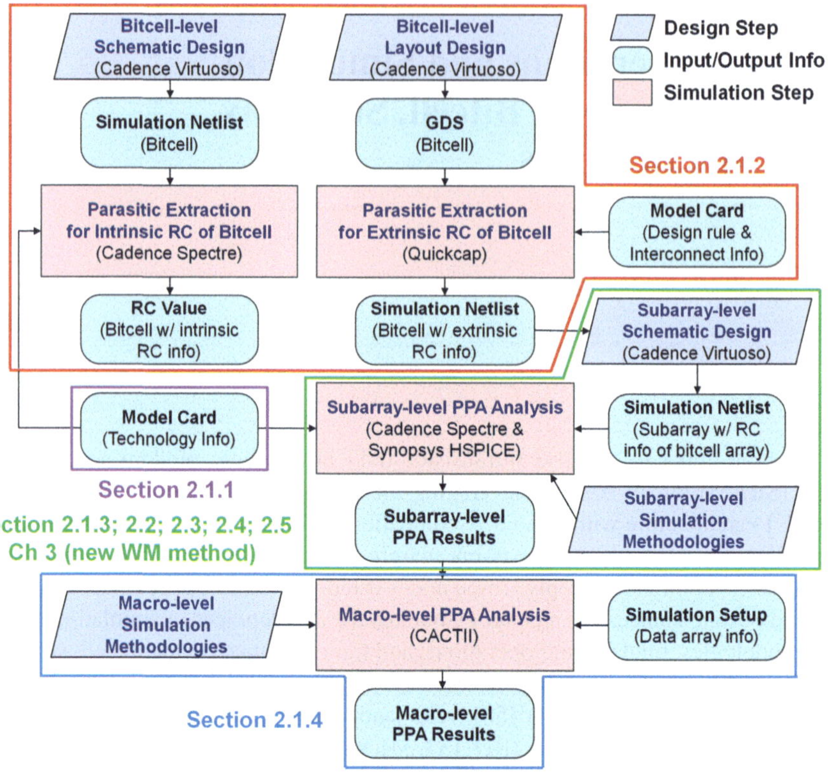

Fig. 2.1 PPA analysis framework in bitcell, subarray, and data-array levels

simulation setups are introduced, along with read assist techniques such as negative cell ground (GND) [66], cell supply voltage (V_{DD}) boost [72], suppressed BL (SBL) [45], and WL underdrive (WLUD) [54]. Section 2.4 covers write operation paths, writability, WD, and WE simulation setups, and write-assist techniques including cell GND boost [72], cell supply voltage (V_{DD}) boost [5], negative bitline (NBL) [38, 39], and boosted wordline (BWL) [53]. The subarray-level read and write simulation metrics are utilized in Chaps. 6, 8, and 9. In resistance-dominated technology nodes, the issue of write failure holds more prominence in comparison to read failure due to increased bitline (BL) resistance (R_{BL}) [32]. The conventional WM methodologies [10, 13, 16, 17, 36, 59, 62, 70, 71] fall short in predicting the time-dependent impacts stemming from parasitic resistance (R), capacitance (C), and pass-gate (PG) devices [32]. To bridge this gap, Chap. 3 proposes a novel WM methodology specifically designed for deeply scaled nodes [32], building upon the foundational knowledge of write operations introduced in Chap. 2. Chapter 2 is dedicated to clarifying each crucial aspect of the PPA simulation framework, while Chap. 3 provides a more concise and comprehensive understanding of the motivation behind and results of the proposed WM methodology, thereby avoiding any potential distractions. Additionally, Sect. 2.5 discusses the simulation methodology utilized for examining write failures and delves into the sources of failure associated with NBL [38, 39] and BWL [53] write-assist techniques. The adoption of these two write-assist techniques over others stems from their ability to mitigate the stability issues of half-selected bitcells under the assumption of divided techniques [25, 69]. These circuits and criteria form the basis for the write-ability-aware subarray optimization discussed in Chaps. 8 and 9. Finally, Sect. 2.6 concludes this chapter.

2.1 Simulation Setups and Methodologies

2.1.1 Compact Models

This book incorporates compact models including A14 NSFET, A10 FSFET, and A5/A3 CFET, as reported in the previous studies [23, 49, 64]. To ensure fair comparison across various technologies, each compact model adopts a consistent effective off current (I_{off}) assumption, which stands at approximately 0.02 nA, 0.2 nA, and 2 nA for both NMOS and PMOS devices with high, standard, and low threshold voltages (V_{T}), respectively. These existing compact models were systematically constructed, as described in prior works [7, 23, 49]. Ballistic, electrostatic, and stress effects were independently modeled, drawing upon extensive TCAD and hardware data, particularly for A14 technology. Furthermore, the compact model accurately reproduces the current-voltage (IV) TCAD curves, incorporating the correct threshold voltage and the aforementioned effects.

2.1.2 Bitcell-Level Setups

2.1.2.1 Circuit and Design Overview

In the context of 6T SRAM bitcell design (Fig. 2.2), the composition consists of two pull-up (PU), two pass-gate (PG), and two PD devices. Conventional SRAM bitcell designs, such as those featuring Fin, NS [32, 35], FS [64] devices, incorporate four n-type channels for both PG and PD devices and two p-type channels for PU devices. In contrast, sequential (seq) [33, 34] and monolithic (mono) [33, 34] CFET SRAM bitcell designs adopt a configuration featuring four p-type channels for PU and PG devices and two n-type channels for PD devices, as elaborated upon in Chap. 4. It is noteworthy that hybrid CFET SRAM bitcell design incorporates both configurations, depending on the specific fabrication process, as detailed in Sect. 5.4 of Chap. 5. Table 2.1 provides a comprehensive listing of the corresponding fin or nanosheet width (W_{NS}) ratio for PU:PG:PD devices, catering to high density (HD), high-performance (HP), and low-power (LP) applications. This table also includes comparative descriptions of their operations for NMOS PG (N-PG) and PMOS

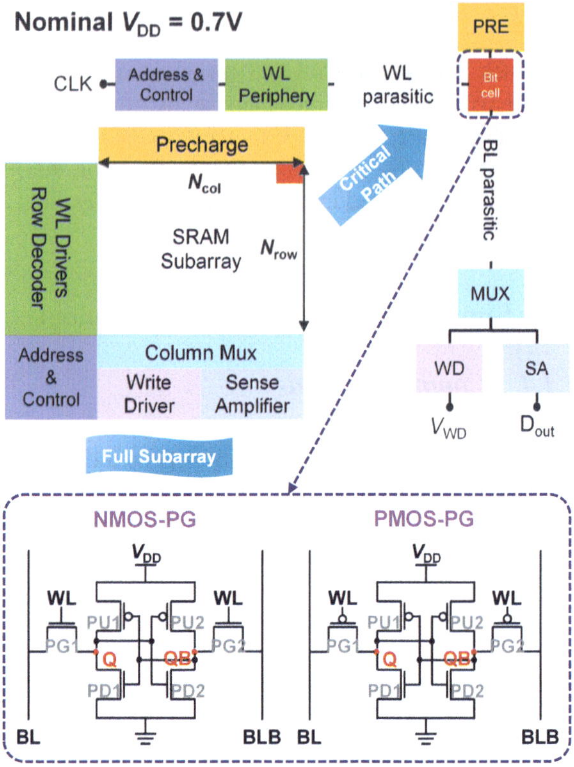

Fig. 2.2 Schematics of the full subarray, critical path, and 6T SRAM bitcell simulation circuits

Table 2.1 Comparison of 6T SRAM bitcell design with n- and p-type PG devices

6T SRAM	NMOS PG (N-PG)	PMOS PG (P-PG)
Transistor type	NSFET, FSFET, and Hybrid CFET	Seq, Mono, and Hybrid CFET
#Fin or W_{NS} ratio of PU:PG:PD	1:1:1 (HD) 1:2:2 (HP) 1:1:2 (LP)	1:1:1 (HD) 2:2:1 (HP) 2:1:1 (LP)
Operation	BL discharging WL active high	BL charging WL active low
Read path	PG, PD, V_{SS}	V_{DD}, PU, PG
Write path	V_{DD}, PU, PG	PG, PD, V_{SS}

PG (P-PG) based 6T SRAM bitcells. For an in-depth exploration of CFET SRAM bitcell design intricacies, please refer to Chaps. 4 and 5. It's essential to note that the simulation circuits exhibit a high degree of similarity between N-PG- and P-PG-based SRAM bitcells, with the primary difference lying in the WL activation setup. Specifically, the WL activation operates on an active high configuration for the N-PG case and an active low configuration for the P-PG case [33, 34].

2.1.2.2 Parasitic Extraction (PEX) of SRAM Bitcell

As technology progresses towards deeply scaled nodes, the resistance and capacitance (RC) within SRAM bitcells increase, which may lead to suboptimized PP outcomes at subarray and macro levels [33, 34]. Thus, conducting parasitic extraction (PEX) is imperative, with a specific focus on assessing the RC impacts in PP analysis. It is noteworthy that the simulation netlist exclusively incorporates the RC of SRAM bitcells, given the dominant influence of the SRAM bitcell array on the overall subarray-level PP results. Additionally, it is crucial to highlight that post-layout simulation is not considered in this study owing to the absence of a customized full subarray layout design and presently unavailable tape-out capabilities for nodes beyond 3 nm.

After designing the SRAM bitcell layout, the Cadence Virtuoso Layout software [9] is utilized to generate the graphic data stream (GDS) file. This GDS file, accompanied by the necessary device and interconnect models, serves as input for the QuickCap simulation tool [56]. The primary output files comprise the spice netlist and RC summary for the SRAM bitcell. Integrating this spice netlist into the subarray circuit netlist enables further subarray-level PP analysis. A comprehensive understanding of capacitance components necessitates consideration of three main categories: front-end-of-line (FEOL), middle-end-of-line (MEOL), and back-end-of-line (BEOL) capacitances. FEOL capacitance comprises intrinsic and extrinsic components. Intrinsic capacitance arises from the device itself, reflecting its inherent characteristics. Extrinsic capacitance includes fringe and overlap effects. To facilitate subarray-level PP analysis, the total capacitance is segmented into two distinct parts: (1) intrinsic FEOL and (2) extrinsic FEOL, MEOL, and BEOL. The former (FEOL intrinsic capacitance) and the latter (FEOL extrinsic + MEOL + BEOL capacitance) are extracted using Cadence Virtuoso Spectre [8] and QuickCap [56] simulation software, respectively. Furthermore, the resistance determination relies on both QuickCap simulation results and calibrated interconnect models as documented in [12]. The outcomes of the RC analysis for each SRAM bitcell design alternative of this book are presented in Fig. 6.1 of Chap. 6.

2.1.3 Subarray-Level Setups

2.1.3.1 Circuit and Design Overview

Figure 2.2 illustrates the critical path and full subarray simulation circuits of the SRAM subarray, including elements such as address/timing control, precharge circuits, row decoder, column multiplexer, sense amplifiers, write drivers, and the bitcell array. Detailed circuit diagrams are provided in Appendix A.1. The critical path circuit primarily facilitates the evaluation of access time (RD and WD) and WM outcomes contingent upon diverse input design parameters (e.g., RC changes in

BL and WL). Conversely, the full subarray circuit serves predominantly for energy (RE and WE) assessment and the attainment of more precise timing results. The PPA analysis focuses on the worst-case bitcell (identified within the red box), positioned farthest from the row decoder, sense amplifier, and write driver. The subsequent sections will provide a comprehensive explanation of the subarray-level metrics along with their respective operational principles.

2.1.3.2 Monte Carlo (MC) Simulation

MC simulations play an essential role in ultra-scaled SRAM design, ensuring functionality and comprehensively capturing multiple simulation outcomes stemming from increased process variability [1, 20, 68]. The A14 compact model integrates process variation and device mismatch by incorporating calibrated device characteristics and parasitic circuits. However, the maturity of silicon data and process flow development remains insufficient for A5 and A3 CFET technologies concerning MC simulations. Furthermore, adapting the bitcell simulation netlist generated by the Quickcap simulation tool [56] necessitates alterations to the bitcell layout design to precisely accommodate the physical variation in device geometry. The repetition of bitcell layout design for a billion iterations, as required for a 6-sigma (6σ) simulation, is unfeasible. Thus, V_T sweeping is performed based on a random Gaussian statistical distribution with a 6σ range for MC simulation, aiming to identify the worst-case scenarios within the scope of this book. While further analysis of temperature and voltage holds significance, such exploration is earmarked for future endeavours.

2.1.4 Macro-Level Setups

2.1.4.1 Circuit and Design Overview

The PPA analysis of SRAM macro is significantly influenced by the delay and energy consumption associated with the extensive inter-subarray interconnects (e.g., H-tree and E-tree) [42–44]. Hence, there is a crucial need to develop a comprehensive simulation framework at the macro level, which has been already established through collaborative studies documented in [42–44].[1] Figure 2.3 illustrates the schematic of the simulation circuit designed for the SRAM macro, highlighting various access pin positions and interconnect technologies: side pin (SP) access with H-tree interconnects (SP-H), centre pin (CP) access with H-tree interconnects (CP-H), side pin access with E-tree interconnects (SP-E), and centre pin access

[1] Acknowledgment is given to Zhenlin Pei for the Macro-level PPA analysis framework and inter-subarray interconnect advancements.

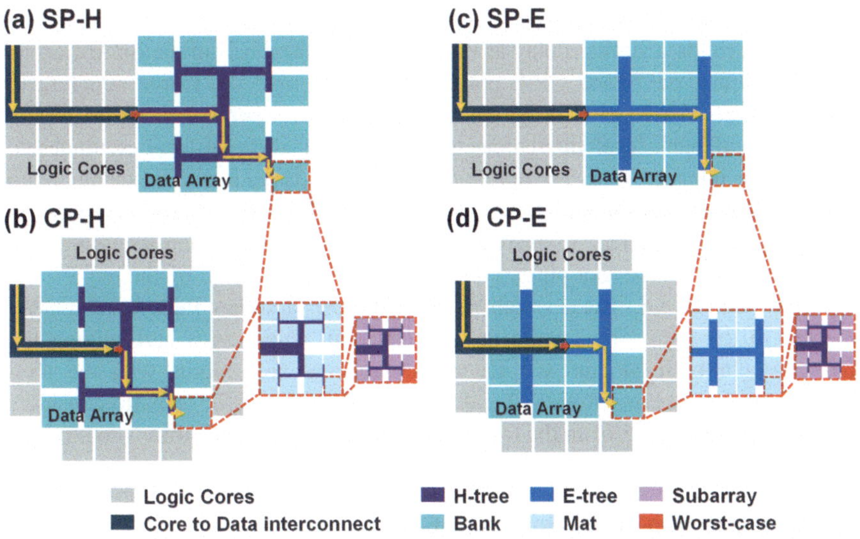

Fig. 2.3 Schematic view of logic cores and SRAM data array consisting of bank, mat, and subarrays based on (**a**) SP-H, (**b**) CP-H, (**c**) SP-E, and (**d**) CP-E technologies

with E-tree interconnects (CP-E). In this hierarchical macro architecture, a data array comprises multiple identical banks, each equipped with its address and data bus, enabling concurrent access. Each bank consists of multiple identical subbanks, with only one subbank being activated per access. Each subbank further consists of multiple identical mats. During access, all mats within a subbank are activated, with each mat storing a portion of the accessed word in the bank. Each mat comprises several identical subarrays that share a common central predecoder and decoder. Each subarray encompasses a bitcell array along with its own peripheral circuitry as depicted in Fig. 2.2. Notably, the current macro architecture lacks the inclusion of the tag array necessary for cache design, as it does not significantly contribute to the PPA in the cache [22]. It is suggested that future investigations prioritize the integration of the tag array to finalize the cache structure.

The macro architecture utilizes H-tree and E-tree distribution networks as request and reply paths across various hierarchy levels, ensuring uniform access to all mats [40, 60]. Signal gating employs multi-stage buffers with a 2-input NAND gate as the input stage [60], with sizing and gate number determined at each H/E-tree node following the methodology outlined in [2]. Throughout operation, the logic cores generate multiple input signals including clock, address, chip enable, chip select, write enable, and data information. These input signals are carried by the reply path from the edge of the data array to the mats, ultimately reaching the subarrays. Conversely, the output data produced by the subarrays are transmitted through the reply path from the mats to the edge of the data array [40, 60].

In SP-H and CP-H configurations, the H-tree architecture spans all three hierarchy levels—array, bank, and mat—facilitating straightforward balance through simple control logic [44]. However, due to symmetry, accessing a cell right beside the root pin yields a similar delay as reaching the farthest cell in the SRAM array. On the contrary, SP-E and CP-E configurations leverage the E-tree structure to reduce average wire length, further enhancing array-level performance [44]. While the array- and bank-level interconnects adopt E-tree, the mat-level continues to utilize H-tree interconnects. The horizontal interconnects per hierarchy segment into vertical interconnects shared by every two columns of banks or mats. Asymmetric routing advantages surface in decreasing interconnect length when accessing data physically closer to the root pin or bank/mat inputs (highlighted by the red arrow). From the perspective of access pin position, the SP assumes logic cores placed adjacent to the data array, while the CP envisions logic cores surrounding the data array. In both SP and CP scenarios, the core area aligns with the total subarray area [44]. For wire length calculations, the worst-case scenario considers interconnects extending from the logic cores' corners to the data array's root pin.

2.1.4.2 CACTI Simulation Framework

The PPA analysis of the SRAM macro is conducted using the CACTI simulator [3, 60], supported by SPICE simulation and referenced data from commercial caches in Intel's 65-nm L3 cache and Sun's SPARC 90-nm L2 cache [60]. Following the evaluation of subarray-level PPA results, these data are integrated into the CACTI simulator [34, 42–44]. The architecture of the data array, logic cores, and interconnects is designed to facilitate PPA optimization through the adjustment of user-defined metrics, such as access pin position, interconnect technology, and data capacity. To model the delay of inter-array and inter-subarray interconnects, we adopt methods from prior research [3, 43] to optimize repeater size and spacing for a balance between delay and energy consumption. This process involves utilizing Synopsys HSPICE [57] and RAPHAEL [58] to extract device-level and interconnect parameters, leveraging copper (Cu) material, A14 compact model [23], and interconnect design rules from imec. The interconnect delay model incorporates repeater insertion for logic cores-to-data-array access and inter-subarray interconnects. By adjusting a width scaling factor that allows variation in the standard interconnect width, its impact on array-level PPA can be assessed. To streamline the parameter sweeping process and avoid generating extensive customized layouts, a simplified RC extraction method for repeaters is adopted. Unlike the Quickcap simulation flow utilized for SRAM bitcells (Sect. 2.1.2.2), this approach relies on classic delay equations for repeaters and wire [43], involving parameters such as gate and drain capacitance, on current (I_{on}), temperature-dependent leakage current, and V_T, simulated using Cadence Spectre [8]. While this approach may overlook extrinsic capacitance inherent in the repeaters, the observed trends at the macro level remain reliable given the consistent simulation flow and model assumptions. However, it is advisable for future research to provide realistic RC data for repeaters, inter-array, and inter-subarray interconnects, aligning with more comprehensive model setups and experimental data for calibration purposes. It is important to highlight that the PPA outcomes presented in Chaps. 7 and 9 primarily focus on read operations. This preference stems from the general observation that read delays and energy consumption typically exceed those associated with write operations. Further calibration of the CACTI simulation framework with silicon data in deeply scaled nodes is recommended, which is currently unavailable.

2.2 Data Retention, Leakage, and Hold Operation

2.2.1 Data Retention

Data retention stands as a critical metric within memory systems, encompassing technologies such as dynamic random-access memory (DRAM) and SRAM, exerting a direct influence on leakage power consumption [19, 21, 65]. This metric defines the ability of these memory technologies to uphold stored data over specified durations, even in the absence of active power or refresh operations [19]. In the realm of DRAM, data retention signifies its proficiency in preserving data accuracy between successive refresh cycles. Interruptions in the refresh process or prolonged power discontinuity compromise the charge within DRAM cells, resulting in data loss. Notably, DRAM demonstrates low standby power dissipation but incurs heightened active power consumption due to its intrinsic reliance on periodic refresh cycles. Conversely, data retention in SRAM characterizes the capacity of memory cells to retain stored data when continuously powered, correlating with an upsurge in leakage power dissipation [65]. Unlike DRAM, SRAM operates without the need for periodic refreshing, maintaining its content solely while powered, with data loss occurring upon power removal.

2.2.2 Leakage

In the context of advanced CMOS technology scaling, a notable increase in the leakage current of SRAM bitcells is apparent [21]. This escalation primarily arises from multiple sources delineated in Fig. 2.4, including subthreshold, gate, and junction leakages [21, 31]. Subthreshold leakage manifests as the persistent current flow through transistors, even when the gate is ostensibly turned off, there is still a faint current due to the weak inversion layer of the channel region. Conversely, gate leakage results from the thinning of gate oxide layers, facilitating electron tunneling and subsequent current flow. Junction leakage emerges due to imperfections within the semiconductor material, fostering unintended current pathways across junctions. The pronounced SRAM leakage poses a substantial challenge in low-power design, significantly impacting power dissipation [21, 31], thereby influencing the overall energy efficiency, stability, and reliability of SRAM-based systems.

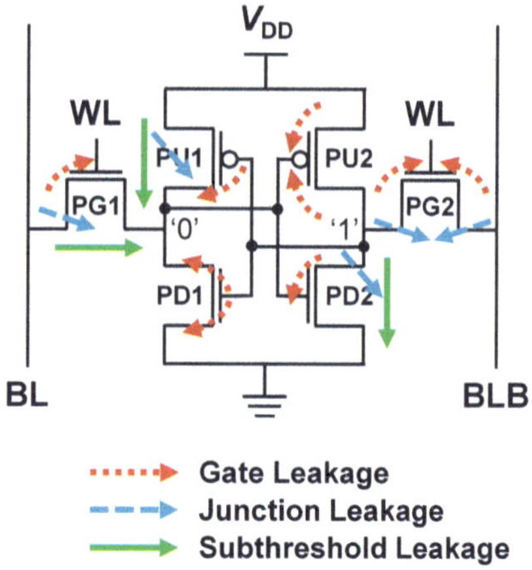

Fig. 2.4 Standby leakage currents in 6T SRAM bitcell

Strategies to mitigate this leakage include the implementation of high V_T methodologies [31] and the minimization of standby V_{DD} to its threshold limit, known as the data retention voltage (DRV), during idle mode [46]. In this book, all circuits are designed with high V_T (HVT) for SRAM bitcells and standard V_T (SVT) for peripheral logics [32] (Sect. 2.1.1). It is important to note that the IV characteristics of the actual devices are influenced by process-determined V_T, which remains undisclosed across the various technology nodes examined in this study (A14 to A3) due to the lack of silicon data. As a result, the primary focus of this study is not specifically on optimizing leakage, but rather on assessing the most favourable PPA outcomes based on consistent leakage characteristics. This approach ensures fair comparisons that are not affected by V_T variations, consistent with methodologies employed in prior studies [23, 49]. To assess the static current of an SRAM bitcell, a direct current (DC) simulation is conducted with WL deactivated, and BL and BLB precharged to V_{DD}. In the case of the A14 SRAM bitcell, the static current measures 2.121 μA, yielding a standby leakage of V_{DD} (0.7 V) × 2.121 μA, equivalent to 1.4847 μW.

2.2.3 *Hold Static Noise Margin (HSNM)*

As on-chip SRAM arrays typically occupy a substantial portion of system-on-chip (SoC), they wield considerable influence over a chip's PPA characteristics. Given the volatile nature of SRAM memory, data retention necessitates the continuous application of power, resulting in significant leakage consumption as just discussed in Sect. 2.2.2. Thus, idle modes are often operated at reduced V_{DD} levels to curtail SRAM standby power [48]. Furthermore, V_{DD} reduction is a common strategy to scale dynamic power consumption in older technology nodes, although it must be

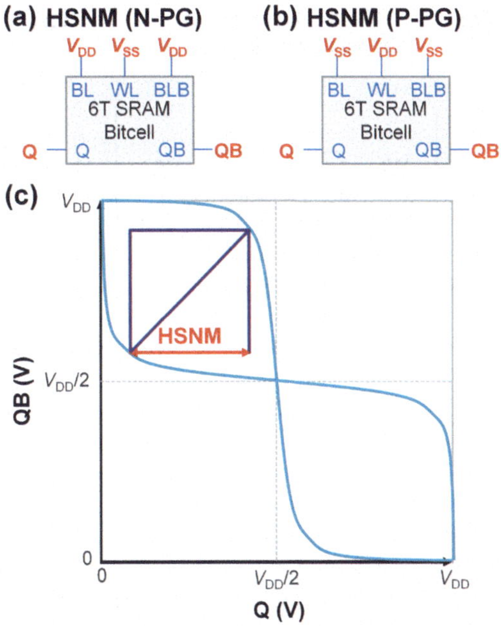

Fig. 2.5 Schematics of HSNM simulation circuits for (**a**) N-PG and (**b**) P-PG-based 6T SRAM bitcell with (**c**) VTC curves

carefully controlled to ensure data retention reliability. Otherwise, excessive V_{DD} scaling can elevate the bit error rate.

In alignment with prior investigations [29], the lower limit for V_{DD} should be established at the thermal voltage, denoted as kT/q (k: Boltzmann's constant; T: temperature; q: electronic charge). This constraint arises from the inescapable presence of thermal noise. Therefore, simulations of hold static noise margin (HSNM) are imperative to verify the data retention ability of SRAM bitcell. Figure 2.5 presents the voltage transfer characteristics (VTC) under specific conditions of 6T SRAM bitcell: when the WL is turned off, and BL and BLB are either pre-charged to supply voltage (V_{DD}) or pre-discharged to ground voltage (V_{SS}) depends on the N-PG and P-PG cases, respectively. The HSNM values are derived from these VTC and are defined as the side length of the largest square that can be fitted inside the butterfly curves [41, 50].

2.3 Read Operation and Read-Assist Techniques

During the read operation in SRAM bitcells, data access occurs through BL and BLB. Crucially, these stored data must remain undisturbed—any change from '0' to '1' during this process characterizes a destructive read operation [52]. Figure 2.6 and Table 2.2 outline the read paths, internal node conditions (Q and QB), as well as BL and BLB voltages for both N-PG and P-PG-based SRAM bitcells. The text primarily

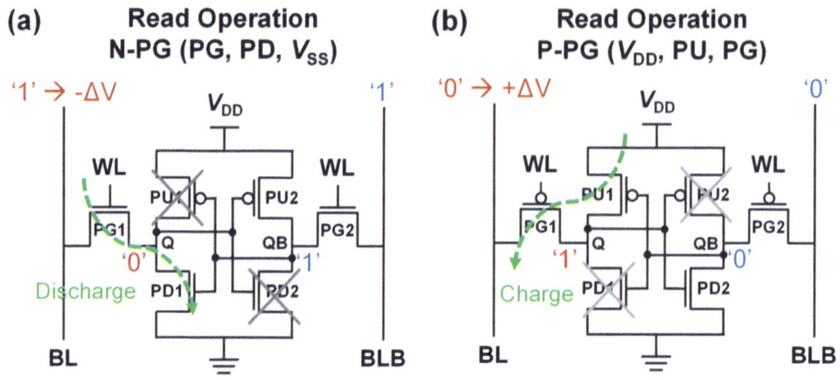

Fig. 2.6 Schematics of read operation path for (**a**) N-PG and (**b**) P-PG-based 6T SRAM bitcells

Table 2.2 Read operation steps and bias conditions for N-PG- and P-PG-based 6T SRAM bitcells

Read operation for N-PG-based SRAM Bitcell (path: PG, PD, V_{SS})		
1	Set initial condition	Q = 0; QB = V_{DD}
2	Pre-charging BL & BLB to ground (WL is off)	$V_{BL} = V_{BLB} = V_{DD}$; $V_{WL} = 0$
3	WL activation	$V_{WL} = V_{DD}$
4	Discharging BL to 0 (through PG and PD devices)	$V_{BL} = 0$; $V_{BLB} = V_{DD}$; $V_{WL} = V_{DD}$
5	Sensing voltage difference ($-\Delta V = 150\,\text{mV}$) between BL & BLB, then activate sense amplifiers	
6	Output data '1' & '0'	
7	WL deactivation	$V_{WL} = 0$
Read operation for P-PG-based SRAM Bitcell (path: V_{DD}, PU, PG)		
1	Set initial condition	Q = V_{DD}; QB = 0
2	Pre-discharging BL & BLB to ground (WL is off)	$V_{BL} = V_{BLB} = 0$; $V_{WL} = V_{DD}$
3	WL activation	$V_{WL} = 0$
4	Charging BL to V_{DD} (through PU and PG devices)	$V_{BL} = V_{DD}$; $V_{BLB} = 0$; $V_{WL} = 0$
5	Sensing voltage difference ($+\Delta V = 150\,\text{mV}$) between BL & BLB, then activate sense amplifiers	
6	Output data '1' & '0'	
7	WL deactivation	$V_{WL} = V_{DD}$

elaborates on the read operation in the N-PG-based SRAM bitcell, focusing on the BL discharging process for simplicity, while acknowledging the similarity in P-PG counterparts that shift to BL charging.

In this context, assuming an initial state of '0' for Q and '1' for QB, the read operation initiates when PD1 and PU2 devices are activated and WL is off. Sequentially: (1) BL and BLB are pre-charged to V_{DD}; (2) WL activation; (3) BL discharges from V_{DD} to V_{DD}-ΔV while BLB remains at V_{DD}; (4) data output via the sense amplifier; (5) WL deactivation. The ratio between PG1 and PD1 significantly influences overall read stability and speed when BL discharges via PG1 and PD1 [52]. In a HD SRAM bitcell design, a 1:1:1 ratio of PU:PG:PD achieves minimal bitcell area. However, in LP SRAM bitcell configurations where improved read stability is essential, adopting a 1:1:2 ratio becomes imperative. This stronger PD device in comparison to PG device serves to prevent data flipping during the current flow from BL to Q node [52]. This section will delve into the methodologies for extracting read operation metrics, which are used in Chap. 6. Moreover, a concise overview of common read-assist techniques will be provided.

2.3.1 Read Static Noise Margin (RSNM)

Lowering the V_{DD} of the memory system is regarded as the primary approach for minimizing dynamic energy consumption with technological scaling [19, 65]. However, V_{DD} scaling faces constraints not only from the thermal limit, akin to HSNM (Sect. 2.2.3), but also from increased process variation, imposing a strict criterion for ensuring successful read operations. Consequently, verifying the read functionality and minimum operating voltage (V_{min}) of an SRAM array necessitates computer-based simulations of read stability prior to fabrication. Several methods were reported including read static noise margin (RSNM) [50], N-curve [17, 63], and dynamic approaches [30, 61], etc. RSNM has been found to overestimate failure rates, as the maximum square side of the butterfly curve is defined by half of V_{DD}, thereby imposing an upper limit on the read stability value as V_{DD} scales [17, 61]. N-curve methodology aims to provide both voltage (static voltage noise margin, SVNM) and current (static current noise margin, SINM) information, with SINM capable of discerning the impact of transistor size on stability [17]. Additionally, N-curve can alleviate the limitations induced by V_{DD} scaling observed in RSNM [17]. Dynamic methods aim to capture the transient and dynamic behaviours of SRAM operation, offering read stability values that better align with silicon failures compared to static methods [61, 72].

Among all the approaches, RSNM yields the most pessimistic value due to the lack of consideration for time-dependent effects and the limited maximum square side [14, 17, 61]. However, adopting a more pessimistic methodology could provide a safer and conservative design boundary, particularly considering the immature process variability and device mismatch calibration employed in deeply scaled nodes (Sect. 2.1.1). Therefore, RSNM is utilized to extract the read stability to ensure read stability in this book. It is recommended that future studies focus on validating the most suitable read stability method based on mature silicon data and compact models.

In Fig. 2.7, the VTC specific to the 6T SRAM bitcell is showcased under distinct conditions: WL activation, with BL and BLB pre-charged to V_{DD} or pre-discharged to V_{SS}, contingent upon the type of PG devices utilized. The RSNM values, derived from these VTC, represent the side length of the largest square fitting within the butterfly curves, offering a crucial measure [41, 50].

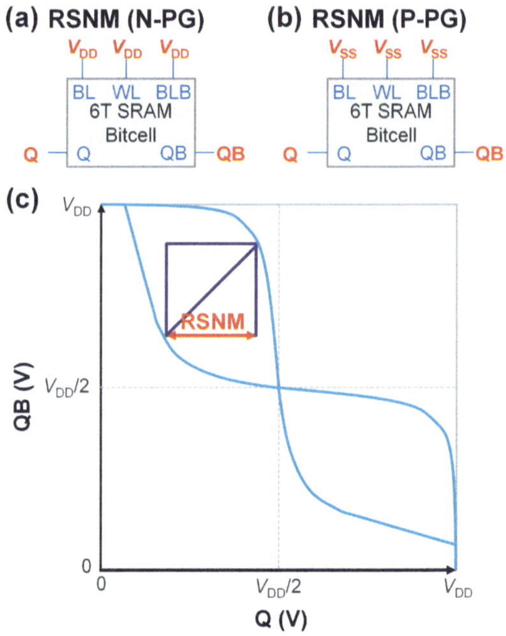

Fig. 2.7 Schematics of RSNM simulation circuits for (**a**) N-PG and (**b**) P-PG-based 6T SRAM bitcell with (**c**) VTC curves

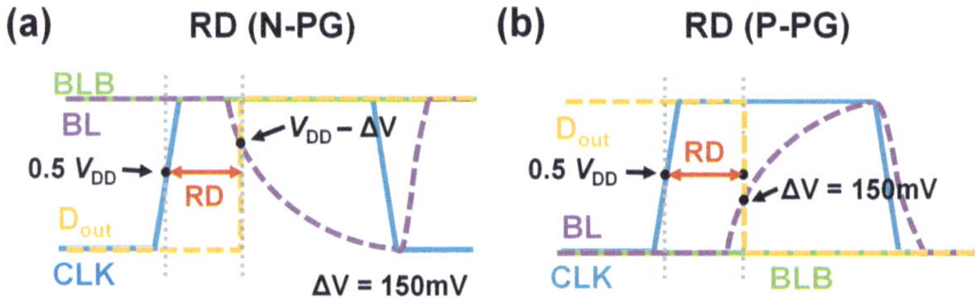

Fig. 2.8 Schematics of RD and RE waveform for (**a**) N-PG and (**b**) P-PG-based 6T SRAM bitcells

2.3.2 "Read Delay (RD)" and "Read Energy (RE)"

Figure 2.8 illustrates the waveform pertaining to the read operation in both N-PG and P-PG-based SRAM subarrays. The RD is precisely defined as the duration from half of the clock (CLK) signal to half of the data output (D_{out}) from the sense amplifier, specifically when reaching a 150-mV voltage difference between BLs [34]. The RE is calculated by integrating the active power of the entire subarray circuit across the operational duration. Notably, the sense amplifier activates solely when the voltage difference between BL and BLB (ΔV) reaches 150 mV for low power consumption. This 150 mV is a conservative assumption considering the amplifier's offset reported in existing work [6]. To confirm this assumption's validity under a mature variability model for nodes below A14, future analysis should involve a comprehensive MC simulation (Sect. 2.1.3.2).

2.3.3 *Read Assist Techniques*

As discussed in Sect. 2.3.1, the determination of V_{min} for SRAM bitcells during read operation relies on the read stability methodology. Scaling V_{DD} to reduce power consumption poses a challenge as it conflicts with the Vmin restriction, potentially jeopardizing read functionality [1, 65]. Therefore, an effective strategy to enhance this Vmin value is to integrate external read-assist circuitry [51, 67, 72]. This section introduces four common read-assist techniques for N-PG-based SRAM bitcell: negative cell GND [66], cell V_{DD} boost [72], suppressed BL (SBL) [45], and WL underdrive (WLUD) [54] (Fig. 2.9). The P-PG-based SRAM bitcell should follow a similar concept, employing an opposing method of voltage adjustment.

2.3.3.1 Negative Cell Ground (GND)

Lowering the voltage of V_{SS} pulls one of the internal nodes '0' below GDN in an SRAM bitcell [66]. It stands as the most effective read-assist technique by reinforcing the PD and PG devices through an increase in their gate-to-source voltage (V_{GS}) [66], thereby enhancing read stability and speed. Nevertheless,

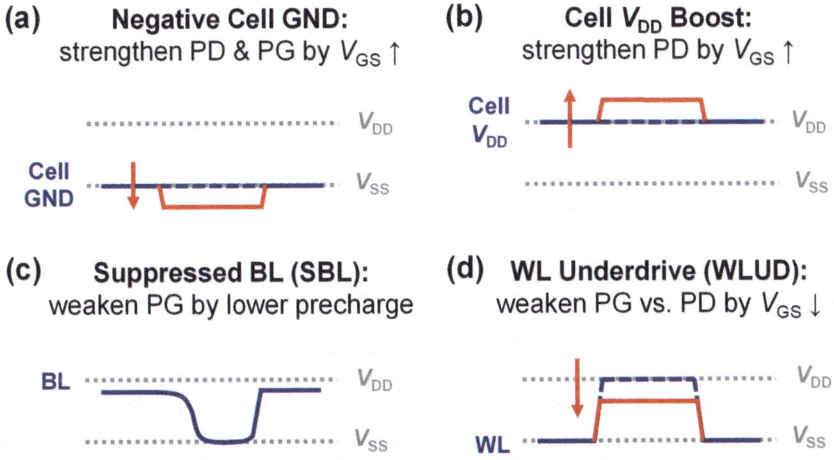

Fig. 2.9 SRAM read-assist techniques. (**a**) Negative cell GND: strengthen PD and PG devices by increasing V_{GS}. (**b**) Cell V_{DD} boost: strengthen PD devices by increasing V_{GS}. (**c**) Suppressed BL (SBL): weaken PG devices by lowering BL precharge level. (**d**) WL underdrive (WLUD): weaken PG devices by decreasing V_{GS}

this approach leads to unexpected data flips in half-selected bitcells within the selected column (deactivated WL but activated BL) and incurs high area and energy costs [72].

2.3.3.2 Cell Supply Voltage (V_{DD}) Boost

Increasing the voltage of V_{DD} elevates one of the internal nodes to '1' above the initial V_{DD} level within an SRAM bitcell [72]. This enhancement bolsters the strength of PD devices by augmenting their V_{GS}, resulting in better read stability. However, it does not affect PG devices, rendering it a less impactful read-assist technique [72]. This technique should be avoided during write operations as it degrades writability due to increased contention current.

2.3.3.3 Suppressed Bitline (SBL)

Pre-charging BL and BLB to a reduced level of V_{DD} (around 70% of V_{DD}) weakens the strength of PG devices in comparison to PD devices, consequently enhancing read stability due to the decreased read disturbance [45]. However, this adjustment might marginally reduce read speed by lowering the drain-to-source voltage (V_{DS}) on the PG devices [72]. Moreover, this technique becomes less effective in low V_{DD} technology [72].

2.3.3.4 Wordline Underdrive (WLUD)

Lowering the WL voltage during read operations weakens the PG devices compared to the PD device by decreasing the V_{GS}. This approach can enhance read stability by mitigating read disturbance. However, this technique bears an inherent timing penalty as it diminishes both read and write speeds of the selected bitcell due to reduced PG devices' strength. Additionally, this scheme induces noise injection in half-selected bitcells [54, 65].

2.3.3.5 Brief Summary

This section provides a brief summary of the advantages and disadvantages of four read-assist techniques. Negative cell GDN offers significant improvements in read stability but comes with high energy overhead and stability issues for half-selected bit cells within the same column [72]. V_{DD} boost is less effective for enhancing read stability and has a higher concern of device aging effect [18, 72]. SBL offers better power, speed, and PPA characteristics compared to WLUD but is less efficient in low V_{DD} conditions [55, 72]. WLUD is the most effective technique in low V_{DD} scenarios with the smallest area overhead, although it may introduce stability issues for half-selected bit cells within the same row [55, 72]. Among these options, the row-based WLUD technique appears to be a more popular design choice in the industry, often combined with other column-based write-assist techniques [11, 27]. It is recommended to analyse the overall effectiveness of read-assist techniques in ultra-scaled SRAM subarrays. However, potential conflicts may arise between certain read- and write-assist techniques, necessitating careful circuit design balance. For example, implementing read-assist circuits aims to enhance read stability of half-selected bitcells during write operations. However, the utilization of column-based read-assist circuits (such as cell V_{DD} boost, negative cell GND, and SBL) necessitates simultaneous control of signals across multiple unselected columns during write operation, resulting in significant power consumption [24]. Therefore, it is preferable to apply hybrid-assist techniques incorporating row-based read-assist WLUD and column-based write-assist like NBL or V_{DD} collapse to address stability issues while maintaining write margin and speed [24, 27, 54, 67]. However, these hybrid-assist methods introduce trade-offs among power, performance, and area costs, necessitating comprehensive optimization efforts to achieve overall PPA enhancement in future work.

2.4 Write Operation and Write-Assit Techniques

During the write operation in SRAM bitcells, the stored data are transitioned from its previous value to a new value ('0' to '1' or vice versa), demanding a fast and dependable data-flipping process [52]. Figure 2.10 and Table 2.3 delineate the write paths, internal node conditions (Q and QB), and BL and BLB voltages for both N-PG and P-PG-based SRAM bitcells. The explanation primarily focuses on the write

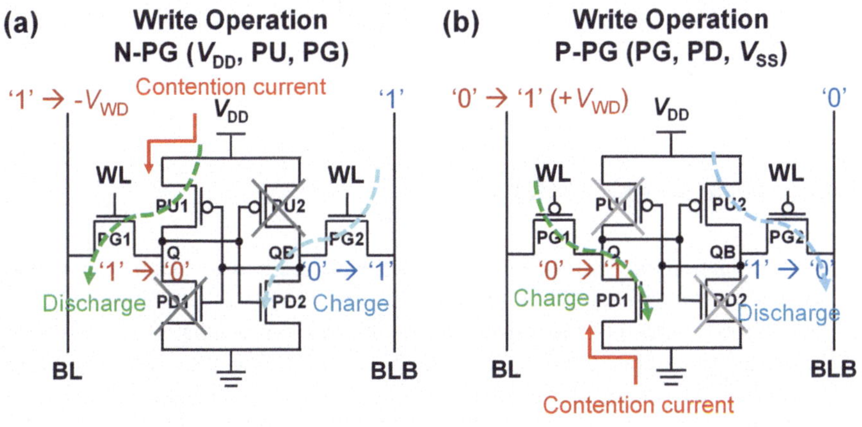

Fig. 2.10 Schematics of write operation path for (**a**) N-PG and (**b**) P-PG-based 6T SRAM bitcells

Table 2.3 Write operation steps and bias conditions for N-PG- and P-PG-based 6T SRAM bitcells

Write operation for N-PG-based SRAM Bitcell (path: V_{DD}, PU, PG)		
1	Set initial condition	Q = V_{DD}; QB = 0
2	Pre-charging BL & BLB to V_{DD} (WL is off)	V_{BL} = V_{BLB} = V_{DD}; V_{WL}=0
3	Asserting data to BL & BLB from the write driver during WL activation	V_{WL} = V_{DD}; V_{WD} < 0 for V_{BL}
4	Internal node data flipping	Q = 0; QB = V_{DD}
5	WL deactivation	V_{WL} = 0
Write operation for P-PG-based SRAM Bitcell (path: PG, PD, V_{SS})		
1	Set initial condition	Q = 0; QB = V_{DD}
2	Pre-discharging BL & BLB to ground (WL is off)	V_{BL} = V_{BLB} = 0; V_{WL} = V_{DD}
3	Asserting data to BL & BLB from the write driver during WL activation	V_{WL} = 0 V_{WD} >V_{DD} for V_{BL}
4	Internal node data flipping	Q = V_{DD}; QB = 0
5	WL deactivation	V_{WL} = V_{DD}

operation in the N-PG-based SRAM bitcell, mirroring the P-PG counterparts but with an inverse write path like read operation.

Assuming an initial state of '1' for Q and '0' for QB, the write operation is instigated when PU1 and PD2 devices are activated but WL is deactivated. The sequence involves: (1) pre-charging BL and BLB to V_{DD}; (2) activating WL; (3) Asserting data to BL and BLB using the write driver—here, employing the NBL write-assist technique [38, 39] with the write driver voltage (V_{WD}) below ground applied on BL; (4) data transition (Sect. 2.4.3.3); (5) deactivating WL. The ratio between PU1 and PG1 significantly impacts the overall write margin and speed when BL discharges through PU1 and PG1 [52]. Generally, PG needs to be stronger than PU to ensure BL voltage overwrites the internal node Q. Therefore, a 1:2:2 ratio for PU:PG:PD becomes imperative in a HP SRAM bitcell design. This section will explore the methodologies employed to extract write operation metrics, essential for application in Chap. 6.

2.4.1 *Write Margin (WM)*

The determination of V_{min} is not solely dependent on read stability but also on writability, which is significantly influenced by increased variability and R_{BL} as technology scales [1]. Therefore, the constraint on scaling V_{DD} to reduce dynamic energy requires a thorough examination of the WM value for identifying V_{min}. Several WM metrics have been proposed including both static and dynamic methodologies [10, 13, 16, 16, 17, 17, 36, 59, 59, 62, 62, 70, 71, 71], with the comparison described in Chap. 3. However, as technology advances towards sub-10 nm nodes, the significantly increased R_{BL} presents challenges to existing WM

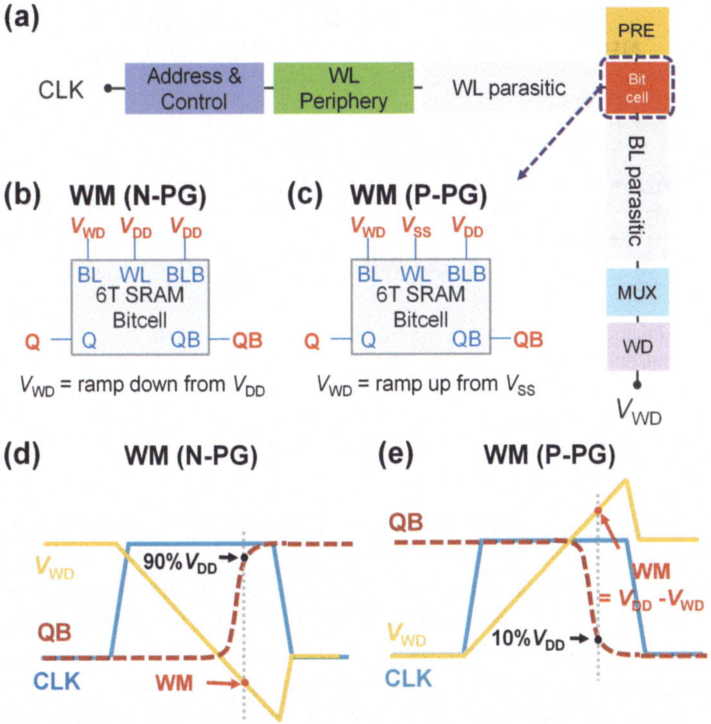

Fig. 2.11 Schematics of WM (**a**) simulation circuits and waveform for (**b**)(**d**) N-PG and (**c**)(**e**) P-PG-based 6T SRAM bitcells

methodologies. These methodologies fail to accurately predict the comprehensive impact of parasitic R, C, and PG leakage, resulting in an overly optimistic estimation of WM values [32]. Thus, an extended WM methodology tailored and validated specifically for nodes below 2 nm is proposed [32], detailed extensively in the forthcoming Chap. 3.

Figure 2.11 illustrates the simulation circuit, bias conditions, and waveform of this proposed WM methodology [32]. Employing an artificial write-assist V_{WD} [32], a negative ramp-down or positive ramp-up slope (S_{BL}) is applied to BL for N-PG- and P-PG-based SRAM bitcells, respectively, ensuring successful write operations without specific peripheral circuits. Furthermore, the S_{BL} is determined by the operational speed, which varies depending on distinct applications, allowing for a more accurate extraction of time-dependencies [32]. In the N-PG scenario, WM is defined as the V_{WD} value confirming data flip (90% of V_{DD}) [32]. While in the P-PG counterpart, WM is V_{DD} minus the minimum V_{WD} confirming data flip (10% of V_{DD}) [34].

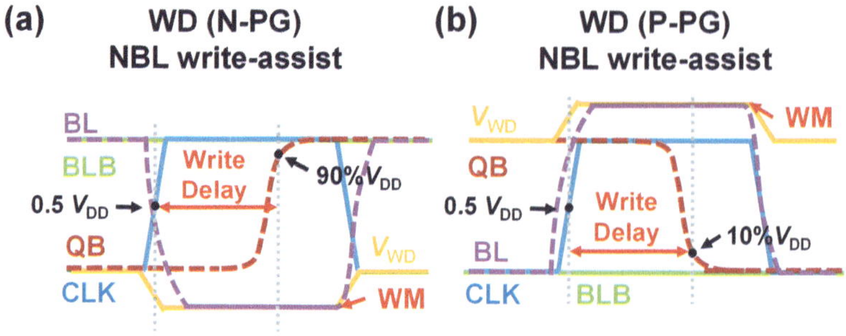

Fig. 2.12 Schematics of WD and WE waveform for (**a**) N-PG and (**b**) P-PG-based 6T SRAM bitcells with conceptual NBL write-assist technique assumption

2.4.2 "Write Delay (WD)" and "Write Energy (WE)"

Figure 2.12 exhibits the waveform of the write operation in both N-PG- and P-PG-based SRAM bitcells. Following the determination of the WM value, a constant V_{WD} equivalent to WM has been assumed for the worst-case simulations involving write delay (WD) and write energy (WE). WD denotes the duration from half of the CLK signal to the confirmation of data flip. On the other hand, WE is computed by integrating the active power of the full subarray circuit throughout its operational time. Notably, a more negative V_{WD} indicates the need for stronger write assistance, which has the potential to reduce write time but may escalate energy consumption. For a more comprehensive study on write operations, future research should explore specific write-assist circuits and realistic voltage waveforms. However, these aspects are not the primary focus of this study.

2.4.3 Write Assist Techniques

As discussed in Sects. 2.3.3.5 and 2.4.1, the V_{min} is determined by both read stability and writability due to the conflicts between them. Therefore, addressing the degradation in WM resulting from V_{DD} scaling, implementation of read-assist techniques, and the heightened R_{BL} in scaled nodes necessitates the adoption of write-assist techniques. Various write-assist techniques have been documented for N-PG-based SRAM bitcell, including cell GND boost [72], cell supply voltage (V_{DD}) boost [5], negative bitline (NBL) [38, 39], and boosted wordline (BWL) [53] (Fig. 2.13). This section will provide a concise overview of these four techniques. The P-PG-based SRAM bitcell should mirror these approaches but adjust voltages conversely.

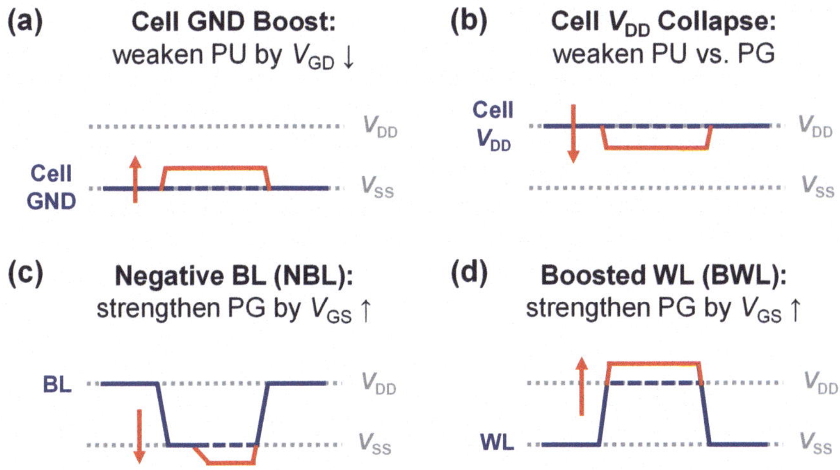

Fig. 2.13 SRAM write assist techniques. (**a**) Cell GND boost: weaken PU devices by decreasing V_{GS}. (**b**) Cell V_{DD} collapse: weaken PU devices. (**c**) Negative BL (NBL): strengthen PG devices by increasing V_{GS}. (**d**) Boosted WL (BWL): strength PG devices by increasing V_{GS}

2.4.3.1 Cell Ground (GND) Boost

Elevating the voltage of V_{SS} raises one of the internal nodes to '0' above the initial GND level within an SRAM bitcell [72], thereby weakening the PU devices due to a decrease in gate-to-drain voltage (V_{GD}). Consequently, as the PU devices and the cross-coupled inverter weaken, the write ability is enhanced. However, the effectiveness of this technique plateaus at around 30% higher than the GND level when the PD device pulls up the internal node. This saturation occurs because the V_T drop for NMOS device limits its ability to pull up the internal node to a maximum of V_{DD}-V_T [72]. It is important to note that the read stability of the half-selected bitcell within the same column on the selected row deteriorates. Moreover, this technique not only makes the circuit design complex but also causes high dynamic power overhead due to shared power rails among all the SRAM bitcells.

2.4.3.2 Cell Supply Voltage (V_{DD}) Collapse

Lowering the cell V_{DD} of an SRAM bitcell weakens the PU devices in comparison to the PG devices, diminishing contention current and enhancing write ability [5]. However, write failures may arise if the PU devices become excessively weak to pull up the internal node on the low-node side [72]. Despite its smaller area overhead compared to other write-assist techniques, this approach leads to degradation in both read and hold stability of the half-selected bitcell on the same column due to the shared power rails of every SRAM bitcell. Moreover, it exhibits reduced efficiency by decreasing the trip-point and gain of the cross-coupled inverter.

2.4.3.3 Negative Bitline (NBL)

Reducing the BL to a negative voltage below GND augments the PG devices by elevating V_{GS} and and V_{DS}, thereby enhancing write capability [38, 39]. This method stands out as the most effective write-assist technique [72], exerting no impact on the cross-coupled inverter. Additionally, it preserves half-select stability in the selected row, thanks to its column-based control characteristics. However, caution must be exercised for the half-selected bitcell in the unassessed row and selected column, as unintended PG leakage due to an excessively negative BL voltage might flip high-node data [32, 37]. To mitigate the risks associated with PG leakage, data retention, and reliability, meticulous design of the NBL voltage (V_{NBL}) level on BL is imperative, as elaborated in the subsequent section. Furthermore, the adoption of a more potent NBL technique necessitates a higher boost capacitance value, potentially incurring significant dynamic power consumption overhead [38, 39]. Thus, a comprehensive examination encompassing PPA, V_{NBL}, and capacitor size holds paramount importance.

2.4.3.4 Boosted Wordline (BWL)

Raising the WL voltage above the initial V_{DD} voltage enhances write ability by elevating the V_{GS} [53]. This technique efficiently augments the strength of both PG devices. However, while it accelerates read speed by injecting more charges into the internal node, it degrades the read stability of half-selected bitcells on the selected row due to increased read disturbance [72]. To counter this, employing an additional read-assist technique alongside the BWL write-assist technique becomes imperative to resolve the degraded read stability issue. Alternatively, delaying the BWL pulse until partway through the WL activation can mitigate the stability problem of half-selected bitcells, as they would have already begun reading and the BL voltage would more closely match the voltage level of the internal nodes [53].

2.4.3.5 Brief Summary

This section presents a succinct overview of the benefits and drawbacks associated with the four distinct write-assist techniques. Cell GND boost demonstrates limited effectiveness in enhancing writability, along with stability issues such as half-selected bitcells within the same column and considerable area and energy overhead [72]. V_{DD} collapse proves to be inferior to NBL and BWL at higher V_{DD} ranges [67], with limitations arising from the presence of half-selected stability issue within the same column and the degradation of WM and write speed [18]. Its efficacy is primarily observed within the lower V_{DD} range [67]. NBL emerges as the most effective option, offering superior write speed, particularly within the small R_{BL} range [18, 55, 72]. Nonetheless, NBL faces half-selected hold stability concerns, while improvements in WM reach a saturation point with increasing

R_{BL}, a condition irrecoverable by timing delay [55]. BWL proves advantageous when considering read speed and larger R_{BL} ranges due to the recovery of timing delay [53]. However, it significantly compromises half-select read stability within the same row [55, 67]. Given the inherent stability issues associated with half-selected bitcells across all write-assist techniques, a strategic approach involving the utilization of mixed assist techniques is imperative to achieve V_{min} enhancement, as discussed in Sect. 2.3.3.5 [55]. Industry reports highlight four common hybrid assist techniques:

1. WLUD (row-based read-assist technique) + NBL (column-based write-assist technique): This combination eliminates stability issues with half-selected bitcells and restores writability and write speed [11].
2. WLUD (row-based read-assist technique) + V_{DD} collapse (column-based write-assist technique): This pairing eliminates stability issues with half-selected bitcells and enhances writability and write speed [26, 27].
3. SBL (column-based read-assist technique) + NBL (column-based write-assist technique): This combination maintains optimized read stability and writability [45].
4. WLUD (row-based read-assist technique) + BWL (row-based write-assist technique): Utilizing timing intervals, this approach improves read stability and consecutively recovers WM [4, 55].

A thorough analysis to identify the most suitable hybrid assist technique for deeply scaled nodes should be conducted in future research endeavours.

2.5 Write Failure Examination Framework

The increasing R_{BL} observed in deeply scaled nodes highlights the critical need to address write failure prevention. Furthermore, the effectiveness of write-assist techniques in recovering the WM is becoming increasingly limited. This underscores the importance of establishing a clear correlation between the WM and write-assist techniques, as the discussion on write failure criteria expands from Sect. 2.4. However, the lack of silicon data validation for functional SRAM bitcell designs in NS, FS, and CFET technology poses challenges in confirming the write failure probability. Consequently, it is essential to develop a design strategy that relies heavily on computer-based simulation to effectively mitigate potential write failure risks. This section outlines the simulation methodologies and identifies sources of write failure based on the N-PG-based SRAM subarray and associated write-assist peripheral designs. These methodologies lay the foundation for defining write failure criteria and set the stage for subsequent discussions in Chap. 8.

2.5.1 Specialized Write-Assist Circuit Design and Simulation Methodology

Although stability can be ensured by staying away from the thermal limit, it is also essential to maintain sufficient HSNM and RSNM in the half-selected bitcell while applying the write-assist techniques. The latter three write-assist techniques suffer from read and hold stability degradation, which is crucial for scaled SRAM bitcell. However, the stability issue of BWL can be resolved by implementing the divided wordline (DWL) technique [69] due to the gated local WL structure. Hence, the NBL [38, 39] and BWL [53] techniques are selected as the reference points for examining write failure in this section.

As discussed in the write operation methodology, the more negative value of the WM represents the necessity of stronger write-assist techniques and higher write failure risk [32]. To provide a realistic impact from the write-assist technique, the most common NBL [38, 39] and BWL [53] techniques are considered. In Figs. 2.14 and 2.15, the circuit schematics and simulation waveforms for the NBL and BWL designs pertinent to the WD are illustrated. Both NBL and BWL are assumed to be driven by the boost capacitor (C_{boost}) for generating the negative voltage on BL (V_{NBL}) and positive voltage (larger than V_{DD}) on WL (V_{BWL}) in this work.

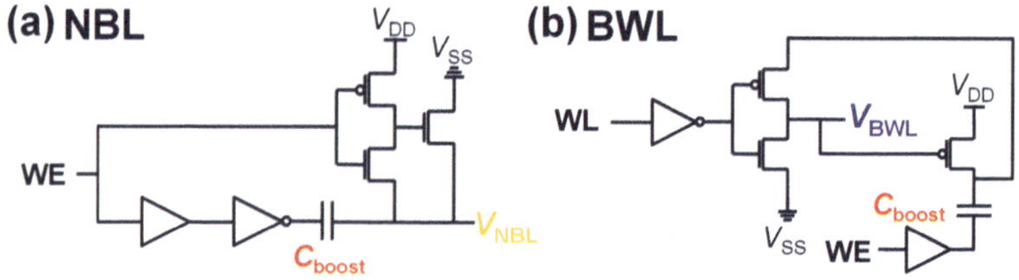

Fig. 2.14 Realistic circuit schematics of (**a**) NBL and (**b**) BWL write-assist techniques, which use C_{boost} as V_{NBL} and V_{BWL} generator. (WE: write enable)

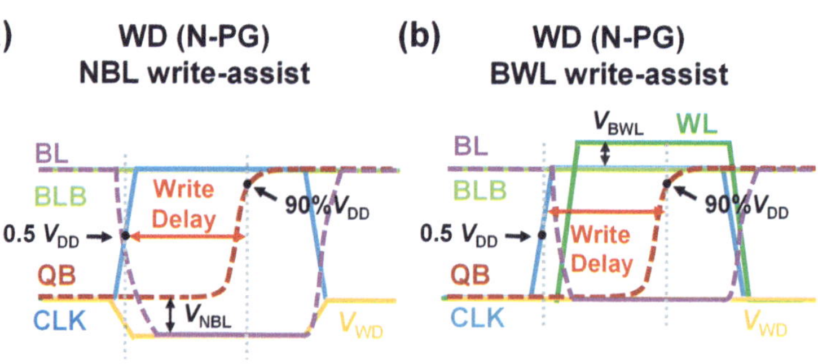

Fig. 2.15 Schematics of WD and WE waveform for N-PG-based 6T SRAM bitcell with conceptual (**a**) NBL and (**b**) BWL write-assist technique assumption

The lower or upper limit of V_{NBL} and V_{BWL} produced by C_{boost} is captured to be as the reference point of the WM constraint. Whenever the simulated WM value exceeds the maximum C_{boost}-generated voltage, the write operation may fail. Cross-checking both the requisite WM value and practical write-assist limitation can guarantee a successful write operation.

To crosscheck the functionality of the write operation, two simulation scenarios are employed. This involves the design of two types of WD circuits: conceptual and realistic, each tailored to specific objectives. The former aims to test the source of write failure attributed to PG leakage, while the latter targets the limited C_{boost}-generated write assist voltage. The subsequent discussion outlines their respective aims, with detailed results presented in Chap. 8.

2.5.1.1 Conceptual Write-Assist Circuits (PG Leakage)

The goal is to investigate the limit of simulated WM induced by PG leakage, achieved by varying specific parameters like N_{row} and N_{col}. To do the sweeping easily, V_{NBL} and V_{BWL} are set as the obtained WM value (as described in Sect. 2.4.1). The more negative of WM, the larger $|V_{NBL}|$ and $|V_{BWL}|$ are required. If V_{NBL} or V_{BWL} is over-boosted, the unintentionally activated PG device will create a leakage path from the internal node to BL [32]. PG leakage impedes write-assist voltage from becoming larger and thereby causes a longer write time that results in write failure eventually. Hence, the upper limit of subarray size that survives the write failure caused by PG leakage can be obtained by this test bench.

2.5.1.2 Realistic Write-Assist Circuits (Limited C_{boost}-Generated Write-Assist Voltage)

The goal is to find the limit of C_{boost}-generated V_{NBL} and V_{BWL}. An ideal capacitor is assumed for this C_{boost} in NBL and BWL write-assist circuits, which can simply provide the impact of varying C_{boost} on V_{NBL} and V_{BWL} and its corresponding WD. As the C_{boost} increases, the $|V_{NBL}|$ or $|V_{BWL}|$ increases and the WD decreases. However, the WD improvement will eventually saturate when the C_{boost} is sufficiently large. In general, it is suggested not to go beyond this threshold C_{boost} ($C_{boost\text{-}th}$) to avoid the cost of larger power consumption and area overhead of the write-assist circuits. Therefore, testing the $C_{boost\text{-}th}$ and maximum allowed $|V_{NBL}|$ and $|V_{BWL}|$ ($V_{NBL\text{-}max}$ and $V_{BWL\text{-}max}$) are crucial. The upper limit of subarray size can also be crosschecked by these threshold write-assist voltage values.

2.5.2 *Other Write Failure Sources*

2.5.2.1 Reliability

While the simulated $V_{NBL\text{-}max}$ and $V_{BWL\text{-}max}$ could potentially denote specific WM values as the write failure boundary following the aforementioned simulation methodologies, the maximum allowable operating gate voltage ($V_{GS\text{-}max}$) may not afford this voltage range. Bias temperature instability (BTI) imposes a restriction on $V_{GS\text{-}max}$ for reliable device operation. The NBL and BWL write-assist techniques transiently augment the gate voltage (V_{GS}) of the NMOS PG transistors in SRAM bitcells, raising concerns regarding positive BTI (PBTI) reliability. Initially, PBTI was a significant reliability concern for n-channel devices in early high-k metal gate (HKMG) technologies, attributed to intensified trapping of channel electrons in high-k dielectric defects [28]. However, notable advancements through gate stack optimization in recent technology nodes have resulted in negligible PBTI V_T shifts at low operating voltages [47].

For a cutting-edge replacement metal gate HKMG stack featuring an equivalent oxide thickness of 0.8 nm and V_T of 0.35 V, we anticipate a $V_{GS\text{-}max}$ of 1 V under bias conditions [47]. This projected value is estimated to induce a 50 mV V_T shift (BTI failure criterion) after a continuous 10-year operation at an elevated temperature of 125°C, based on imec experimental data [15]. Notably, this suggests the potential utilization of voltages surpassing the nominal V_{DD} of 0.7 V in the targeted A14 technology. Additionally, it's vital to highlight that the proposed write-assist techniques momentarily subject only a fraction of the total SRAM bitcell count in the array to excessive V_{GS}. Thus, during a 10-year continuous chip operation, each cell experiences this excessive V_{GS} solely for a fraction of the chip's operating time, calculated as 10 years multiplied by the number of active cells at any given time, divided by the total number of cells within the array.

For instance, in Chap. 8, the worst-case ratio of active cells to the subarray size stands at 32 over 128×128. Hence, in the proposed implementation and under a worst-case scenario of continuous chip operation, each cell would experience overstress for a maximum of approximately 171 hours across the entire 10-year chip lifetime. Factoring in this reduced duty time, the projected $V_{GS\text{-}max}$ for the target device is 1.2 V. This projection offers a margin of 0.5 V above the nominal V_{DD} (0.7 V) of A14 technology for the write-assist implementation. However, a more comprehensive reliability analysis should be pursued in separate work, given that it is not the primary focus here.

2.5.2.2 Substrate Current in NBL Write-Assist Technique

Apart from reliability concerns, subjecting the devices to an overdriven ground towards a negative value based on the NBL technique may result in substrate current (wherein the diode activates between the n+ source and p-well region). This diode's

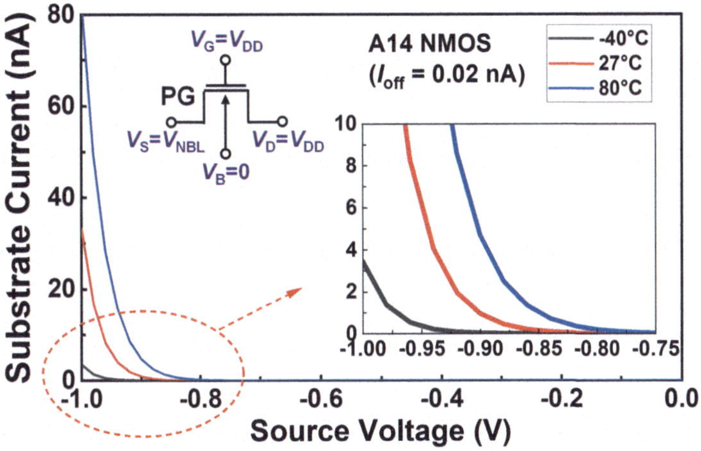

Fig. 2.16 Substrate current versus source voltage when applying NBL from 0 to -1 V on the source of A14 NMOS PG device. (V_G: gate voltage; V_S: source voltage; V_D: drain voltage; V_B: bulk voltage)

turn-on voltage, verified by TCAD simulation at 80 °C based on A14 NS device (as depicted in Fig. 2.16), measures approximately 800 mV. Consequently, this is equivalent to a relatively relaxed WM constraint of -0.8 V.

2.5.3 *Brief Summary*

The methodology for examining write failures aims to determine a specific WM value for identifying the write failure boundary, in conjunction with the corresponding write-assist techniques. This approach is particularly utilized for optimizing subarray size with consideration of sufficient writability, as discussed in Chap. 8. The criteria for determining write failure take into account several factors including PG leakage, limitations on write-assist voltage for both NBL and BWL circuits, reliability concerns, and substrate current in NBL technique. The first test seeks to establish the relationship between PG leakage and the maximum allowable subarray size for safe write operations. Subsequently, the second test is conducted to ascertain the practical limitations of write-assist voltage, which is then translated into the targeted WM value necessary for successful write operations. Based on the findings from these tests and the targeted WM value, the upper limits of subarray sizes are adjusted accordingly. Moreover, to ensure stability and prevent substrate current in critical nodes such as A14, the WM value should not exceed -500 mV. This constraint is essential for maintaining the functionality and reliability of the overall system.

2.6 Conclusions

This chapter presents the simulation setups, operational principles, and methodologies for both read and write metrics across SRAM bitcell, subarray, and macro levels. Figure 2.1 provides a comprehensive overview of the PPA analysis framework, along with associated sections and chapters in this dissertation. The bitcell-level framework includes setups of 6T SRAM with two types of PG devices and the PEX methodology using the Quickcap tool specifically for a single bitcell (Sect. 2.1). Detailed information on the extracted RC data for each bitcell is provided in Sect. 6.2 of Chap. 6. The compact model, bitcell netlist including RC information, and the subarray netlist serve as inputs for subarray-level simulation. Table 2.4 summarizes subarray-level metrics, including HSNM, RSNM, WM, RD, WD, RE, and WE, as delineated in Sects. 2.2, 2.3, and 2.4. Chapters 3, 6, 8, and 9 elaborate on this subarray-level analysis using Cadence Virtuoso and Synopsys HSPICE. The chosen read stability and writability methods are considered more conservative metrics compared to other existing methods due to increased variability in deeply scaled nodes. In resistance-dominated technology nodes, write failures gain prominence over read failures owing to increased R_{BL} resistance [32]. A

Table 2.4 Evaluation terms and methodologies for N-PG-based SRAM bitcell and subarray simulation. (CLK: clock signal; SA: sense amplifier; WD: write driver; V_{WD}: write driver voltage; WL: wordline; V_{WL}: WL voltage; BL/BLB: bitlines; C_{boost}: boost capacitor; V_{NBL}: negative BL voltage; V_{BWL}: boosted WL voltage)

Parameters	Definition
Worst-case bitcell	The bitcell is located furthest away from row decoder, WD, and SA
Hold static noise margin (HSNM)	The side length of the largest square can be fitted inside the butterfly curve when WL = 0; BL/BLB = V_{DD}
Read static noise margin (RSNM)	The side length of the largest square can be fitted inside the butterfly curve when WL = BL/BLB = V_{DD}
Read delay (RD)	Half of CLK to half of the output of SA
Write margin (WM)	VWD which confirms data flip (90% of V_{DD}) of the worst-case bitcell
Write delay (WD) with conceptual write-assist circuits	Half of CLK to data flips (90% of V_{DD})
	NBL: set WM as constant V_{WD} (V_{NBL}). BWL: set V_{DD} (when WM <0) or V_{DD}—WM (when WM $\geqq 0$) as constant V_{WL} (V_{BWL})
Write delay (WD) with realistic write-assist circuits	Half of CLK to data flips (90% of V_{DD})
BWL: V_{BWL} ($>V_{DD}$) generated by C_{boost}	NBL: V_{NBL} ($<$) generated by C_{boost}
Read/write energy	Integration of power over the read/write time

proper WM methodology and write failure examination are necessary for subarray optimization with the ability to predict write failure. However, existing WM metrics may not adequately account for time-dependent effects of BL parasitics, which could lead to overly optimistic results (Sect. 2.4 and Chap. 3). Therefore, this chapter introduces the motivation and provides a brief overview of the proposed WM methodology, with detailed explanations reserved for Chap. 3. Additionally, the development of a write failure prediction model solely reliant on computer-based simulations becomes crucial due to the lack of silicon data for SRAM subarrays in A14 to A3. The simulation methodologies and associated concerns regarding write failure sources, such as PG leakage, limited write-assist voltage, maximum operating voltage, and substrate current, are discussed in Sect. 2.5. A comprehensive analysis of the usage of this framework to assess write failure risks based on A14 SRAM subarray optimization is provided in Chap. 8. At the macro level, the simulation setups include considerations for inter-subarray interconnects, access pin positions, and macro architecture hierarchy, along with the utilization of the CACTI simulation framework. Following functionality verification in both read and write operations at the subarray level, the subarray-level PPA data serve as input for deriving macro-level metrics in Chaps. 7 and 9.

References

1. Aburahma, M., & Anis, M. (2012). *Nanometer variation-tolerant SRAM: Circuits and statistical design for yield* (pp. 1–170). ISBN: 978-14-61417-48-4. https://doi.org/10.1007/978-1-4614-1749-1
2. Amrutur, B. S., & Horowitz, M. A. (2001). Fast low-power decoders for RAMs. In: *IEEE Journal of Solid-State Circuits, 36*(10), 1506–1515. https://doi.org/10.1109/4.953479
3. Balasubramonian, R., et al. (2017). CACTI 7: New tools for interconnect exploration in innovative off-chip memories. *ACM Transactions on Architecture and Code Optimization, 14*(2). ISSN: 1544-3566. https://doi.org/10.1145/3085572
4. Bhargava, M., et al. (2014). Low VMIN 20nm embedded SRAM with multivoltage wordline control based read and write assist techniques. In *2014 Symposium on VLSI Circuits Digest of Technical Papers* (pp. 1–2). https://doi.org/10.1109/VLSIC.2014.6858412
5. Bhavnagarwala, A., et al. (2007). A Sub-600mV, fluctuation tolerant 65nm CMOS SRAM array with dynamic cell biasing. In *2007 IEEE Symposium on VLSI Circuits* (pp. 78–79). https://doi.org/10.1109/VLSIC.2007.4342773
6. Biswas, A., et al. (2017). An offset-cancelling four-phase voltage sense amplifier for resistive memories in 14nm CMOS. In *2017 IEEE Custom Integrated Circuits Conference (CICC)* (pp. 1–4). https://doi.org/10.1109/CICC.2017.7993689
7. Bufler, F. M., et al. (2020). Monte Carlo comparison of n-type and p-type nanosheets with FinFETs: Effect of the number of sheets. *IEEE Transactions on Electron Devices, 67*(11), 4701–4704. https://doi.org/10.1109/TED.2020.3024267
8. Cadence® (2021). *Spectre®Circuit Simulator*. https://www.cadence.com/en_US/home/tools/custom-ic-analog-rf-design/circuit-simulation.html
9. Cadence®(2021). *Virtuoso®Layout Suite*. https://www.cadence.com/en_US/home/tools/customic-analog-rf-design/layout-design/virtuoso-layout-suite.html

10. Carmona, C., Torrens, G., & Alorda, B. (2014). SRAM write margin cell estimation using wordline modulation and read/write operations. In *2014 5th European Workshop on CMOS Variability (VARI)* (pp. 1–6). https://doi.org/10.1109/VARI.2014.6957077
11. Chang, J., et al. (2013). A 20nm 112Mb SRAM in high-metal-gate with assist circuitry for low-leakage and low-VMIN applications. In *2013 IEEE International Solid-State Circuits Conference Digest of Technical Papers* (pp. 316–317). https://doi.org/10.1109/ISSCC.2013.6487750
12. Ciofi, I., et al. (2019). RC benefits of advanced metallization options. *IEEE Transactions on Electron Devices, 66*(5), 2339–2345. https://doi.org/10.1109/TED.2019.2902031
13. Dong, W., Li, P., & Huang, G. M. (2008). SRAM dynamic stability: Theory, variability and analysis. In *2008 IEEE/ACM International Conference on Computer-Aided Design* (pp. 378–385). https://doi.org/10.1109/ICCAD.2008.4681601
14. Elthakeb, A. T., et al. (2015). Analysis and optimization for dynamic read stability in 28nm SRAM bitcells. In *2015 IEEE International Symposium on Circuits and Systems (ISCAS)* (pp. 1414–1417). https://doi.org/10.1109/ISCAS.2015.7168908
15. Franco, J., et al. (2022). Low thermal budget PBTI and NBTI reliability solutions for multi-Vth CMOS RMG stacks based on atomic oxygen and hydrogen treatments. In *2022 International Electron Devices Meeting (IEDM)* (pp. 30.4.1–30.4.4). https://doi.org/10.1109/IEDM45625.2022.10019385
16. Gierczynski, N., et al. (2007). A new combined methodology for write-margin extraction of advanced SRAM. In *2007 IEEE International Conference on Microelectronic Test Structures* (pp. 97–100). https://doi.org/10.1109/ICMTS.2007.374463
17. Grossar, E., et al. (2006). Read stability and write-ability analysis of SRAM cells for nanometer technologies. *IEEE Journal of Solid-State Circuits, 41*(11), 2577–2588. https://doi.org/10.1109/JSSC.2006.883344
18. Gupta, M. K., et al. (2017). Device circuit and technology co-optimisation for FinFET based 6T SRAM cells beyond N7. In *2017 47th European Solid-State Device Research Conference (ESSDERC)* (pp. 256–259). https://doi.org/10.1109/ESSDERC.2017.8066640
19. Harris, D., & Harris, S. (2007). *Digital design and computer architecture*. Morgan Kaufmann. ISBN: 01-23-704-97-9.
20. Huynh-Bao, T., et al. (2017). SRAM designs for 5nm node and beyond: Opportunities and challenges. In *2017 IEEE International Conference on IC Design and Technology (ICICDT)* (pp. 1–4). https://doi.org/10.1109/ICICDT.2017.7993502
21. Ishibashi, K., & Osada, K. (2013). *Low power and reliable SRAM memory cell and array design*. Springer Series in Advanced Microelectronics. Springer. ISBN: 978-36-42270-18-5.
22. Ishihara, T., & Fallah, F. (2005). A way memoization technique for reducing power consumption of caches in application specific integrated processors. In *Design, automation and test in europe* (Vol. 1, pp. 358–363). https://doi.org/10.1109/DATE.2005.45
23. Jang, D., et al. (2017). Device exploration of nanosheet transistors for sub-7-nm technology node. *IEEE Transactions on Electron Devices, 64*(6), 2707–2713. https://doi.org/10.1109/TED.2017.2695455
24. Jeong, H., et al. (2015). Offset-compensated cross-coupled PFET bit-line conditioning and selective negative bit-line write assist for high-density low-power SRAM. *IEEE Transactions on Circuits and Systems I: Regular Papers, 62*(4), 1062–1070. https://doi.org/10.1109/TCSI.2015.2388837
25. Karandikar, A., & Parhi, K. K. (1998). Low power SRAM design using hierarchical divided bit-line approach. In *Proceedings International Conference on Computer Design. VLSI in Computers and Processors (Cat. No.98CB36273)* (pp. 82–88). https://doi.org/10.1109/ICCD.1998.727027
26. Karl, E., et al. (2012). A 4.6GHz 162Mb SRAM design in 22nm tri-gate CMOS technology with integrated active VMIN-enhancing assist circuitry. In *2012 IEEE International Solid-State Circuits Conference* (pp. 230–232). https://doi.org/10.1109/ISSCC.2012.6176988

27. Karl, E. et al. (2012). The impact of assist-circuit design for 22nm SRAM and beyond. In *2012 International Electron Devices Meeting* (pp. 25.1.1–24.1.4). https://doi.org/10.1109/IEDM.2012.6479099
28. Kerber, A., & Cartier, E. A. (2009). Reliability challenges for CMOS technology qualifications with hafnium oxide/titanium nitride gate stacks. *IEEE Transactions on Device and Materials Reliability, 9*(2), 147–162. https://doi.org/10.1109/TDMR.2009.2016954
29. Keyes, R. W. (1969). Physical problems and limits in computer logic. *IEEE Spectrum, 6*(5), 36–45. https://doi.org/10.1109/MSPEC.1969.5213809
30. Khalil, D., et al. (2008). Accurate estimation of SRAM dynamic stability. *IEEE Transactions on Very Large Scale Integration (VLSI) Systems, 16*(12), 1639–1647. https://doi.org/10.1109/TVLSI.2008.2001941
31. Kim, N. S., et al. (2003). Leakage current: Moore's law meets static power. In *Computer, 36*(12), 68–75. https://doi.org/10.1109/MC.2003.1250885
32. Liu, H.-H., et al. (2022). Extended methodology to determine SRAM write margin in resistance-dominated technology node. *IEEE Transactions on Electron Devices, 69*(6), 3113–3117. https://doi.org/10.1109/TED.2022.3165738
33. Liu, H.-H., et al. (2023). CFET SRAM DTCO, interconnect guideline, and benchmark for CMOS scaling. *IEEE Transactions on Electron Devices, 70*(3), 883–890. https://doi.org/10.1109/TED.2023.3235701
34. Liu, H.-H., et al. (2023). CFET SRAM with double-sided interconnect design and DTCO benchmark. *IEEE Transactions on Electron Devices, 70*, 1–8. https://doi.org/10.1109/TED.2023.3305322
35. Liu, H.-H., et al. (2023). DTCO of sequential and monolithic CFET SRAM. In R.-H. Kim & N. V. Lafferty (Eds.), *DTCO and computational patterning II* (Vol. 12495, p. 124950Z). International Society for Optics and Photonics. SPIE. https://doi.org/10.1117/12.2657524
36. Makino, H., et al. (2011). Reexamination of SRAM cell write margin definitions in view of predicting the distribution. *IEEE Transactions on Circuits and Systems II: Express Briefs, 58*(4), 230–234. https://doi.org/10.1109/TCSII.2011.2124531
37. Mann, R. W., et al. (2010). Limits of bias based assist methods in nano-scale 6T SRAM. In *2010 11th International Symposium on Quality Electronic Design (ISQED)* (pp. 1–8). https://doi.org/10.1109/ISQED.2010.5450413
38. Mukhopadhyay, S., et al. (2008). Capacitive coupling based transient negative bit-line voltage (Tran-NBL) scheme for improving write-ability of SRAM design in nanometer technologies. In *2008 IEEE International Symposium on Circuits and Systems (ISCAS)* (pp. 384–387). https://doi.org/10.1109/ISCAS.2008.4541435
39. Mukhopadhyay, S., et al. (2011). SRAM write-ability improvement with transient negative bit-line voltage. *IEEE Transactions on Very Large Scale Integration (VLSI) Systems, 19*(1), 24–32. https://doi.org/10.1109/TVLSI.2009.2029114
40. Muralimanohar, N., Balasubramonian, R., & Jouppi, N. (2009). Cacti 6.0: A tool to model large caches. In *HP Laboratories*.
41. Pavlov, A., & Sachdev, M. (2008). *CMOS SRAM circuit design and parametric test in nano-scaled technologies: Process-Aware SRAM design and test* (1st ed.). Springer. ISBN: 14-0-2083-62-9.
42. Pei, Z., et al. (2023). Emerging interconnect exploration for SRAM application using non-conventional H-tree and center-pin access. In *2023 24th International Symposium on Quality Electronic Design (ISQED)* (pp. 1–1). https://doi.org/10.1109/ISQED57927.2023.10129316
43. Pei, Z., et al. (2023). Graphene-based interconnect exploration for large SRAM caches for ultrascaled technology nodes. *IEEE Transactions on Electron Devices, 70*(1), 230–238. https://doi.org/10.1109/TED.2022.3225512
44. Pei, Z., et al. (2023). Technology/memory Co-design and co-optimization using E-tree interconnect. In *Proceedings of the Great Lakes Symposium on VLSI 2023*. GLSVLSI '23 (pp. 159–162). Association for Computing Machinery. ISBN: 979-84-00701-25-2. https://doi.org/10.1145/3583781.3590311

45. Pilo, H., et al. (2011). A 64Mb SRAM in 32nm High-k metal-gate SOI technology with 0.7V operation enabled by stability, write-ability and read-ability enhancements. In *2011 IEEE International Solid-State Circuits Conference* (pp. 254–256). https://doi.org/10.1109/ISSCC.2011.5746307
46. Qin, H., et al. (2003). SRAM leakage suppression by minimizing standby supply voltage. In *International Symposium on Signals, Circuits and Systems. Proceedings, SCS 2003. (Cat. No.03EX720)* (pp. 55–60). https://doi.org/10.1109/ISQED.2004.1283650
47. Rahman, A., et al. (2018). Reliability studies of a 10nm high-performance and low-power CMOS technology featuring 3rd generation FinFET and 5th generation HK/MG. In *2018 IEEE International Reliability Physics Symposium (IRPS)* (pp. 6F.4-1-6F.4–6). https://doi.org/10.1109/IRPS.2018.8353648
48. Rezaei, E., et al. (2020). Fundamental thermal limits on data retention in low-voltage CMOS latches and SRAM. *IEEE Transactions on Device and Materials Reliability, 20*(3), 488–497. https://doi.org/10.1109/TDMR.2020.2996627
49. Schuddinck, P., et al. (2022). PPAC of sheet-based CFET configurations for 4 track design with 16nm metal pitch. In *2022 IEEE Symposium on VLSI Technology and Circuits (VLSI Technology and Circuits)* (pp. 365–366). https://doi.org/10.1109/VLSITechnologyandCir46769.2022.9830492
50. Seevinck, E., List, F. J., & Lohstroh, J. (1987). Static-noise margin analysis of MOS SRAM cells. *IEEE Journal of Solid-State Circuits, 22*(5), 748–754. https://doi.org/10.1109/JSSC.1987.1052809
51. Shafaei, A., & Pedram, M. (2016). Energy-efficient cache memories using a dual-Vt 4T SRAM cell with read-assist techniques. In *2016 Design, Automation Test in Europe Conference Exhibition (DATE)* (pp. 457–462).
52. Shin, C. (2016). *Variation-aware advanced CMOS devices and SRAM* (Vol. 56). ISBN: 978-94-01775-95-3. https://doi.org/10.1007/978-94-017-7597-7
53. Sinangil, M. E., Mair, H., & Chandrakasan, A. P. (2011). A 28nm high-density 6T SRAM with optimized peripheral-assist circuits for operation down to 0.6V. In *2011 IEEE International Solid-State Circuits Conference* (pp. 260–262). https://doi.org/10.1109/ISSCC.2011.5746310
54. Song, T., et al. (2014). 13.2 A 14nm FinFET 128Mb 6T SRAM with VMIN-enhancement techniques for low-power applications. In *2014 IEEE International Solid-State Circuits Conference Digest of Technical Papers (ISSCC)* (pp. 232–233). https://doi.org/10.1109/ISSCC.2014.6757413
55. Song, T., et al. (2017). A 10 nm FinFET 128 Mb SRAM with assist adjustment system for power, performance, and area optimization. *IEEE Journal of Solid-State Circuits, 52*(1), 240–249. https://doi.org/10.1109/JSSC.2016.2609386
56. Synopsys (2020). *QuickCap R-2020.09-SP4*. https://www.synopsys.com/implementation-and-signoff/signoff/quickcap-nx.html
57. Synopsys® (2023). *Synopsys®HSPICE*. https://www.synopsys.com/
58. Synopsys® (2023). *Synopsys®Raphael*. https://www.synopsys.com/manufacturing/tcad/interconnect-simulation/raphael.html
59. Takeda, K., et al. (2006). Redefinition of write margin for next-generation SRAM and write-margin monitoring circuit. In *2006 IEEE International Solid State Circuits Conference - Digest of Technical Papers* (pp. 2602–2611). https://doi.org/10.1109/ISSCC.2006.1696326
60. Thoziyoor, S., et al. (2008). *CACTI 5.1*.
61. Toh, S. O., et al. (2011). Characterization of dynamic SRAM stability in 45 nm CMOS. *IEEE Journal of Solid-State Circuits, 46*(11), 2702–2712. https://doi.org/10.1109/JSSC.2011.2164300
62. Wang, J., Nalam, S., & Calhoun, B. H. (2008). Analyzing static and dynamic write margin for nanometer SRAMs. In *Proceedings of the 2008 International Symposium on Low Power Electronics & Design*. ISLPED '08 (pp. 129–134). Association for Computing Machinery. ISBN: 9781605581095. https://doi.org/10.1145/1393921.1393954

63. Wann, C., et al. (2005). SRAM cell design for stability methodology. In *IEEE VLSI- TSA International Symposium on VLSI Technology, 2005. (VLSI-TSA-Tech).* (pp. 21–22). https://doi.org/10.1109/VTSA.2005.1497065
64. Weckx, P., et al. (2019). Novel forksheet device architecture as ultimate logic scaling device towards 2nm. In *2019 IEEE International Electron Devices Meeting (IEDM)* (pp. 36.5.1–36.5.4). https://doi.org/10.1109/IEDM19573.2019.8993635
65. Weste, N., & Harris, D. (2010). *CMOS VLSI design: A circuits and systems perspective* (4th ed.). Addison-Wesley Publishing. ISBN: 03-2-1547-74-8
66. Yabuuchi, M. et al. (2009). A 45nm 0.6V cross-point 8T SRAM with negative biased read/write assist. In *2009 Symposium on VLSI Circuits* (pp. 158–159).
67. Yahya, F. B., et al. (2015). Combined SRAM read/write assist techniques for near/sub-threshold voltage operation. In *2015 6th Asia Symposium on Quality Electronic Design (ASQED)* (pp. 1–6). https://doi.org/10.1109/ACQED.2015.7273998
68. Yamauchi, H. (2010). A discussion on SRAM circuit design trend in deeper nanometer-scale technologies. *IEEE Transactions on Very Large Scale Integration (VLSI) Systems, 18*(5), 763–774. https://doi.org/10.1109/TVLSI.2009.2016205
69. Yoshimoto, M., et al. (1983). A divided word-line structure in the static RAM and its application to a 64K full CMOS RAM. *IEEE Journal of Solid-State Circuits, 18*(5), 479–485. https://doi.org/10.1109/JSSC.1983.1051981
70. Zhang, B., et al. (2006). Analytical modeling of SRAM dynamic stability. In *2006 IEEE/ACM International Conference on Computer Aided Design* (pp. 315–322). https://doi.org/10.1109/ICCAD.2006.320052
71. Zhang, K., et al. (2006). A 3-GHz 70-mb SRAM in 65-nm CMOS technology with integrated column-based dynamic power supply. *IEEE Journal of Solid-State Circuits, 41*(1), 146–151. https://doi.org/10.1109/JSSC.2005.859025
72. Zimmer, B., et al. (2012). SRAM assist techniques for operation in a wide voltage range in 28-nm CMOS. *IEEE Transactions on Circuits and Systems II: Express Briefs, 59*(12), 853–857. https://doi.org/10.1109/TCSII.2012.2231015

Chapter 3
Extended Write Margin Methodology for SRAM Subarray in Resistance-Dominated Technology Node

In the sub-10 nm node, the increased bitline (BL) resistance of static random-access memory (SRAM) bitcell poses challenges to an efficient write operation. The rise in BL resistance hinders the complete discharge of BL during a write operation. Furthermore, the write ability is undermined by the augmented leakage current observed in half-selected bitcells situated on both the BL and the BL capacitance. This issue becomes particularly pronounced when operating at higher frequencies. The BL parasitics also cause 30% SRAM yield loss for nanosheet (NS) FET SRAM subarray in 14Å-compatible technology (A14) [9]. As indicated in Chap. 2, the current methodology for determining write margin (WM) lacks sufficient consideration of the comprehensive effects of BL parasitics. To enhance the accuracy of write failure prediction beyond existing methods, an extended write-ability methodology for SRAM subarray in advanced technology nodes is proposed in this chapter. This proposed method analyses the time-dependent impacts of BL parasitic resistors (R), capacitors (C), and pass-gate (PG) transistors on WM encompassing the negative bitline (NBL) write-assist technique [14, 15]. For brief overviews of NBL technique, simulation circuit, and simulation methodology, please refer to Chap. 2.

This chapter aims to elaborate the challenges and necessity of employing an extended WM methodology within resistance-dominated technology nodes, specifically focusing on the A14 SRAM technology. Notably, this proposed methodology demonstrates potential applicability in technology nodes beyond A14, owing to the coverage of similar RC values outlined in this chapter. Section 3.1 discusses the new challenges posed by previous WM methods in ultra-scaled nodes and the motivation behind this chapter. Section 3.2 elaborates on the proposed WM methodology, building upon the conventional bitline write margin (BLWM) methods [13, 25, 28]. Section 3.3 demonstrates the impact of BL parasitic PG transistors, parasitic capacitors, overall parasitic circuitry, and different operating frequencies on WM. Section 3.4 compares the conventional BLWM and proposed WM methods. Finally, Sect. 3.5 concludes this chapter.

H.-H. Liu, F. Catthoor, *Circuit-Technology Co-Optimization of SRAM Design in Advanced CMOS Nodes*, https://doi.org/10.1007/978-3-031-76109-6_3

3.1 New Challenges Arising From Deeply Scaled Technology Nodes

Continued geometric scaling of devices in advanced CMOS technology intensifies process variation and bolsters the driver strength of PMOS transistors [4, 15, 16, 22, 23]. These factors are becoming the primary sources of SRAM subarray's write ability deterioration [3, 4, 9, 15, 17]. To enhance the write ability, circuit assist techniques such as NBL have been introduced [4, 15, 22, 23]. However, the effectiveness of NBL remains limited by the rapid increase in BL resistance within sub-10 nm technology nodes due to metal geometric scaling and surface scattering [16, 22]. The modulation of write driver voltage (V_{WD}) through the NBL assist circuit employs a boost capacitor (C_{boost}) [15]. As discussed in Sect. 2.5 of Chap. 2, the necessity for a stronger write-assist technique accompanied by additional area and energy overhead caused by larger C_{boost} becomes inevitable as the WM deteriorates. Moreover, a higher level of degradation in WM signifies an increased risk of write failure and may ultimately break the SRAM functionality even when utilizing write-assist techniques. Hence, identifying a methodology to accurately evaluate WM value is crucial for assessing write failure probability and the consequential impact of write-assist techniques for further design-technology co-optimization (DTCO) analysis of this book.

Several approaches have been proposed to evaluate the static WM of SRAM bitcell, including voltage transfer curves (VTC) [25], sweeping BL [28] and wordline (WL) voltages [6, 24], and N-curve [7]. However, these methodologies often assume an infinite WL pulse width disregarding time dependency, which may underestimate write failures [2]. In response, dynamic WM analysis [5, 27] was introduced to consider dynamic noise sources, finite WL pulse widths, and transient behaviours. Nevertheless, existing static and dynamic methodologies [2, 5–7, 13, 24, 25, 27, 28] only perform simulations at the bitcell level, overlooking the interplay between each signal pulse and BL parasitics within the subarray.

During a typical NBL process, BL discharge coincides with WL activation [3, 4], making the WM a result of BL parasitic impacts and SRAM subarray operational speeds. As depicted in Fig. 3.1, achieving a successful data flip requires higher NBL voltage when facing greater BL parasitic resistance (R_{BL}) and capacitance (C_{BL}). Otherwise, an increased data flip time or write failure may occur. Typically, R_{BL} and C_{BL} range from 10 to 100 Ω/cell and remain below 50 aF/cell, respectively [10, 11, 16]. Therefore, R_{BL} of 20 Ω/cell and 40 Ω/cell, and C_{BL} of 25 aF/cell and 50 aF/cell are taken as references for an initial examination in Fig. 3.1. Notably, the effects of BL parasitic circuits under NBL write process are also determined by the BL ramp-down speed as discussed in Sect. 3.4. These time-dependent effects of BL parasitic components might be negligible in less-scaled nodes, but they become pivotal in advanced resistance-dominated technology nodes as demonstrated in Fig. 3.1. Moreover, they are only observable in the subarray-level simulations, not at the bitcell level. Failing to consider both factors could

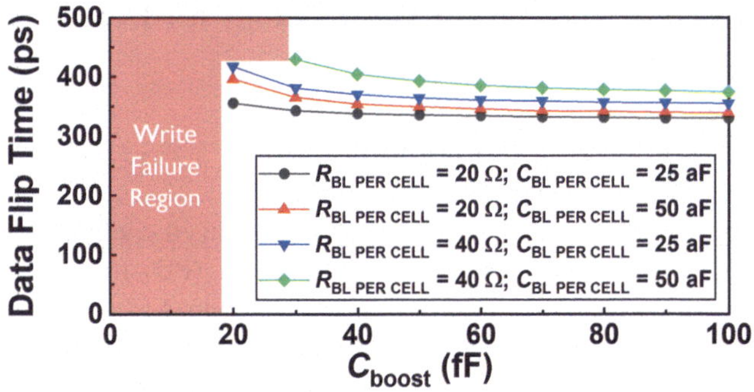

Fig. 3.1 Data flip time with varying C_{boost}, R_{BL}, and C_{BL} of HD SRAM subarray design for a typical NBL write operation. The requirement of NBL voltage (controlled by C_{boost}) for a successful data flip is increased with larger BL parasitics

lead to an overly optimistic estimation of WM values in resistance-dominated nodes. Therefore, accurately predicting the time-dependent impacts of BL parasitics on write operations necessitates the development of a tailored WM methodology specifically designed for deeply scaled nodes.

Takeaways

- **Challenges in Existing WM Methodologies:**
  - **Increased Resistance in Advanced Technology Nodes:** As technology advances, resistance increases significantly. It is crucial to consider the full BL parasitic circuits to accurately assess time-dependent impacts.
  - **Missing Subarray-Level Impacts:** Current WM methodologies often consider bitcell-level simulation setup but neglect the peripheral circuits, particularly the write driver. This omission can obscure the interplay between signal pulses and BL parasitics within the subarray.
- **Motivated Design Strategy:**
  - **Complete Simulation Circuit Setups:** Integrate the complete BL parasitic circuit (R, C, and PG devices) and peripheral circuits (timing control, write driver, etc) into the simulation.
  - **Realistic BL Timing Settings:** Establish an appropriate time-dependent BL discharging setting to accurately reflect realistic NBL write-assist techniques.

3.2 Proposed Write Margin Methodology for Deeply Scaled Technology Nodes

3.2.1 SRAM Bitcell Specification

In this chapter, we adopt a 6-transistor (6T) SRAM bitcell configuration featuring 2 pull-up (PU), 2 pass-gate (PG), and 2 pull-down (PD) transistors for high-density (HD) and high-performance (HP) applications in A14 NSFET technology. SRAM bitcell area scaling represents a common strategy for enhancing power, performance, and area (PPA) metrics across successive technology nodes in system-on-chip (SoC) designs [19–21]. Therefore, the HD option emerges as a favourable choice to attain maximal bitcell area density, particularly for on-chip caches such as the last-level cache (LLC) with large capacities. Furthermore, maintaining sufficiently high speed is crucial for frequently accessed and low-capacity caches such as registers or level 1 (L1) caches. Therefore, the incorporation of both HD and HP options is deemed essential within this chapter.

The HD and HP SRAM bitcell layout designs and their design rules using A14 four-stacked NSFETs are presented in Fig. 3.2 and Table 3.1 [18], excluding consideration of the buried power rail (BPR) technique [16]. Table 3.2 lists the nanosheet width (W_{NS}) ratio (PU:PG:PD), BL and WL back-end-of-line (BEOL) parasitic values extracted from the layout by Calibre tool, and array configuration of HD and HP subarray designs. The HD subarray design features a W_{NS} ratio of 1:1:1

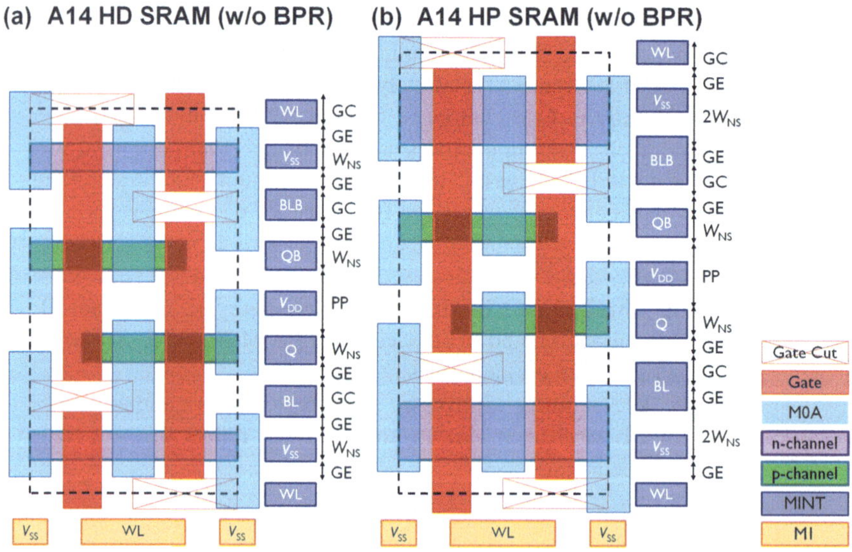

Fig. 3.2 Schematic views of (**a**) HD and (**b**) HP A14 NSFET SRAM bitcell layout designs (without BPR)

Table 3.1 Design rules of A14 NSFET SRAM bitcell layout design [18] (without BPR)

Parameters	Value (nm)
Gate Length (L_g)	15
Metal Pitch (MP)	18
Contacted Poly Pitch (CPP)	42
Gate Extension (GE)	7.5
Gate Cut (GC)	12
P-channel-to-P-channel Separation (PP)	25
Nanosheet Width (W_{NS})	11
Nanosheet Height (H_{NS})	5
Number of NS Stacks (NNS)	4
BL/WL Metal Thickness	27/24
BL Metal Width for HD/HP	12/18.5
WL Metal Width for HD/HP	42/42

Table 3.2 Nanosheet width (W_{NS}) ratio, BL and WL parasitic values, and array configuration of HD and HP A14 NSFET SRAM subarray designs (without BPR)

Type	W_{NS} Ratio	R_{BL}/R_{WL} (Ω/cell)	C_{BL}/C_{WL} (aF/cell)	Configuration
HD	1:1:1	38/20	30/79	256 bits/WL; 256 bits/BL
HP	1:2:2	18/23	33/86	256 bits/WL; 128 bits/BL

with 256-bit WL and BL. On the other hand, the HP subarray design has a W_{NS} ratio of 1:2:2 with 256-bit WL and 128-bit BL [4, 16, 22, 23]. The internal nodes in the SRAM bitcell Q and QB are initialized to store '1' and '0', respectively. This initialization aligns with the write driver enabling the BL path for negative voltage application. Reversing this setting in a symmetric SRAM bitcell design would yield identical results. It is noteworthy that parasitic RC extraction in this chapter is performed using Calibre rather than Quickcap as referenced in Chap. 2. This choice stems from the unavailability of the Quickcap tool during the duration of this research. Therefore, the extracted parasitic RC data (Table 3.2) only contain the BEOL component within this chapter.

3.2.2 Simulation Setup

During write operation, the BL voltage (V_{BL}) is pulled down to a lower voltage to write '0', while BLB remains at the supply voltage (V_{DD}) of 0.7 V to write '1'. Simultaneously, the WL is asserted to activate the PG transistors. To facilitate the write process, tri-state inverters function as the write driver, and the voltage source for lowering BL is directly connected to this write driver. The BL discharge commences when the WL is triggered, allowing us to capture the NBL characteristics detailed in [3, 4]. To avoid write failures, the ramp-down time (t_{BL}) lasts

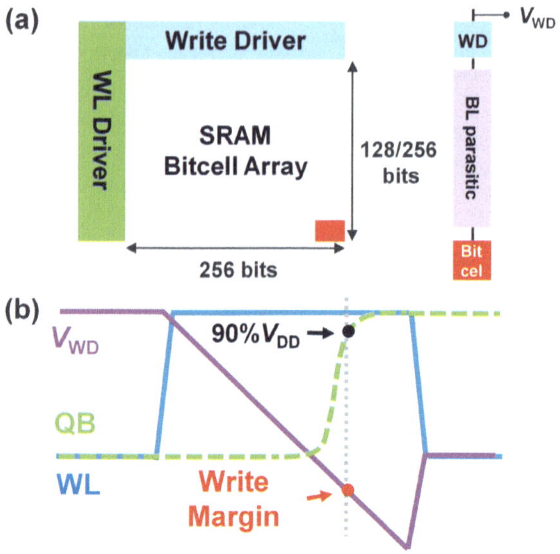

Fig. 3.3 Schematic views of (**a**) the locations of V_{WD} and the worst-case bitcell under test; and (**b**) the waveforms and definition of WM

for the entire WL enable period and follows a certain ramp-down slope (S_{BL} = $-$voltage/t_{BL}). In this chapter, S_{BL} of -2 V/ns is assumed to represent an SRAM macro design operating at 2 GHz frequency for write operations in A14. Notably, the value of 2 GHz is derived from published research [3, 4]. The WM is defined as the maximum V_{WD} which confirms data flip when node '0' reaches 90% of its full swing. This criterion applies to the bitcell situated farthest from both the WL driver and write driver (Fig. 3.3).

Process and mismatch variations are evaluated via Monte Carlo (MC) simulations using Cadence tools. To mitigate extensive runtime demands associated with complete memory and logic circuits, a dedicated SRAM subarray circuit is purpose-built for WM extraction. This specialized simulation circuit includes timing control, the last stage buffer of row decoder, transmission gate, write driver, WL and BL parasitic circuits, and precharge circuits. Furthermore, the device characteristics and BEOL parasitic circuits have been calibrated to ensure simulation accuracy [8]. The simulations focus on mean minus 6-sigma ($\mu - 6\sigma$) points and are conducted at the temperature of 80 °C to ensure the worst-case conditions for write operation. The A14 compact model and MC simulation strategy are detailed in Sect. 2.1 of Chap. 2. The 6σ range is established through the analysis of process variation and device mismatch, calibrated using silicon data and TCAD results [1, 8]. Notably, no alterations to the bitcell design are necessary, as the bitcell simulation netlist is determined by the variations provided in compact model but not Quickcap tool.

Takeaways

- **Proposed WM Methodologies:**
  - Incorporate R, C, and PG devices into simulation circuits to assess the impact of BL parasites.
  - A customized SRAM subarray circuit is designed specifically for WM extraction, with timing control, WL driver, and write driver setup. To keep the simulation time manageable, only circuits on the crucial path that affect WM findings are included.
  - Adopt a novel BL ramp-down timing setup (S_{BL} and t_{BL}) that can adjust to different application speeds.

3.3 Results and Discussion

This section investigates the distinct impacts of BL parasitic components on an SRAM subarray, using A14 as an initial case study in deeply scaled nodes. Section 3.3.1 illustrates how the current flow from internal nodes of half-selected bitcells to BL through inadvertently activated PG devices during BL discharge impedes data flipping, leading to degraded WM. Furthermore, high-frequency operations exacerbate WM degradation due to voltage drops on BL caused by parasitic capacitors, as described in Sect. 3.3.2. Section 3.3.3 underscores the importance of incorporating complete BL parasitic circuits (R, C, and PG devices) into simulations to avoid underestimating the yield of successful write operation. In Sect. 3.3.4, the impact of R and C under varying operating frequencies on WM is presented, emphasizing the significance of proper S_{BL} settings to accurately assess WM across different technologies and applications. Section 3.3.5 concludes that V_{DD} should remain at the nominal value of 0.7 V for the forthcoming analysis in other chapters. It is important to note that these time-dependent aspects also manifest in more advanced technologies such as forksheet (FS) [26] and complementary transistor (CFET) technologies [10–12] presented in this book, owing to their comparable RC values (Sect. 6.2 of Chap. 6) to those discussed in this chapter. To concentrate on the impacts linked between BL parasitics and SRAM bitcell area scaling, the proposed WM methodology at constant SBL is applied across A14 to 3-Å-compatible (A3) SRAM technologies in subsequent chapters. As the SRAM bitcell area shrinks beyond A3, the BL parasitic effects may become more pronounced due to the potential increase in R_{BL}. Future studies should include a comprehensive analysis to confirm the applicability of employing this proposed WM methodology for advanced SRAM technologies beyond the scope of this dissertation.

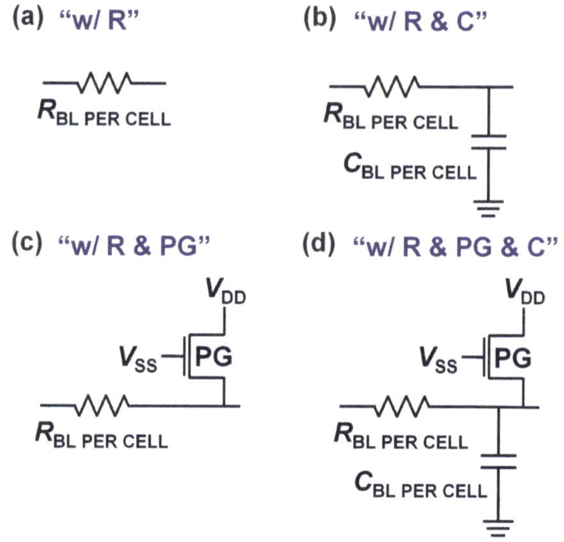

Fig. 3.4 Schematic view of BL parasitic circuits including (**a**) "w/ R", (**b**) "w/ R & C", (**c**) "w/ R & PG", and (**d**) "w/ R & PG & C". To ensure a worst-case analysis, the source voltage of PG transistors is set as V_{DD}

Due to the notably increased R_{BL}, there is a significant voltage drop in the write current, preventing V_{BL} from fully discharging to zero voltage. Hence, it is crucial to consider the effects of R_{BL}. To assess the impacts of different BL parasitic elements on WM, the combinations of resistor (R), PG transistor, and capacitor (C) are categorized into four configurations: "w/ R", "w/ R & PG", "w/ R & C", and "w/ R & PG & C". Figure 3.4 illustrates the schematic view of BL parasitic circuits consisting of BL resistance per bitcell ($R_{BL\ PER\ CELL}$), PG transistor, and BL capacitance per bitcell ($C_{BL\ PER\ CELL}$). In total, there are 128 and 256 segments for HP and HD subarray designs, respectively, within the complete BL parasitic circuits. Notably, the worst-case source voltage for PG transistors is set to V_{DD} to account for the maximum potential leakage current from the internal node to BL in case of unintended PG transistor activation.

3.3.1 Impact of BL Parasitic Pass-Gate (PG) Devices

Applying a negative bias to the falling BL can strengthen PG transistors connected to the internal node Q, which stores a high voltage [4, 15, 22, 23]. This improvement results in faster discharge times, thereby enhancing write ability. However, a larger R_{BL} necessitates a more negative V_{WD} to guarantee successful data flipping. A notable challenge arises from the parasitic PG transistors in half-selected bitcells on the same BL row. These transistors may unintentionally turn on, even when their WL is inactive, provided the gate-to-source voltage (V_{GS}) surpasses their threshold voltage (V_T) as depicted in Fig. 3.5. Thus, the current flowing from the internal node of the half-selected bitcells to the BL interferes with the discharging process.

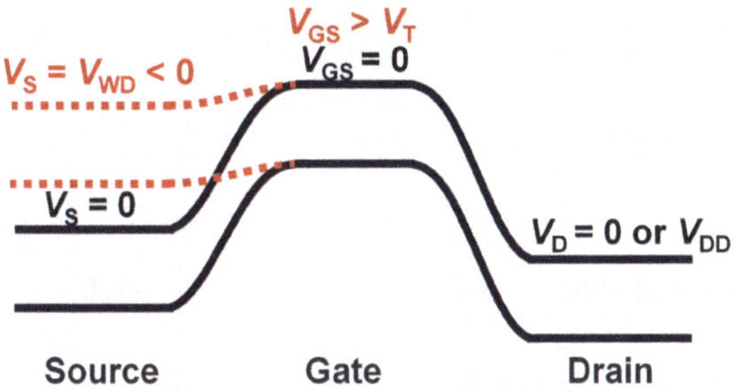

Fig. 3.5 Band diagrams of parasitic NMOS PG transistors of BL in the initial off states at $V_{GS} = 0$ V (solid black lines) and after applying an over-boosted negative V_{WD} (dashed red lines). If $V_{GS} > V_T$, the PG transistors will be turned on and cause current flow from drain to source

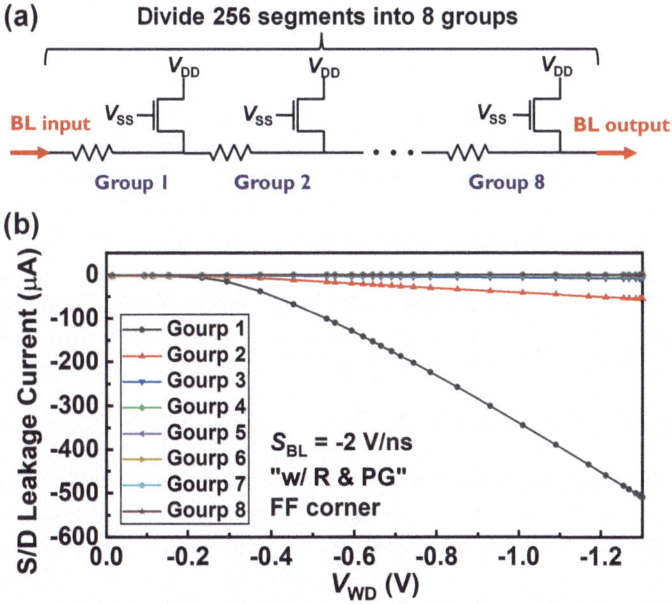

Fig. 3.6 Test circuit of PG leakage in HD subarray design. (**a**) Schematic view of BL parasitic circuits considering "w/ R & PG" using 8 groups to represent 256-bit BL. (**b**) S/D leakage current from PG with ramping down V_{WD} at S_{BL} of -2 V/t_{BL} with t_{BL} of 1 ns

For the evaluation of PG leakage in HD subarray design, tests are conducted at the fast NMOS and fast PMOS (FF) corners. The 256-segment BL parasitic circuits, denoted as "w/ R & PG" are segmented into eight groups, as depicted in Fig. 3.6a. The source-to-drain (S/D) leakage currents of PG transistors increase as V_{WD} ramps down negatively, as demonstrated in Fig. 3.6b. Notably, the amount of leakage current gradually decreases from group 1 to group 8 since the negative voltage input from the BL drops along the BL path. Figure 3.7 illustrates the influence of PG transistors on WM, considering varying $R_{BL\ PER\ CELL}$ and BL parasitics at tBL of

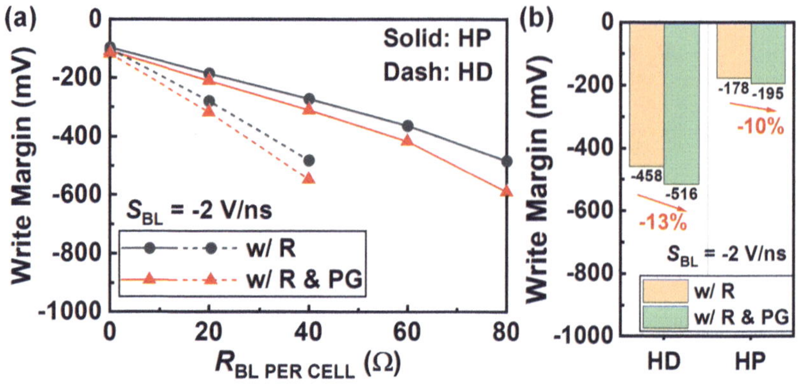

Fig. 3.7 WM with (**a**) varying $R_{BL\ PER\ CELL}$ and (**b**) BL parasitics in HD and HP A14 NSFET SRAM subarray designs considering "w/ R" and "w/ R & PG" BL parasitic circuits at S_{BL} of $-2\ V/t_{BL}$ with t_{BL} of 1 ns

1 ns in both HD and HP A14 NSFET SRAM subarray designs. In Fig. 3.7b, it is evident that the WMs of "w/ R & PG" are 13% and 10% worse than "w/ R" for HD and HP subarray designs, respectively. This occurs because V_{BL} no longer keeps discharging linearly with V_{WD}. Therefore, it is suggested to include PG transistors in the BL parasitic circuits to prevent underestimating the write failures when $|V_{WD}| > |V_T|$. This proposed WM methodology makes a significant contribution by elucidating the impact of PG leakage on WM of an SRAM subarray design.

Takeaways

- **PG Leakage:** Applying a negative bias for assisting write operation may unintentionally turn on PG devices of a half-selected bitcells. This would create a leakage path from the internal node to the BL, impeding the BL discharging process with slower writing speeds and indicating a deteriorated WM.
- **Inclusion of PG Device:** The simulation setup should take into account the PG devices in order to assess the effect of PG leakage on WM outcomes.

3.3.2 Impact of BL Parasitic Capacitors

Parasitic capacitors, when not fully charged during transient states, essentially act as short circuits. Consequently, the current flowing from the write driver to the bitcell also passes through these capacitors, leading to a voltage drop on BL during transient periods. This underscores the importance of considering BL parasitic

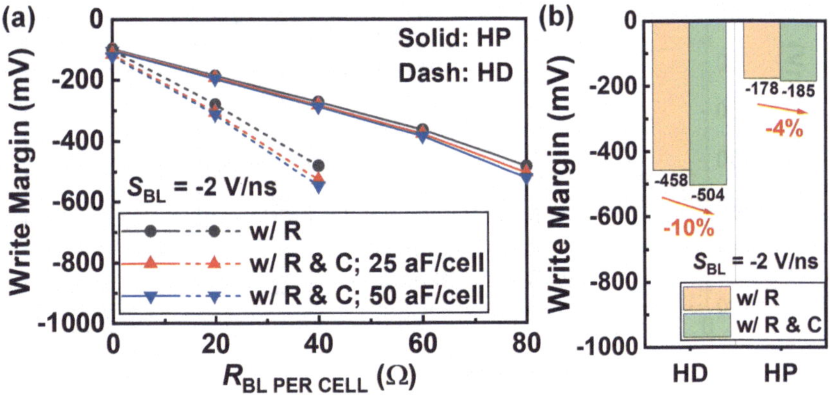

Fig. 3.8 WM with (**a**) varying $R_{BL\ PER\ CELL}$ and $C_{BL\ PER\ CELL}$, and (**b**) BL parasitics in HD and HP A14 NSFET SRAM subarray designs considering "w/ R" and "w/ R & C" BL parasitic circuits at S_{BL} of -2 V/t_{BL} with t_{BL} of 1 ns

capacitance (C_{BL}), particularly in high-frequency operations. As $R_{BL\ PER\ CELL}$ and $C_{BL\ PER\ CELL}$ increase, the internal node requires a more negative V_{WD} to successfully write data within 1 ns t_{BL} (Fig. 3.8a). When examining BL parasitics in A14 NSFET SRAM subarray, it becomes apparent that the WM variation is more significant in the HD subarray design (10%) compared to the HP counterpart (4%) due to the former's smaller write current and larger total R_{BL} (Fig. 3.8b). Note that the $C_{BL\ PER\ CELL}$ has been reduced to less than 50 aF through bitcell area scaling [17]; thus, 50 aF is used to be as an upper limit in the figure.

For a sufficiently long t_{BL} duration, the terminal voltage of parasitic capacitors rises to match the bias voltage and the current can no longer be drawn to the capacitors, rendering them effectively open circuits. Figure 3.9 demonstrates that the WMs in both HD and HP options remain consistent, even with varying $R_{BL\ PER\ CELL}$ and $C_{BL\ PER\ CELL}$, and with BL parasitics in A14 NSFET SRAM subarray design at t_{BL} of 10 μs. Therefore, the consideration of parasitic capacitance is advisable only when t_{BL} is short enough to maintain a transient state and when R_{BL} is significant enough to manifest the capacitors' effects. The subsequent section validates the conditions for the required parasitic capacitance in A14 NSFET SRAM subarray design.

Takeaways

- **Inclusion of BL Capacitors:** When the BL discharges 2 V every 1 ns, i.e., $S_{BL} = -2$ V/t_{BL} and $t_{BL} = 1$ ns, the capacitors will influence the WM. However, the BL capacitor has no effect on WM when BL discharges 2 V per 10 μs. Thus, BL capacitors should be included in the simulation parameters only when the write operation time is short enough.

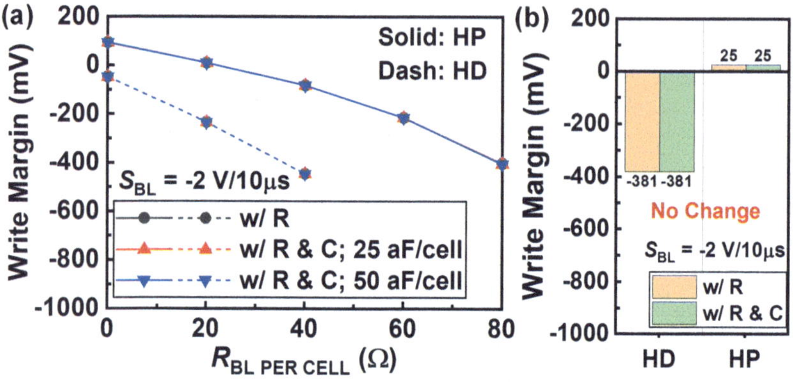

Fig. 3.9 WM with (**a**) varying $R_{\mathrm{BL\ PER\ CELL}}$ and $C_{\mathrm{BL\ PER\ CELL}}$, and (**b**) BL parasitics in HD and HP A14 NSFET SRAM subarray designs considering "w/ R" and "w/ R & C" BL parasitic circuits at S_{BL} of $-2\,\mathrm{V}/t_{\mathrm{BL}}$ with t_{BL} of 10 μs

3.3.3 Impact of Overall BL Parasitic Circuits

Figure 3.10 presents an overview of the collective impact of BL parasitics in both HD and HP subarray designs at t_{BL} of 1 ns. When using conventional BLWM for simulation, the WMs of 0 V are obtained for both HP and HD subarrays designs. This approach appears overly optimistic compared to the proposed method since it ignores all the time-dependent BL parasitics effects. Notably, the 6σ WMs of "w/ R & PG & C" experience a degradation of 17% and 16% compared to "w/ R" for HD and HP A14 NSFET SRAM subarray designs, respectively.

To validate the impact of BL parasitics on real write operations in relation to the WM settings, BL discharge is set to reach zero voltage, referring to the WM of 0 V given by BLWM. As shown in Fig. 3.11, the yield for successful data flipping in HP subarray design remains at 100% disregarding any BL parasitics. However, the yield significantly drops to only 70% when we incorporate all the BL parasitics into the HP subarray design at $S_{\mathrm{BL}} = -2\,\mathrm{V/ns}$. To ensure a realistic yield assessment and avoid overestimation, it is crucial to analyse the impact of BL parasitics on the WM using our extended methodology.

Takeaways

- **Inclusion of Complete BL Parasitic Circuits:** Incorporating whole BL parasitic circuits (R, C, and PG devices) into simulation circuits ensures realistic yield assessment and avoids underestimating the write failure risk.

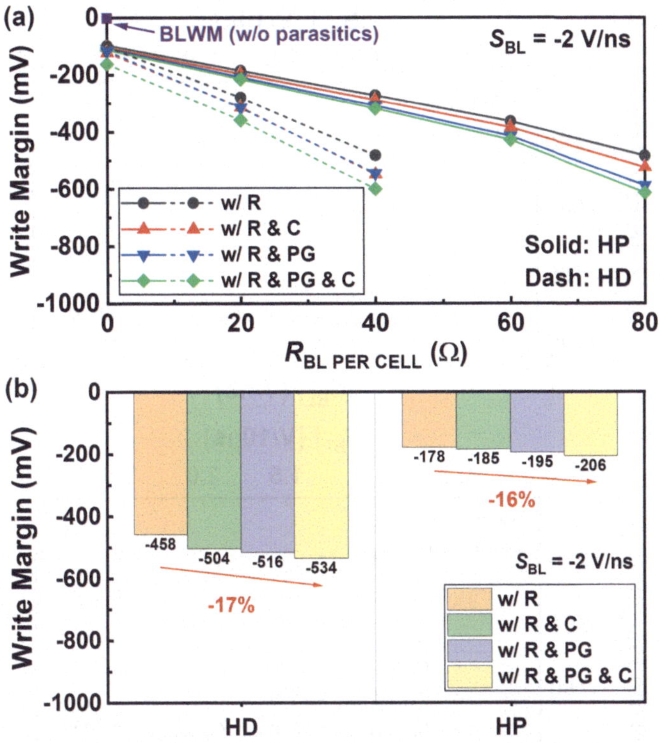

Fig. 3.10 WM with (**a**) varying $R_{\mathrm{BL\ PER\ CELL}}$ and (**b**) BL parasitics in HD and HP A14 NSFET SRAM subarray designs considering all the BL parasitic circuits (50 aF/cell) at S_{BL} of $-2\,\mathrm{V}/t_{\mathrm{BL}}$ with t_{BL} of 1 ns

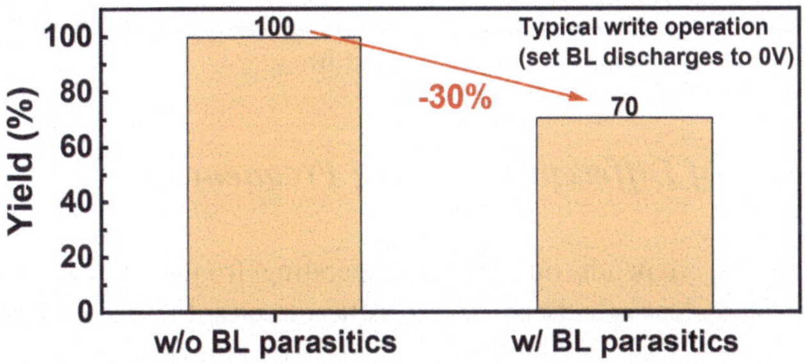

Fig. 3.11 Yield verification simulated in a typical write operation of HP A14 NSFET SRAM subarray design (left column: without BL parasitics; right column: with all the BL parasitics). Yield loss of 30% is obtained when considering the BL parasitics in the simulation

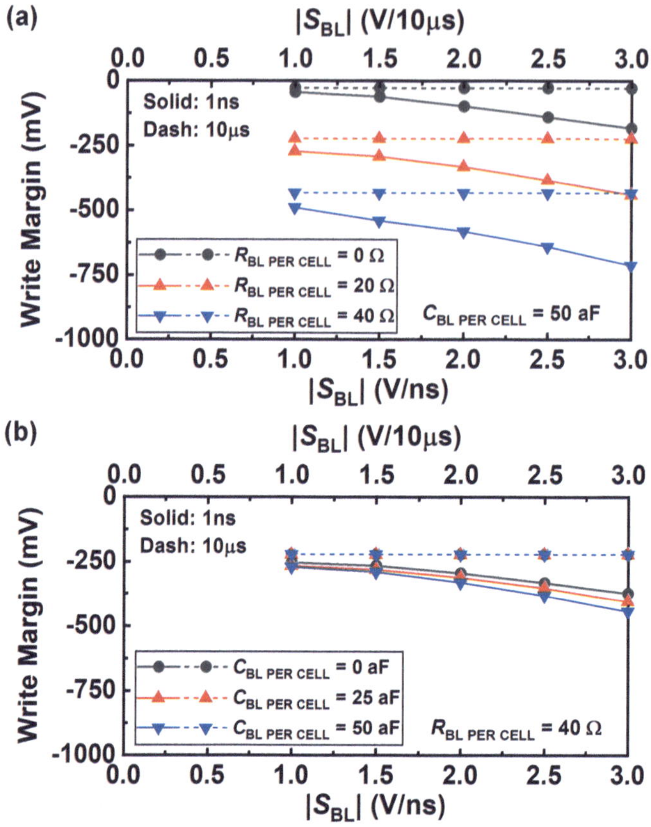

Fig. 3.12 WM with varying (**a**) $R_{\text{BL PER CELL}}$ (at fixed 50 aF/cell) and (**b**) $C_{\text{BL PER CELL}}$ (at fixed 40 Ω/cell) of HD A14 NSFET SRAM subarray designs considering "w/ R & PG & C" BL parasitic circuits at S_{BL} from 0 to $-3\ \text{V}/t_{\text{BL}}$ with t_{BL} of 1 ns and 10 μs

3.3.4 Impact of Different Operating Frequencies

The choice of S_{BL} depends on different operating frequencies, each exhibiting varying time-dependent effects of BL parasitic circuits. Figure 3.12 provides the impacts of S_{BL} at t_{BL} of 1 ns (solid line) and 10 μs (dashed line), while varying $R_{\text{BL PER CELL}}$ and $C_{\text{BL PER CELL}}$. As $|S_{\text{BL}}|$ increases within the frequency range of 1–3 GHz, the influence of R_{BL} and C_{BL} on WM intensifies. As a result, the WM degrades with larger $|S_{\text{BL}}|$. However, at slower frequencies ranging from 100 to 300 KHz, the increased R_{BL} still diminishes the WM. On the contrary, S_{BL} and C_{BL} exhibit negligible impact on the results. This highlights the significance of considering BL parasitics under specific discharging speeds in SRAM subarray design.

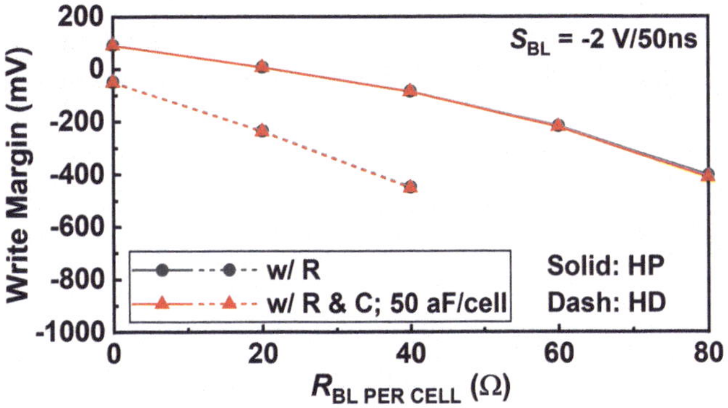

Fig. 3.13 WM with varying $R_{\mathrm{BL\ PER\ CELL}}$ of HD and HP A14 NSFET SRAM subarray designs considering "w/ R" and "w/ R & C (50 aF/cell)" BL parasitic circuits at S_{BL} of $-2\ \mathrm{V}/t_{\mathrm{BL}}$ with t_{BL} of 50 ns

As described in Sect. 3.3.2, it is confirmed that the parasitic capacitance is relevant only at sufficiently high frequencies. To further determine the range of S_{BL} that necessitates the inclusion of parasitic capacitors in BL parasitic circuits, the WMs with varying $R_{\mathrm{BL\ PER\ CELL}}$ at $C_{\mathrm{BL\ PER\ CELL}}$ of 50 aF and t_{BL} of 50 ns in HD and HP subarray designs (Fig. 3.13). As t_{BL} decreases from 10 μs to 1 ns, the reduction in WM is observed, and the variation between "w/ R" and "w/ R & C" becomes more pronounced. While reduced S_{BL} can improve data flipping speed, it comes at the cost of degrading the WM. However, the WM variation becomes negligible when t_{BL} exceeds 50 ns. At each $R_{\mathrm{BL\ PER\ CELL}}$, the difference between "w/ R" and "w/ R & C" is merely up to 1%. Consequently, parasitic C_{BL} can be neglected to reduce simulation time in SRAM macro design if S_{BL} exceeds -2 V/50 ns. In sub-10 nm nodes, the increased R_{BL} amplifies the effect of C_{BL}. Moreover, typical access time for SRAM macro in L1 and L2 cache are below 1 ns [4, 16, 22, 23]. Hence, the maximum S_{BL} is approximately around $-V_{\mathrm{DD}}$ V/ns without NBL. This implies that the parasitic capacitance should be considered for WM calculations in sub-10 nm nodes.

Takeaways

- **Application-Determined BL Discharging Speed:** As $|S_{\mathrm{BL}}|$ increases, the influence of R_{BL} and C_{BL} on WM intensifies. As a result, altering the S_{BL} to match the operation speed is critical for achieving a realistic impact on WM.
- **Criteria for including BL Capacitors:** The impact of BL capacitors on WM becomes significant if the BL discharge period is sufficiently short. For instance, this is evident when the BL discharges 2 V in 50 ns or less.

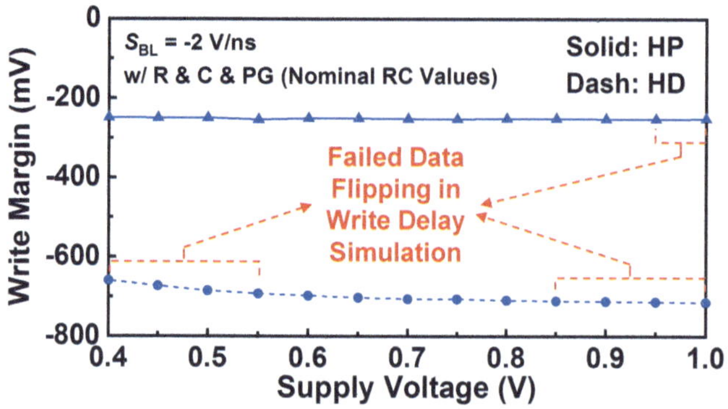

Fig. 3.14 WM with varying V_{DD} of HD and HP A14 NSFET SRAM subarray designs considering "w/ R & C & PG" BL parasitic circuits at S_{BL} of $-2\,V/t_{BL}$ with t_{BL} of 1 ns

3.3.5 Impact of Operating Supply Voltage

Figure 3.14 illustrates the impact of varying V_{DD} on WM for HD and HP subarray designs, incorporating complete BL parasitic components and values from Table 3.2. In the HD subarray, WM improves as V_{DD} scales down due to decreased contention current. As V_{DD} decreases to 0.4 V, WM shows a marginal improvement of 47 mV and 50 mV compared to 0.7 V and 1.0 V, respectively. Moreover, the insignificant difference in WM between each V_{DD} is less than 3 mV in the HP subarray. Consequently, V_{DD} scaling appears redundant when the write path is already optimized using the HP bitcell configuration.

Additionally, the worst-case write delay (WD) simulations at $-40\,°C$ are conducted to ascertain successful data flipping under excessive V_{DD} downscaling or upscaling. Decreasing or increasing V_{DD} leads to increased PG leakage or contention current, respectively, potentially resulting in write failure. As depicted in Fig. 3.14, successful write operations, considering both WM and WD simulations, occur within the V_{DD} range of 0.55–0.85 V for HD subarrays and 0.4–0.95 V for HP subarrays. Expectedly, HP subarrays demonstrate a better minimum V_{DD} for write operations due to their lower contention current. Furthermore, reducing V_{DD} may severely degrade read stability, posing challenges to maintaining SRAM functionality, particularly in deeply scaled nodes. Hence, it is recommended to maintain V_{DD} at the nominal value of 0.7 V for both read and write operations in this book.

Takeaways

- V_{DD} **Scaling is not Recommended:** Decreasing or increasing V_{DD} causes greater PG leakage or contention current, which may result in write failure. Furthermore, lowering V_{DD} may significantly degrade the read stability, making it difficult to maintain SRAM functionality in deeply scaled nodes. It is advised that V_{DD} be kept at the nominal value (0.7 V in this work) for both read and write operations.

3.4 Comparison Between Extended and Conventional BLWM

The primary contribution of the extended WM methodology presented in this chapter lies in its incorporation of complete BL parasitic circuits and a suitable S_{BL} setup mainly for deeply scaled nodes. Figure 3.15 compares the WM obtained through conventional and extended BLWM methodologies, both utilizing the same BL parasitic circuits on an identical SRAM subarray. The extended WM analysis is performed at S_{BL} of −2 V/ns with 50 aF/cell based on HP and HD subarray designs. The results from BLWM continue to exhibit an overly optimistic outlook, primarily because they disregard transient behaviours. Figure 3.15 illustrates that the conventional approach has overestimated the WM values by 112 mV, 317 mV, and 393 mV at 0 Ω/cell, 40 Ω/cell, and 80 Ω/cell in the HP subarray, respectively, compared to those in the proposed method. In the HD case, the conventional BLWM yields optimistically higher WM of 162 mV, 300 mV, and 283 mV at 0 Ω/cell, 20 Ω/cell, and 40 Ω/cell, respectively, compared to the proposed method. As a result, the extended WM methodology considering a realistic ramp-down slope setting is strongly recommended to avoid underestimating the negative impact of BL parasitics on WM. This approach proves essential for addressing significant time-dependent BL parasitic effects, especially in resistance-dominated nodes.

3.5 Conclusions

This chapter introduces an extended WM characterization methodology based on A14 SRAM subarray design in advanced BEOL resistance-dominated technology nodes. The implementation of BL ramp-down slope (S_{BL}) and time (t_{BL}) has been introduced in this proposed WM method to adapt to varying operating speeds. We demonstrated that the unintentionally activated PG transistors also degrade the WMs by 13% (HD) and 10% (HP) at S_{BL} of −2 V/ns. Furthermore, the WM variations caused by the parasitic capacitors are 10% (HD) and 4% (HP) at S_{BL} of −2 V/ns.

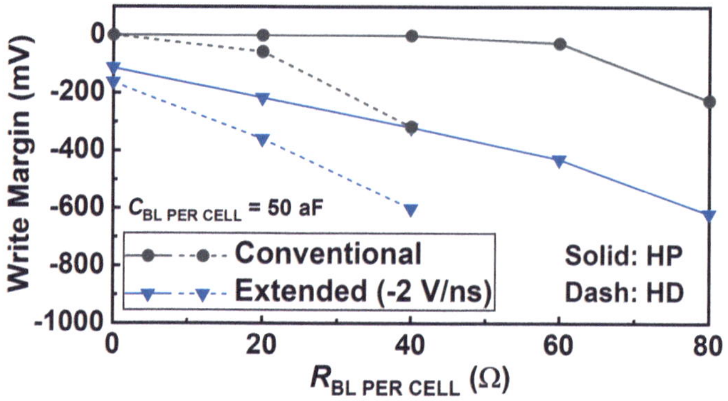

Fig. 3.15 WM comparison between conventional and extended (S_{BL} of -2 V/t_{BL} with t_{BL} of 1 ns) methods with varying $R_{BL\ PER\ CELL}$ of HD and HP A14 NSFET SRAM subarray designs. Note that both methods consider the same "w/ R & PG & C (50 aF/cell)" BL parasitic circuits

The time-dependent effects of BL parasitic circuits differ with SRAM subarray operating speeds, which can be referred to S_{BL} settings. Our results highlight the necessity of considering the complete BL parasitic circuits, accompanied by a proper BL ramp-down speed setup, for write-ability extraction in sub-10 nm nodes. Otherwise, the yield can be overestimated by 30% (in HP A14 NSFET SRAM subarray design) if WM is determined without accounting for time-dependent BL parasitic. The proposed WM methodology will be utilized in all book.

References

1. Bufler, F. M., et al. (2020). Monte Carlo comparison of n-type and p-type nanosheets with FinFETs: Effect of the number of sheets. *IEEE Transactions on Electron Devices, 67*(11), 4701–4704. https://doi.org/10.1109/TED.2020.3024267
2. Carmona, C., Torrens, G., & Alorda, B. (2014). SRAM write margin cell estimation using wordline modulation and read/write operations. In *2014 5th European Workshop on CMOS Variability (VARI)* (pp. 1–6). https://doi.org/10.1109/VARI.2014.6957077
3. Chang, J., et al. (2013). A 20nm 112Mb SRAM in high-metal-gate with assist circuitry for low-leakage and low-VMIN applications. In *2013 IEEE International Solid-State Circuits Conference Digest of Technical Papers* (pp. 316–317). https://doi.org/10.1109/ISSCC.2013.6487750
4. Chang, J., et al. (2017). 12.1 A 7nm 256Mb SRAM in high-k metal-gate FinFET technology with write-assist circuitry for low-VMIN applications. In *2017 IEEE International Solid-State Circuits Conference (ISSCC)* (pp. 206–207). https://doi.org/10.1109/ISSCC.2017.7870333
5. Dong, W., Li, P., & Huang, G. M. (2008). SRAM dynamic stability: Theory, variability and analysis. In *2008 IEEE/ACM International Conference on Computer-Aided Design* (pp. 378–385). https://doi.org/10.1109/ICCAD.2008.4681601
6. Gierczynski, N., et al. (2007). A new combined methodology for write-margin extraction of advanced SRAM. In *2007 IEEE International Conference on Microelectronic Test Structures* (pp. 97–100). https://doi.org/10.1109/ICMTS.2007.374463

7. Grossar, E., et al. (2006). Read stability and write-ability analysis of SRAM cells for nanometer technologies. *IEEE Journal of Solid-State Circuits, 41*(11), 2577–2588. https://doi.org/10.1109/JSSC.2006.883344
8. Jang, D., et al. (2017). Device exploration of nanosheet transistors for sub-7-nm technology node. *IEEE Transactions on Electron Devices, 64*(6), 2707–2713. https://doi.org/10.1109/TED.2017.2695455
9. Liu, H.-H., et al. (2022). Extended methodology to determine SRAM write margin in resistance-dominated technology node. *IEEE Transactions on Electron Devices, 69*(6), 3113–3117. https://doi.org/10.1109/TED.2022.3165738
10. Liu, H.-H., et al. (2023). CFET SRAM DTCO, interconnect guideline, and benchmark for CMOS scaling. *IEEE Transactions on Electron Devices, 70*(3), 883–890. https://doi.org/10.1109/TED.2023.3235701
11. Liu, H.-H., et al. (2023). CFET SRAM with double-sided interconnect design and DTCO benchmark. *IEEE Transactions on Electron Devices, 70*, 1–8. https://doi.org/10.1109/TED.2023.3305322
12. Liu, H.-H., et al. (2023). DTCO of sequential and monolithic CFET SRAM. In R.-H. Kim & N. V. Lafferty (Eds.), *DTCO and computational patterning II* (Vol. 12495, p. 124950Z). International Society for Optics and Photonics. SPIE. https://doi.org/10.1117/12.2657524
13. Makino, H., et al. (2011). Reexamination of SRAM cell write margin definitions in view of predicting the distribution. *IEEE Transactions on Circuits and Systems II: Express Briefs, 58*(4), 230–234. https://doi.org/10.1109/TCSII.2011.2124531
14. Mukhopadhyay, S., et al. (2008). Capacitive coupling based transient negative bit-line voltage (Tran-NBL) scheme for improving write-ability of SRAM design in nanometer technologies. In *2008 IEEE International Symposium on Circuits and Systems (ISCAS)* (pp. 384–387). https://doi.org/10.1109/ISCAS.2008.4541435
15. Mukhopadhyay, S., et al. (2011). SRAM write-ability improvement with transient negative bit-line voltage. *IEEE Transactions on Very Large Scale Integration (VLSI) Systems 19*(1), 24–32. https://doi.org/10.1109/TVLSI.2009.2029114
16. Salahuddin, S., et al. (2020). Buried power SRAM DTCO and system-level bench-marking in N3. In *2020 IEEE Symposium on VLSI Technology* (pp. 1–2). https://doi.org/10.1109/VLSITechnology18217.2020.9265076
17. Salahuddin, S. M., et al. (2019). SRAM with buried power distribution to improve write margin and performance in advanced technology nodes. *IEEE Electron Device Letters, 40*(8), 1261–1264. https://doi.org/10.1109/LED.2019.2921209
18. Samavedam, S. B., et al. (2020). Future logic scaling: Towards atomic channels and deconstructed chips. In *2020 IEEE International Electron Devices Meeting (IEDM)* (pp. 1.1.1–1.1.10). https://doi.org/10.1109/IEDM13553.2020.9372023
19. Singh, T., et al. (2017). 3.2 Zen: A next-generation high-performance ×86 core. In *2017 IEEE International Solid-State Circuits Conference (ISSCC)* (pp. 52–53). https://doi.org/10.1109/ISSCC.2017.7870256
20. Singh, T., et al. (2018). Zen: An energy-efficient high-performance 86 core. *IEEE Journal of Solid-State Circuits, 53*(1), 102–114. https://doi.org/10.1109/JSSC.2017.2752839
21. Singh, T., et al. (2020). 2.1 Zen 2: The AMD 7nm energy-efficient high-performance x86-64 microprocessor core. In *2020 IEEE International Solid-State Circuits Conference—(ISSCC)* (pp. 42–44). https://doi.org/10.1109/ISSCC19947.2020.9063113
22. Song, T., et al. (2017). A 10 nm FinFET 128 Mb SRAM with assist adjustment system for power, performance, and area optimization. *IEEE Journal of Solid-State Circuits, 52*(1), 240–249. https://doi.org/10.1109/JSSC.2016.2609386
23. Song, T., et al. (2018). A 7nm FinFET SRAM using EUV lithography with dual write-driver-assist circuitry for low-voltage applications. In *2018 IEEE International Solid - State Circuits Conference—(ISSCC)* (pp. 198–200). https://doi.org/10.1109/ISSCC.2018.8310252
24. Takeda, K., et al. (2006). Redefinition of write margin for next-generation SRAM and write-margin monitoring circuit. In *2006 IEEE International Solid State Circuits Conference - Digest of Technical Papers* (pp. 2602–2611). https://doi.org/10.1109/ISSCC.2006.1696326

25. Wang, J., Nalam, S., & Calhoun, B. H. (2008). Analyzing static and dynamic write margin for nanometer SRAMs. In *Proceedings of the 2008 International Symposium on Low Power Electronics & Design*. ISLPED '08 (pp. 129–134). Association for Computing Machinery. ISBN: 978-16-05581-09-5. https://doi.org/10.1145/1393921.1393954
26. Weckx, P., et al. (2019). Novel forksheet device architecture as ultimate logic scaling device towards 2nm. In *2019 IEEE International Electron Devices Meeting (IEDM)* (pp. 36.5.1–36.5.4). https://doi.org/10.1109/IEDM19573.2019.8993635
27. Zhang, B., et al. (2006). Analytical modeling of SRAM dynamic stability. In *2006 IEEE/ACM International Conference on Computer Aided Design* (pp. 315–322). https://doi.org/10.1109/ICCAD.2006.320052
28. Zhang, K., et al. (2006). A 3-GHz 70-mb SRAM in 65-nm CMOS technology with integrated column-based dynamic power supply. *IEEE Journal of Solid-State Circuits, 41*(1), 146–151. https://doi.org/10.1109/JSSC.2005.859025

Part II
SRAM Bitcell Scaling Roadmap Towards CFETs

Chapter 4
Sequential and Monolithic CFET SRAM Bitcell Designs with Stacked Fin, NS, and FS Channels Towards A3

Static random-access memory (SRAM) stands as a fundamental building block within complementary metal-oxide-semiconductor (CMOS) integrated circuits (ICs). It often dominates a significant portion of the total power, performance, and area (PPA) in various system-on-chip (SoC) designs [22–24]. The growing demand for larger last-level cache (LLC) sizes and improved on-die SRAM density exerts substantial pressure on delay and energy consumption metrics. However, realizing an economically attractive SRAM area scaling solution necessitates comprehensive enhancements in the overall PPA of evolving SoCs [24]. Consequently, this chapter delves into the exploration of bitcell area scaling design possibilities, particularly towards 5-Å-compatible technology (A5) and 3-Å-compatible technology (A3).

The complementary field-effect transistor (CFET) [18, 20] emerges as a highly promising candidate to continue CMOS technology scaling in sub-3-nm nodes. Thanks to its vertically stacked architecture, CFET possesses the potential to scale standard cell and SRAM bitcell area beyond traditional scaling [18]. This stacked architecture of CFET eliminates the need for PMOS and NMOS (PN) separation, resulting in remarkable cell height (H_{cell}) scaling of up to 55% and 40% in A5 CFET SRAM bitcell as compared to its 14-Å-compatible (A14) nanosheet (NS) [9] and 10-Å-compatible (A10) forksheet (FS) [32] technology counterparts, respectively. However, the middle-end-of-line (MEOL) routing design becomes challenging with aggressive bitcell area scaling. Therefore, corresponding routing and process integration options are introduced as basic design references for A5 CFET SRAM bitcells.

In this chapter, we explore the challenges encountered in further scaling the bitcell area beyond A5, particularly focusing on the limitations posed by substantial gate-cut (GC) requirements and the bottleneck in GC scaling [12]. This challenge is explained through the gate stack process flow for fabricating a functional vertically stacked gate-all-around (GAA) NS-based CFET device [18]. This process involves multiple steps, including Si/SiGe epitaxial layer deposition for the active region, shallow trench isolation (STI) formation, and selective removal of sacrificial SiGe

H.-H. Liu, F. Catthoor, *Circuit-Technology Co-Optimization of SRAM Design in Advanced CMOS Nodes*, https://doi.org/10.1007/978-3-031-76109-6_4

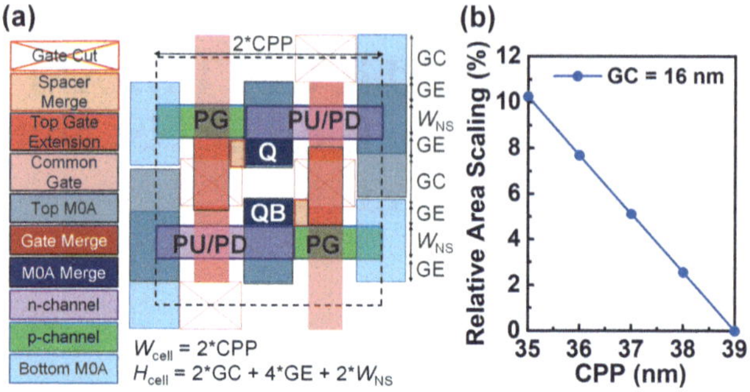

Fig. 4.1 (**a**) HD A5 Mono NS-on-NS CFET SRAM bitcell layout design (W_{cell}: W_{cell}; H_{cell}: cell height; CPP: contacted poly pitch; GC: gate-cut; GE: gate extension; W_{NS}: NS width). (**b**) Relative bitcell area scaling with varying CPP at fixed GC

layers to reveal the channels. Conventionally, the gate stack is constructed using a dual work function replacement metal gate (RMG) module [30], followed by an etching step to separate the gates (i.e., GC) and actives (active cut) for individual transistors [11]. The dimension of GC, i.e., the tip-to-tip spacing between gates orthogonal to the channel, directly impacts the H_{cell} in SRAM bitcell design (Fig. 4.1a) [12]. GC scaling relies on the precision of available patterning techniques and the selectivity of the etch process [11]. Moreover, a larger GC dimension (16 nm) is required in stacked monolithic CFET architecture due to the high aspect ratio (AR) process [3, 12]. If the GC remains unchanged due to unscalable challenges, the relative bitcell area scaling can only achieve marginal gains in cell width (W_{cell}) direction (less than 10%) by scaling the contacted poly pitch (CPP) (Fig. 4.1b). To address this ineffective scaling issue, we propose using dielectric isolation wall (DIW) instead of GC to separate gates in A3 CFET SRAM bitcell design as a scaling booster [12]. We take advantage of the gap created by spacer deposition after the fin reveal process. This gap can be much narrower than the one produced by etching. By filling these gaps with DIW that has a potentially smaller dimension (7 nm) compared to conventional etching methods, gates can be separated by the already present DIW without the need for additional etching steps [12]. Consequently, the tip-to-tip spacing between gates can be reduced, no longer constrained by gate patterning techniques. Our findings demonstrate that replacing GCs with DIW yields up to 17% bitcell area scaling for A3 bitcells compared to their A5 counterparts in this chapter. However, aggressive bitcell area scaling still introduces routing complexity and limits the potential node-to-node power and performance (PP) gains [12, 14]. Hence, this chapter provides essential insights into interconnect design and process integration options to overcome these challenges and achieve subarray- and macro-level PPA enhancements (discussed in Chaps. 6 and 7) for both A5 and A3 sequential (seq) and monolithic (mono) CFET SRAM bitcell designs. The following chapter will explore alternative CFET options for A3 SRAM bitcell design.

Section 4.1 discusses previous challenges in Fin, NS, and FSFET SRAM bitcell designs, highlighting the saturation of bitcell scaling from N2 to A7. To break through this scaling bottleneck, Sect. 4.2 introduces the basic concepts of seq and mono CFET, known for their area scaling benefits due to their stacked architecture. Section 4.3 outlines SRAM bitcell specifications and design rules. Section 4.4 provides a detailed description of A5 high-density (HD) seq and mono CFET SRAM bitcells, with a focus on common design features for all flavours including HD, high-performance (HP), and low-power (LP) configurations. Section 4.5 introduces major challenges and potential design strategies for continuous bitcell scaling. This section shows that the bitcell area scaling potential is mainly constrained by a significant GC requirement, especially evident in A5 mono CFET SRAM bitcells. Therefore, Sect. 4.6 introduces DIW to replace the GC as an alternative design option for HD seq and mono CFET SRAM bitcells towards A3. In addition to HD bitcell, Sect. 4.7 covers HP and LP CFET SRAM bitcell designs. Finally, Sect. 4.8 concludes this chapter. The proposed bitcell layout designs are presented in Appendices A.2 and A.3.

4.1 Previous Challenges in Fin, NS, and FSFET SRAM Bitcell Designs

Figure 4.2 summarizes a comprehensive overview of the HD 6-transistor (6T) SRAM bitcell area scaling benchmarks across different companies, namely imec [12–14, 19, 32], Samsung Electronics [26–28], Intel [6, 16], and TSMC [1, 2, 4]. The corresponding key design rules from imec [14], including CPP and metal pitch (MP), are also included for reference. When transitioning from Fin to NS technology, NSFET offers advantages in effective width per footprint and

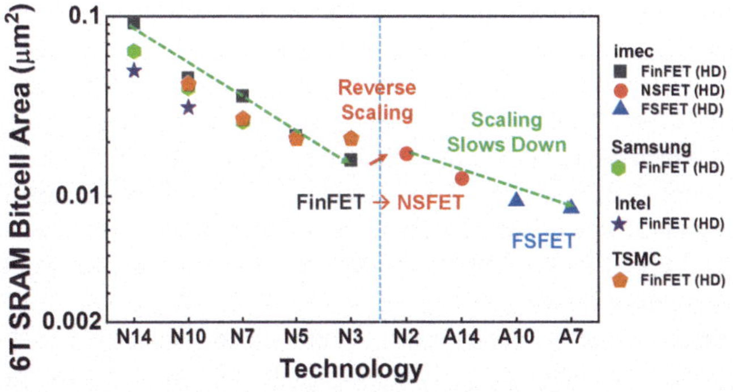

imec Design Rules		
Tech	CPP	MP
N14	90	64
N10	64	44
N7	56	40
N5	48	28
N3	45	22
N2	45	21
A14	42	18
A10	42	18
A7	39	16

Fig. 4.2 HD 6T SRAM bitcell area scaling benchmarks across different companies, namely imec [12–14, 19, 32], Samsung Electronics [26–28], Intel [6, 16], and TSMC [1, 2, 4], and corresponding imec's key design rules from N14 to A7 [14]

channel controllability. However, the 2-nm-compatible technology (N2) NSFET SRAM bitcell faces challenges in reverse scaling due to the identical CPP but wider channel width compared to 3-nm-compatible technology (N3) FinFET counterpart. To achieve area scaling compared to N3 FinFET SRAM bitcell, NSFET counterpart requires more advanced design rules, denoted as A14, with reduced CPP and MP.

The incorporation of an additional FS dielectric wall process in A10 significantly reduces the need for PMOS and NMOS (PN) separation [32]. Consequently, A10 FSFET SRAM bitcell attains a 25% bitcell area scaling advantage over A14 NSFET counterpart, even when using the same CPP and MP. However, the bitcell area scaling of 7-Å-compatible technology (A7) FSFET SRAM bitcell slows down when only CPP and MP are reduced. This bottleneck motivates the development of next-generation transistor types to further sustain SRAM bitcell scaling.

Takeaways

- **Ineffective Scaling Beyond N3:** In comparison to FinFET counterparts, the scaling of SRAM bitcell designs based on NSFET and FSFET technology is slowing down.
- **Motivated Design Strategy:** An alternative transistor structure is needed to continue advancing the bitcell area.

4.2 Next-Generation Devices for SRAM Bitcell Designs: Complementary Transistor (CFET)

To enable further scaling by eliminating the need of PN separation, CFET is introduced in A5 and A3 SRAM bitcell designs [3, 18, 21]. A5 CFET SRAM bitcell [12, 14] successfully achieves up to 40% bitcell area scaling compared to FSFET counterpart due to its stacked architecture. Beyond the standard A5 bitcell structure, introducing DIW as GC replacement and self-aligned M0A/gate merge brings up to 29% area scaling in A3 CFET SRAM bitcell design [12]. However, the dramatic bitcell area scaling poses challenges in designing the MEOL routing and process flow for all types of SRAM bitcells. As a result, this chapter proposes corresponding interconnect and process integration guidelines for A5 and A3 stacked Fin, NS, and FS CFET SRAM bitcell design, all within the context of design technology co-optimization (DTCO).

Seq and mono CFET are the two prominent integration methods for CFET, each with distinct process characteristics. This section briefly summarizes the key differences between seq and mono CFET, leveraging 3D and cross-sectional views of stacked Fin, NS, and FS CFETs in Figs. 4.3 and 4.4. Detailed process and performance comparisons for these CFET options can be found in previous reports [3, 18, 21, 29]. Specific process considerations for seq and mono CFET

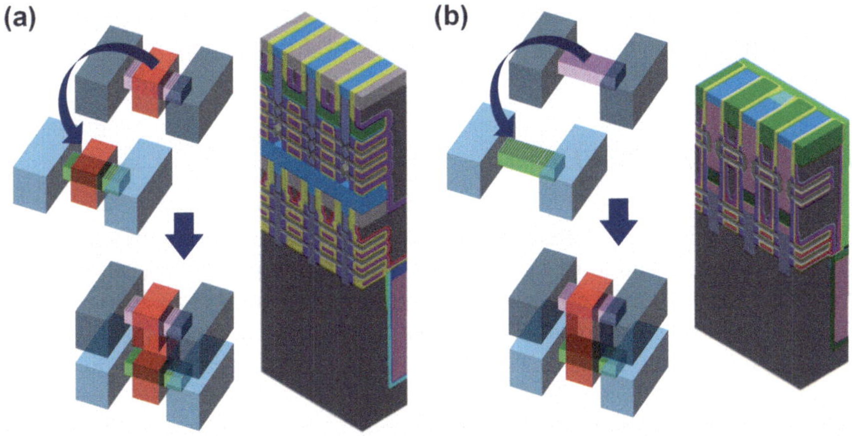

Fig. 4.3 3D views of (**a**) seq and (**b**) mono NS-on-NS CFET by stacking NMOS devices on top of the PMOS devices [3]

Fig. 4.4 Cross-sectional views of the gate for seq/mono (**a**)/(**d**) Fin-on-Fin, (**b**)/(**e**) NS-on-NS, and (**c**)/(**f**) FS-on-FS CFETs

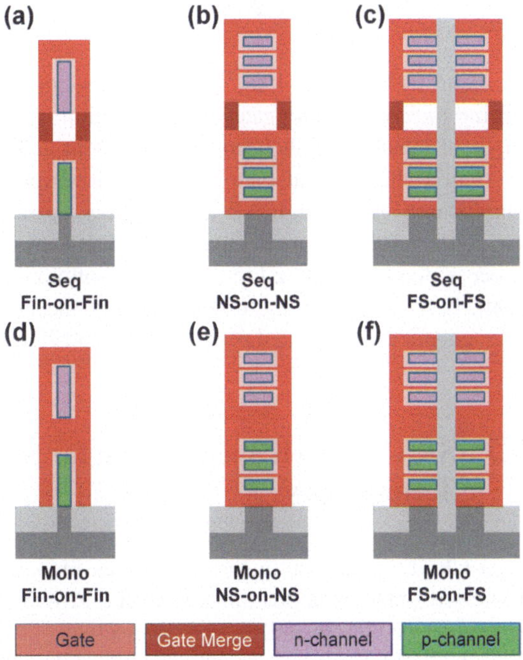

SRAM bitcell designs are discussed in subsequent sections, while an overview of all CFET types (seq, mono, and hybrid [13]) is provided in Chap. 5.

4.2.1 Sequential CFET

In seq CFET (Figs. 4.3a, 4.4a, b, c), the top and bottom devices are initially fabricated separately, followed by layer transfer and bonding processes. Consequently, seq CFET involves independent gate pickup, a higher number of masks due to double active and gate formation steps and the risk of misalignment between the top and bottom tiers. Notably, it requires an additional gate merging process to connect the top and bottom gates in CFET SRAM bitcell design due to separate top and bottom tiers. Furthermore, the requirement for void-free wafer bonding at transistor-level scales is a fundamental difficulty in the seq situation.

4.2.2 Monolithic CFET

In mono CFET (Figs. 4.3b, 4.4d, e, f), the top and bottom tiers are tied together during the channel and gate stack formation. Thus, mono CFET offers the advantages of inherent top-bottom self-alignment and a reduced number of masks due to shared channel and gate modules. However, the tied structure necessitates a significantly larger high AR process flow, thereby leading to the need for larger GC requirements compared to seq CFET. It is worth noting that a greater number of process steps are required to protect the bottom tier while the top transistor is being built. Furthermore, in the mono scenario, the main challenges would be multiple fills, chemical mechanical planarization, and etch-back processes.

4.3 SRAM Bitcell Specification for Seq and Mono CFET Technologies

In a CFET configuration that prioritizes high stress on the p-channel, which is preferred for logic applications [21], the PMOS and NMOS devices are typically placed on the bottom and top tiers, respectively [20]. Unlike conventional 6T SRAM bitcell, a CFET SRAM bitcell consists of four PMOS and two NMOS devices (Fig. 4.5). The fin or nanosheet width ratio for pull-up (PU): pass-gate (PG): pull-down (PD) devices is 1:1:1, 2:2:1, and 2:1:1 for HD [1], HP [4], and LP [5] PMOS PG-based SRAM bitcell designs, respectively. Table 4.1 provides details on the 6T SRAM bitcell design rules and other parameters in A5 and A3.

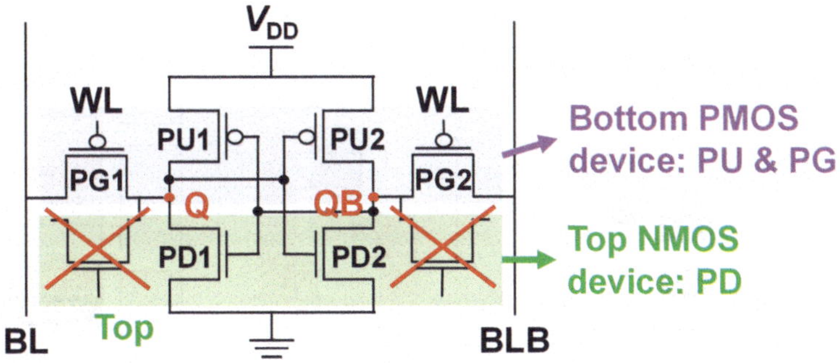

Fig. 4.5 Schematic view of 6T CFET SRAM bitcell design

Table 4.1 6T SRAM bitcell design rules and other parameters in A5 and A3

Parameters	A5	A3
Transistor Type	Fin-, NS-, FS- based Seq & Mono CFET	
Gate Length (L_g)	12 nm	
Metal Pitch (MP)	16 nm	
Contacted Poly Pitch (CPP)	39 nm	
Fin/NS Pitch (FP/NSP)	16 nm	
M0A/MINT Spacing	7 nm	
M1 Spacing	10 nm	
Fin/Nanosheet Width (W_{fin}/W_{NS})	5/11 nm	
Fin/Nanosheet Height (H_{fin}/H_{NS})	35/5 nm	
Number of Stacked Nanosheet (*NNS*)	2	3
Gate-Cut (GC)	12 nm (Seq) 16 nm (Mono)	
Gate Extension (GE)	9 nm (Seq) 7.5 nm (Mono)	9 nm
Gate to Internal Node Contact (GI) Gate to Internal Node Extension (GIE) Gate to Internal Node Peak (GIP)	9 nm	–
Isolation Wall Width (ISO)	7 nm	
Pass-gate (PG) Devices	PMOS	

4.4 HD Standard Seq and Mono CFET SRAM Bitcell Designs in A5

Achieving SRAM bitcell area scaling always comes with the challenge of developing suitable interconnect routing designs. Existing routing solutions have primarily focused on creating additional space within the frontside MEOL MINT layer to

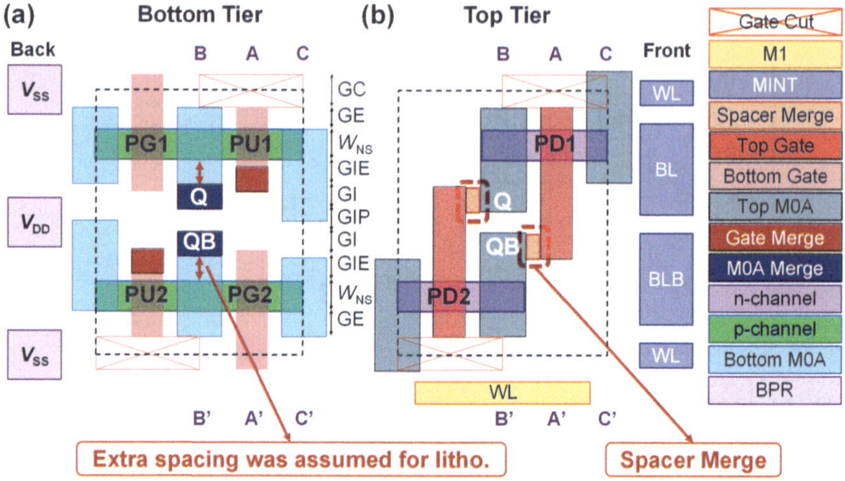

Fig. 4.6 Top views of (**a**) bottom and (**b**) top tiers for HD A5 seq NS-on-NS CFET SRAM bitcell layout design. (W_{BL} = 33.5 nm; W_{WL} = 65 nm; W_{cell} = 78 nm; H_{cell} = 97 nm)

widen bitlines (BL/BLB). For instance, in the context of A14 NS and A10 FSFET SRAM bitcell designs discussed in Chap. 1, the buried power rail (BPR) technique was used to relocate power rails from the frontside to the backside back-end-of-line (BEOL) [7, 19]. However, relying solely on the BPR technique is no longer sufficient to address the routing challenges posed by CFET SRAM bitcell design. Therefore, previous works introduce the spacer merge technique, which enlarges BL width (W_{BL}) by moving two metal rails responsible for cross-coupled formation (Q and QB) from BEOL to the MEOL [12, 14, 18]. However, these techniques are insufficient for A3 bitcells due to the extensively squeezed H_{cell} [12]. Hence, we propose to rearrange the V_{SS} rail to the frontside and connected with a supervia to avoid using extra MINT rails in A3 CFET SRAM bitcell design [12]. This approach prevents the surplus V_{SS} rails from exceeding the maximum allowed count of MINT metal layers at the cost of narrower WL width (W_{WL}) [12]. In this section, the process features of standard A5 seq and mono CFET SRAM bitcell designs are presented. Details of interconnect solutions and corresponding process flow developments for alternative A3 seq and mono CFET SRAM bitcell designs will be discussed in the following section. To simplify, the concepts of required process integration are introduced using HD CFET SRAM bitcell designs.

4.4.1 HD A5 Seq NS-on-NS CFET SRAM Bitcell Design

Figures 4.6 and 4.7 illustrate the top and cross-sectional views, respectively, of the HD A5 seq NS-on-NS CFET SRAM bitcell layout design. In this A5 design,

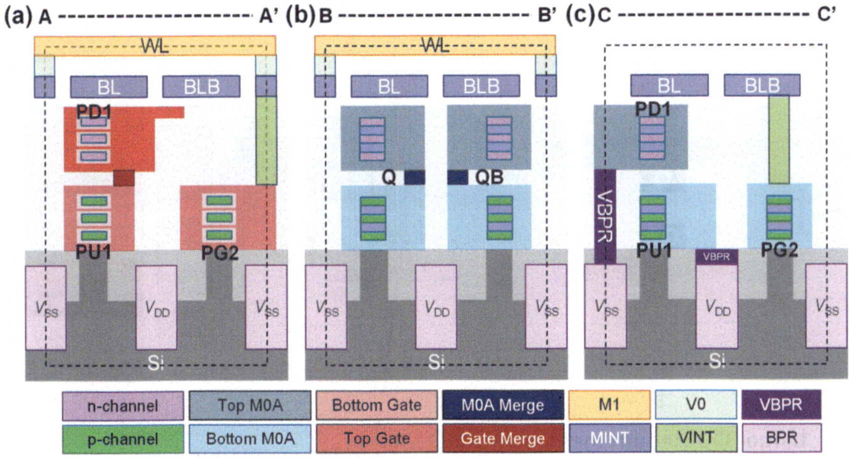

Fig. 4.7 Cross-sectional views along (**a**) gate (A-A'), (**b**) internal nodes (B-B'; across WL), and (**c**) edge (C-C'; across V_{DD} and V_{SS}) for HD A5 seq NS-on-NS CFET SRAM bitcell layout design

additional spacings away from the active region, i.e., gate to internal node extension (GIE), become necessary for maintaining sufficient lithography window (Figs. 4.6a and 4.7b). These spacings are vital during the lithography of top-bottom non-self-aligned M0A (dark blue) and gate (dark red) merges due to the separate fabrication of the top and bottom tiers. Moreover, the imprecise lithography process may pose a risk of damaging the peripheral surroundings. Thus, these extra spacings represent a significant obstacle to achieving further H_{cell} scaling in A5 seq CFET SRAM bitcell design.

4.4.2 HD A5 Mono NS-on-NS CFET SRAM Bitcell Design

Figures 4.8 and 4.9 depict the top and cross-sectional views, respectively, of the HD A5 mono NS-on-NS CFET SRAM bitcell layout design. In this A5 design, a larger GC requirement (16 nm) compared to the sequential counterparts (12 nm) should be considered due to the high AR manner [3]. This GC assumption may limit the node-to-node H_{cell} scaling as it occupies up to 44% of the H_{cell}. However, mono CFET SRAM bitcell design utilizes a self-aligned approach due to the tied top and bottom tiers. This allows the M0A merge to abut directly to the active region, eliminating the extra spacing (GIE) observed in seq CFET SRAM bitcell design shown in Fig. 4.9a. As a result, the final bitcell area of mono CFET SRAM bitcell design is up to 15% smaller than its seq counterparts, albeit at the cost of narrower BL.

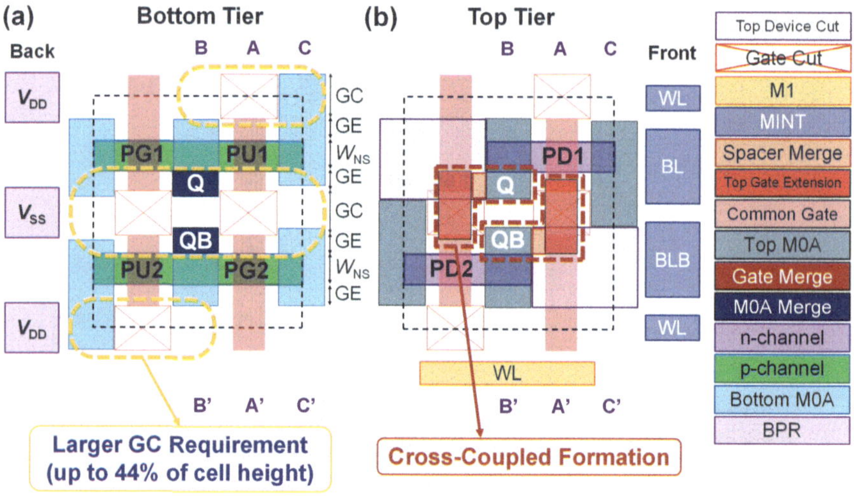

Fig. 4.8 Top views of (**a**) bottom and (**b**) top tiers for HD A5 mono NS-on-NS CFET SRAM bitcell layout design. (W_{BL} = 27 nm; W_{WL} = 65 nm; W_{cell} = 78 nm; H_{cell} = 84 nm)

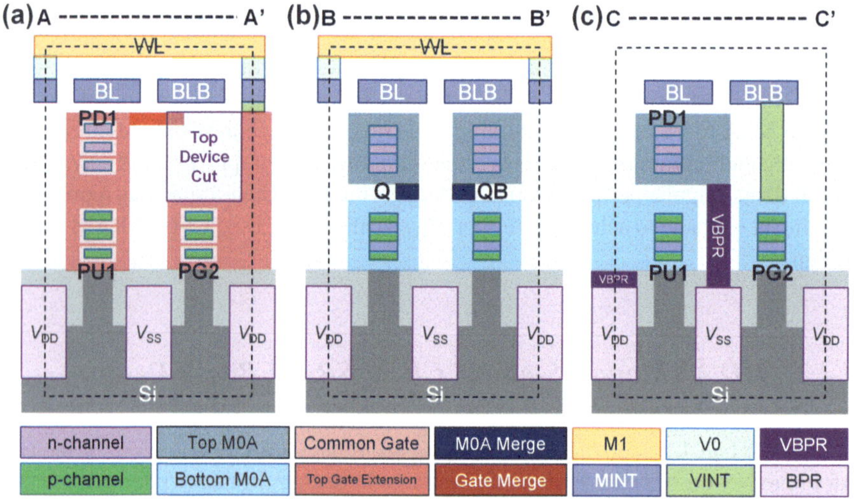

Fig. 4.9 Cross-sectional views along (**a**) gate (A-A'), (**b**) internal nodes (B-B'; across WL), and (**c**) edge (C-C'; across V_{DD} and V_{SS}) for HD A5 mono NS-on-NS CFET SRAM bitcell layout design

4.4.3 Major Challenges and Highlight Process Features in HD A5 Seq and Mono CFET SRAM Bitcell Designs

4.4.3.1 Parasitic Top Device Removal

In a CFET SRAM bitcell, PU (PMOS) and PD (NMOS) devices are perfectly matched with the CFET architecture. However, this CFET stack also unintentionally forms redundant transistors on top of the bottom PG (PMOS) devices, which need to be removed for proper functionality [18]. As a result, developing a patterning process to eliminate these top devices from the PG stacks is essential to maintain SRAM functionality. In seq CFET SRAM bitcell design, the removal of these parasitic top devices is relatively straightforward. Active and gate removal processes can be applied to the top tier without affecting the bottom devices, thanks to the independent top-bottom fabrication feature. Conversely, the process of removing parasitic top devices in mono CFET SRAM bitcell design is more complex due to its inherently interconnected top-bottom architecture. However, with a controlled time-etch method, it is possible to remove these parasitic top devices (region of top device cut marked in purple box in Fig. 4.9b). The resulting space on top of the PG devices can then be utilized for forming cross-coupled connections without increasing the H_{cell} [12, 14, 18], as discussed in the following section.

4.4.3.2 Cross-Coupled Formation

To construct the cross-coupled feedback structure in a CFET SRAM bitcell, each inverter's gate should extend to merge with the drain of the other inverter (Figs. 4.6, 4.7, 4.8, 4.9). This connection for cross-coupled inverters is established by merging the spacing between the gate and drain after removing the gate spacer locally [18]. Figures 4.6b and 4.8b highlight the location of the gate-to-drain merge (marked in brown frame), known as the spacer merge technique, which is a specific solution implemented in CFET SRAM bitcell design to address one of the routing challenges [18]. Instead of using two metal tracks in the BEOL, these cross-coupled connections are achieved through gate-to-drain connections in the MEOL, reducing complexity and saving space.

4.4.4 MEOL Routing Design in A5 Seq and Mono CFET SRAM Bitcell Designs

In A5 CFET SRAM bitcell design, V_{DD} and V_{SS} connections are established from PU and PD devices, respectively, to BPRs through vias (VBPR), as illustrated in Figs. 4.7 and 4.9. It's crucial to ensure sufficient spacing between VBPR and the active region in this configuration. On the frontside BEOL, only signal lines (BL and WL) are connected to PG devices because power rails and cross-coupled formation utilize BPR and spacer merge techniques, respectively (Figs. 4.6, 4.7, 4.8, 4.9) [12, 14, 18]. Therefore, A5 CFET SRAM bitcell has wider W_{BL} compared to A10 FSFET counterpart resulting in remarkable R_{BL} reduction (Fig. 6.1 in Chap. 6).

Takeaways

- **A5 Seq CFET SRAM Bitcell Designs:** These designs have a higher H_{cell} compared to their monolithic counterparts due to the additional spacing required between the active region and the top-bottom merge region.
- **A5 Mono CFET SRAM Bitcell Designs:** These designs necessitate an extra step to remove redundant top devices situated above the bottom PG devices. Additionally, the higher AR process increases the GC requirement significantly.

4.5 Potential Design Directions and Their Major Challenges in Continuous Scaling SRAM Bitcells Beyond A5

4.5.1 Essential Strategy for Limited Routing Space: Innovative Routing Solutions

The aggressive scaling enabled by stacked architecture leads to insufficient routing space in both the frontside and backside BEOL layers. As illustrated in Fig. 4.9c, the spacing between the VBPR and the active region is already constrained by the most compact design rules applicable to A5 technology. If the bitcell area continues to scale down, this spacing will become inadequate to accommodate more than one VBPR, which is necessary for connecting both V_{SS} and V_{DD} from the frontend device to the backside BEOL. For more detailed information, please refer to Sect. 4.6.5. Moreover, the number of metal tracks at MINT layer is constrained by the H_{cell} and MP. If the MP stops scaling while the H_{cell} continues to scale, the number of metal tracks may even decrease. These discussions indicate that additional innovative approaches in routing design are crucial when further scaling the bitcell area in advanced technology nodes.

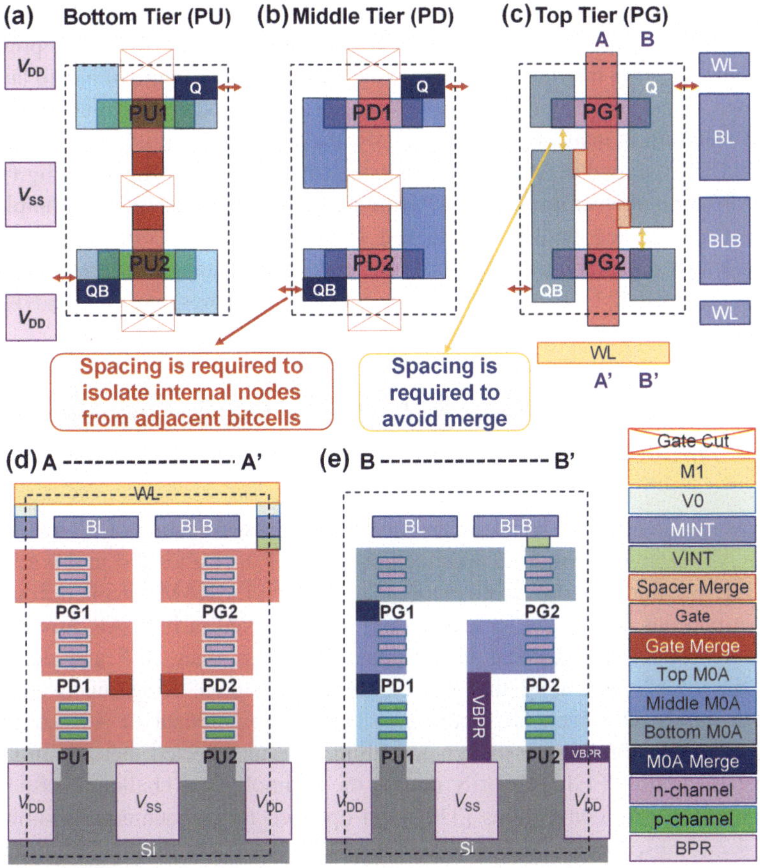

Fig. 4.10 Top views of (**a**) bottom, (**b**) middle, and (**c**) top tiers; Cross-sectional views along (**d**) gate (A-A'), (**e**) internal nodes (B-B') for HD triple-stacked seq NS-on-NS CFET SRAM bitcell layout design. ($W_{BL} = 33.5$ nm; $W_{WL} = 52$ nm; $W_{cell} = 65$ nm; $H_{cell} = 97$ nm)

4.5.2 Suboptimal Strategy for Device Scaling: Triple-Stacked Architecture

Figure 4.10 illustrates that triple-stacked SRAM bitcell designs using seq fabrication approach may offer potential area scaling benefits, dependent on the achievable design compactness in W_{cell} direction. While the H_{cell} dimension remains the same as that of the A5 seq counterpart, it is larger than that of the mono counterpart. Consequently, estimated bitcell area reductions are 17% and 4% when compared to the seq and mono designs, respectively. This scaling benefit should degrade to less than 10% with the use of a mono fabrication approach due to larger GC requirement. The primary factor contributing to bitcell area scaling is the reduction in the W_{cell} dimension, which results in a narrower WL and may potentially cause speed degradation. Critical spacings are required to isolate internal nodes between (1) adjacent cells and (2) adjacent source and drain epitaxy layers. Unlike previous A5 designs, placing internal nodes at the side rather than the centre of bitcell layouts will impact data integrity during read, write, and hold operations in neighbouring cells if isolation is inadequate. Additionally, the gate-to-gate pitch, larger than the original CPP assumed in A5, may necessitate extra layout adjustments to align gates across different devices and circuits. Furthermore, the increased process complexity and costs associated with more stacked layers and masks could outweigh the marginal area gains achieved by triple-stacked architecture. Therefore, exploring triple-stacked SRAM bitcell designs for marginal area savings may not be advisable.

4.5.3 Recommended Strategy for Device Scaling: Reducing Tip-to-Tip Gate Spacing

Previous research [3, 18] has demonstrated the successful scaling of SRAM bitcell area to A5 through the stacked architecture of CFET, as elaborated in this section. However, we find out that scaling the H_{cell} of CFET SRAM bitcell beyond A5 faces limitations due to the increasing GC requirements in high AR process [3]. Figure 4.11 demonstrates that GC occupancy is less than 20% and decreases from N3 to A7, indicating that the GC issue is not receiving much attention in these nodes. However, GC occupancy becomes significant in A5 due to the high AR process and aggressive H_{cell} scaling. The unexpected increase in GC is specific to mono CFET, primarily due to the technological challenge associated with common gate connections [3, 12], as opposed to its counterparts shown in Table 4.2 [8, 14, 15, 17, 21, 29, 32]. Figure 4.8a illustrates that the GC (marked in orange circle) occupies up to 44% of H_{cell} in A5 CFET SRAM bitcell design. Limited GC scaling occurs since the separation of gates through etching happens

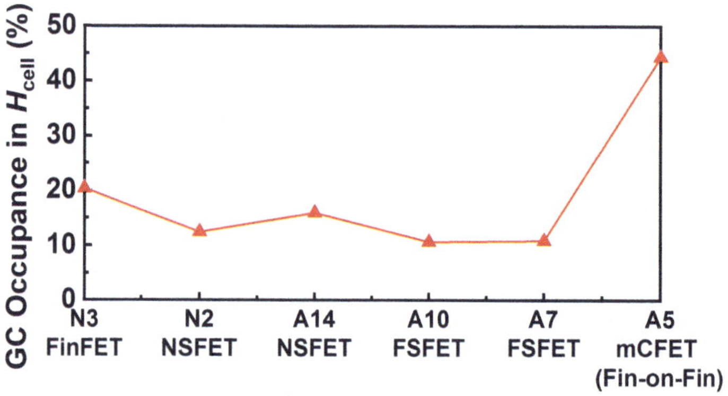

Fig. 4.11 GC occupancy percentage in H_{cell} for SRAM bitcell designs from N3 to A5

Table 4.2 GC scaling trend from A14 to A3. GC assumption increases in mono CFET due to the common gate connection of the top and bottom tiers

Technology	A14	A10	A5	A3
Gate-Cut (GC)	12 nm	12 nm	12 nm (seq) 16 nm (mono)	

after active and gate stack formations. Hence, this tip-to-tip spacing between gates depends on the patterning and etching techniques [11]. Moreover, as demonstrated in Fig. 4.1b, the relative bitcell area can only achieve minimal increases (less than 10%) in the W_{cell} direction by scaling the CPP if the GC stays fixed as a result of unscalable challenges. This indicates that as CPP (and even MP) scaling become less effective, the issue of limited H_{cell} scaling due to large GC occupancy becomes dominant.

Takeaways

- **Bitcell Scaling Challenges:** When scaling parameters such as CPP and MP lose their effectiveness in advanced nodes, the problem of limited GC scaling and routing space becomes more acute and restricts bitcell scalability.
- **Motivated Design Strategy:** Develop an alternative bitcell architecture to continuously scale down the area.
- **Non-Recommended Approach:** The triple-stacked bitcell design provides marginal scaling benefits but complicates the process and increases costs due to a higher number of masks and non-aligned gate-to-gate pitch with logic circuits.
- **Recommended Approach:** An alternative process flow for gate separation should be explored, specifically in CFET SRAM bitcell designs, to avoid limited pattern capability and etch control in the GC process.

4.6 HD Alternative Seq and Mono CFET SRAM Bitcell Designs in A3

This section discusses the primary contribution of this chapter, focusing on scaling the SRAM bitcell towards A3 through innovative process flow development for gate separation [11, 12]. To address the scaling bottleneck arising from the large GC requirement, we propose replacing GC with DIW in A3 CFET SRAM bitcell design [12]. By fabricating the DIW in the gaps created by the spacer deposition, which occurs right after the fin reveal process and before the gate stack formation, the gates can be separated using the existing DIW without the need for additional

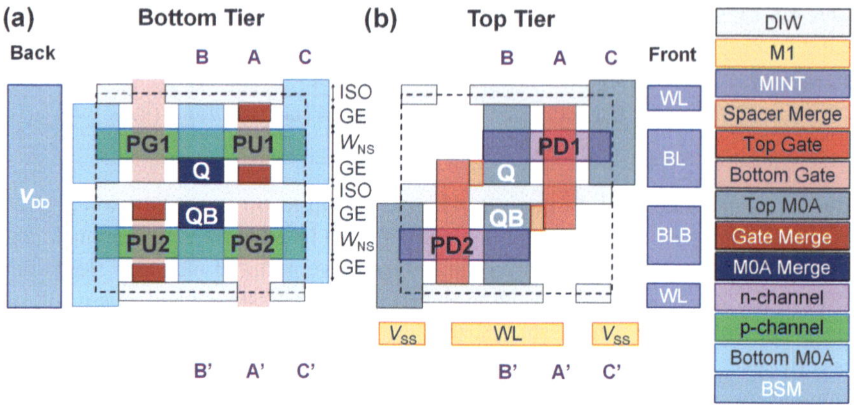

Fig. 4.12 Top views of (**a**) bottom and (**b**) top tiers for HD A3 seq NS-on-NS CFET SRAM bitcell layout design. (W_{BL} = 21 nm; W_{WL} = 41 nm; W_{cell} = 78 nm; H_{cell} = 72 nm)

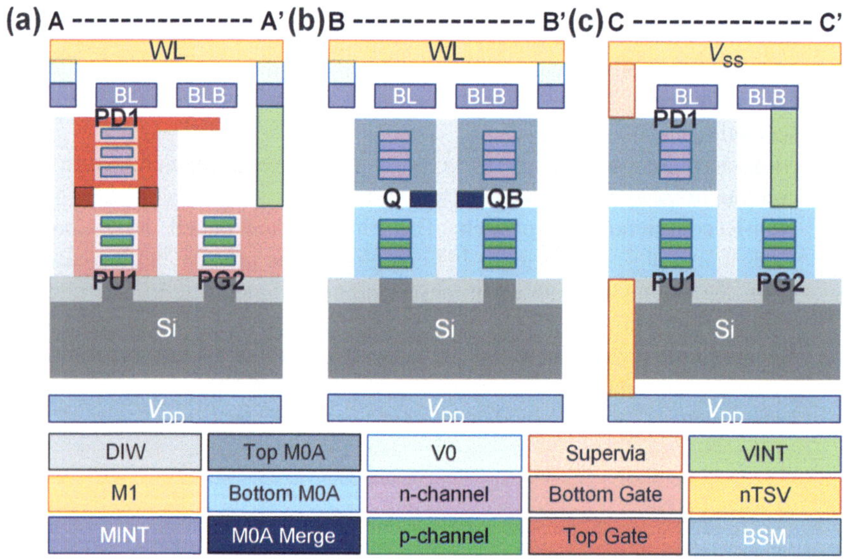

Fig. 4.13 Cross-sectional views along the (**a**) gate (A-A'), (**b**) internal nodes (B-B'; across WL), and (**c**) edge (C-C'; across V_{DD} and V_{SS}) for HD A3 seq NS-on-NS CFET SRAM bitcell layout design

etching steps [10, 11]. The essential innovation in the DIW technique is to bypass the etch procedure for gate separation and instead use the gap between the spacers to form required isolation. It allows for reduced tip-to-tip spacing between gates (7 nm) compared to GC method [12]. Therefore, this proposed design choice, i.e., DIW before gate stack, enables aggressive SRAM H_{cell} scaling from A5 to A3. In this section, seq (Figs. 4.12 and 4.13) and mono (Figs. 4.12 and 4.13) CFET SRAM bitcell design options are introduced based on the DIW process, highlighting the new routing challenges and their corresponding solutions. Note that the identical bitcell layout design concept is applicable to stacked Fin, NS, and FS cases by simply replacing the NS with Fin or FS.

4.6.1 HD A3 Seq NS-on-NS CFET SRAM Bitcell Design

Figures 4.12 and 4.13 depict the top and cross-sectional views, respectively, of the HD A3 seq NS-on-NS CFET SRAM bitcell layout design. Due to the less GC requirement in seq CFET SRAM bitcell designs, the scalability of H_{cell} in stacked Fin and NS-based channels is constrained to 8% and 7%, respectively, through the substitution of GC with the DIW process. It is important to emphasize that the scaling of A5 seq CFET SRAM bitcell designs is limited by the additional spacing away from the active region (GIE) (Sect. 4.4.1). Therefore, we also propose the utilization of self-aligned M0A (dark blue) and gate (dark red) merging processes when transitioning from A5 to A3 designs. This approach allows for another 21% and 19% H_{cell} scaling for stacked Fin and NS CFET SRAM bitcell design, respectively, by eliminating the extra spacings required in the A5 counterpart (Fig. 4.6).

4.6.2 HD A3 Mono NS-on-NS CFET SRAM Bitcell Design

Figures 4.14 and 4.15 display the top and cross-sectional views, respectively, of the HD A3 mono NS-on-NS CFET SRAM bitcell layout design. In this A3 design, the processes involving self-aligned M0A and gate merges, as well as top device removal, are identical to those in A5 counterparts. The key distinction is the adoption of DIW as a replacement for GC, which leads to a 17% and 14% reduction in H_{cell} for stacked Fin and NS CFET SRAM bitcell design, respectively.

4.6.3 Coventor Process Flow

Figure 4.16 presents the proposed process flow for A3 mono NS-on-NS CFET SRAM bitcell design by Coventor simulation,[1] giving a brief overview of the fabrication sequence. This flow also highlights the remaining challenges in the process integration stage, including DIW formation, removal of redundant top devices, and cross-coupled contacts formation, which require future attention. Both the bottom

[1] Acknowledgment is given to BT Chan for the Coventor process flow advancements.

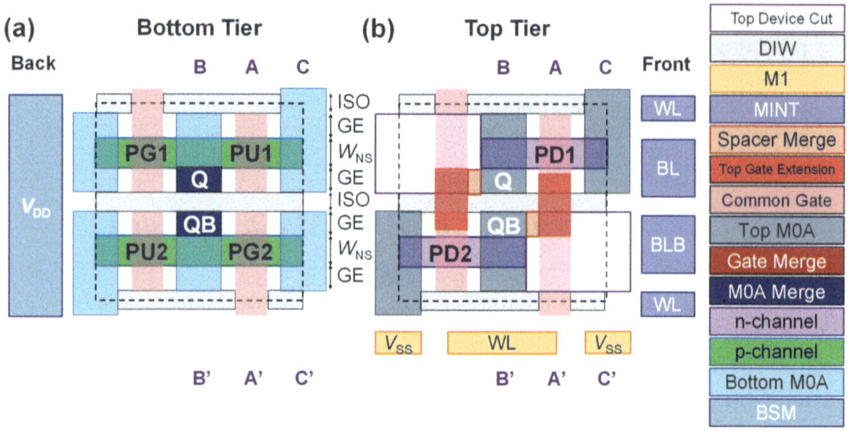

Fig. 4.14 Top views of (**a**) bottom and (**b**) top tiers for HD A3 mono NS-on-NS CFET SRAM bitcell layout design. ($W_{BL} = 21$ nm; $W_{WL} = 41$ nm; $W_{cell} = 78$ nm; $H_{cell} = 72$ nm)

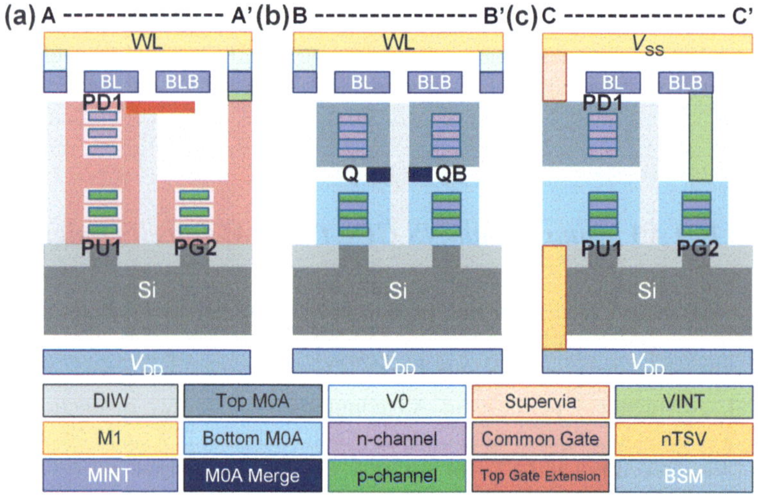

Fig. 4.15 Cross-sectional views along the (**a**) gate (A-A'), (**b**) internal nodes (B-B'; across WL), and (**c**) edge (C-C'; across V_{DD} and V_{SS}) for HD A3 mono NS-on-NS CFET SRAM bitcell layout design

PMOS and top NMOS devices are fabricated with vertically stacked lateral NS, followed by shallow trench isolation (STI) formation (Fig. 4.16a). Optionally, BPR backside connections can be fabricated after fin/NS reveal. Afterward, a spacer material layer is conformally deposited on the fin/NS structures, creating a narrow gap between them. The gap is subsequently filled with a dielectric substance to form DIW (Fig. 4.16b). Subsequently, gate patterning, gate and inner spacer formation, and bottom source/drain (S/D) epitaxy processes are performed (Fig. 4.16c–e). An

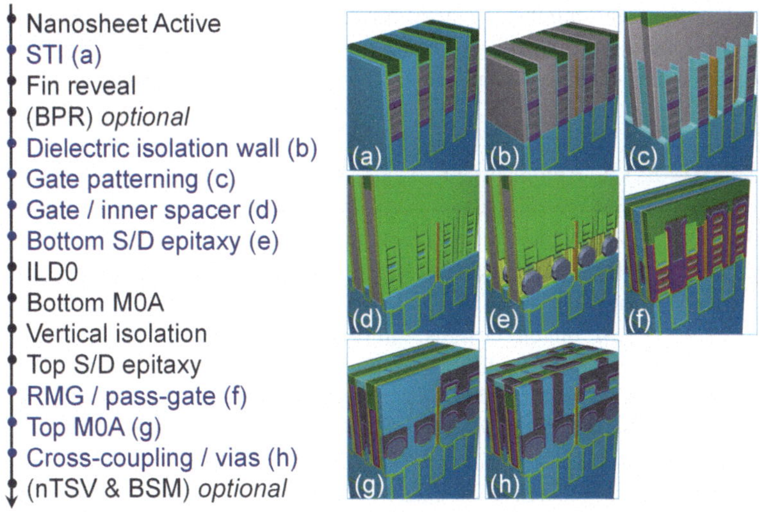

Fig. 4.16 A3 mono NS-on-NS CFET SRAM bitcell design process flow description with Coventor simulation

interlayer dielectric (ILD0) is deposited, followed by bottom S/D metal (M0A) formation. To isolate the bottom and top tiers, vertical isolation is established before top S/D epitaxy. After the replacement metal gate (RMG) process, redundant top PG devices are eliminated to create the bottom PG device (Fig. 4.16f), and top S/D M0A is formed (Fig. 4.16g). The final frontside process involves cross-coupling inverter (L-shaped region) and vias formation (Fig. 4.16h). Optionally, nTSV and BSM can be fabricated after the frontside process as backside PDN. Note that the concept of fabricating DIW before gate patterning remains consistent with seq counterparts.

4.6.4 Major Challenges and Highlight Process Features in HD A3 Seq and Mono CFET SRAM Bitcell Designs

4.6.4.1 M0A Merge and Gate Merge Process Assumptions

The top and bottom M0A merge (dark blue) and gate merge (dark red) in A5 seq CFET SRAM bitcell design (Figs. 4.6 and 4.7) are non-self-aligned, requiring additional spacings from the active region during lithography. The imprecise lithography poses a risk of damage to peripheral surroundings. To eliminate the risk of damage and reduce up to 21% of H_{cell} occupied by these spacings, the self-aligned process should be implemented for area scaling in A3. Thus, M0A merge and gate merge are proposed to be abutted to the active region in A3 seq CFET SRAM bitcell design (Fig. 4.12a) at the cost of routing challenges. However, reducing the H_{cell} below A5 design poses challenges in routing space allocation for maintaining the same number of metal tracks and adhering to metal design rules, even when applying the existing spacer merge and BPR techniques. Section 4.6.5 will present interconnect solutions aimed at overcoming these routing challenges.

4.6.4.2 DIW as GC Replacement

To address the challenge of limited H_{cell} reduction due to substantial GC dimension requirements, the DIW is introduced before the gate stack formation unlike the conventional GC method (Figs. 4.12, 4.13, 4.14, 4.15). In the case of seq and mono CFET SRAM bitcell designs, litho-defined GCs are 12 nm and 16 nm, respectively. By substituting these litho-defined GCs with the DIW, the tip-to-tip gate spacing is reduced to 7 nm (critical dimension in A3). This reduction is achieved by establishing the DIW before gate patterning. Although DIW effectively reduces the tip-to-tip gate spacing from A5 to A3, the alteration in the gate process sequence may increase the gate extension (GE) assumption. In A5 design, single work function is assumed with 2 nm gate oxide during gate stack formation, followed by the GC process. However, A3 design incorporates the DIW process, necessitating the fabrication of the gate stack between two edges (channel and DIW), leading to the creation of double work function formations with 4 nm gate oxide. Consequently, the GE increases from 7.5vnm (A5 mono assumption) to 9 nm (A3) to accommodate the required double work function assumptions. Ultimately, H_{cell} scaling of up to 8% and 17% are obtained for seq and mono CFET SRAM bitcell designs, respectively, coming purely from DIW implementation. The routing challenges are resolved with the following interconnect design (Sect. 4.6.5) and process flow exploration (Sect. 4.6.3). While the DIW shows promise as an effective scaling factor beyond A5, there are still unresolved challenges in finding a suitable material that can endure subsequent processing steps and in ensuring precise process control to fill the narrow wall trench. In comparison, the GC process after gate stack formation seems more feasible and has fewer sustainability concerns. Therefore, to ascertain if it would be beneficial to continue scaling the SRAM bitcell area using the DIW approach, the PPA evaluation will be carried out in Chaps. 6 and 7.

4.6.5 MEOL and Backside Routing Challenge and Solution

Figure 4.17a shows the power delivery network (PDN) solution for A5 CFET SRAM bitcell design. Both the supply (V_{DD}) and ground (V_{SS}) voltages are connected to the backside metal (BSM) through buried power rail (BPR) and via BPR (VBPR) [7]. However, the aggressive H_{cell} scaling from A5 to A3 introduces routing challenges. A3 design has insufficient space between the active region and VBPR (V_{SP}) at MEOL to accommodate another BPR for V_{SS} connection at the backside BEOL, as illustrated in Fig. 4.17b. Consequently, V_{SS} power rail should be placed at the frontside BEOL in A3 CFET SRAM bitcell design. In A3 design,

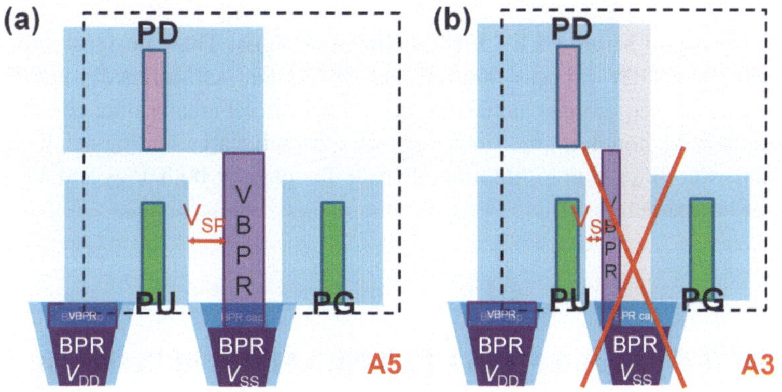

Fig. 4.17 Schematic views of (**a**) previous PDN solution (BPRs are implemented on both V_{DD} and V_{SS}) in A5 and (**b**) PDN challenge (BPR is only applied for V_{DD} since there is insufficient V_{SP} to place BPR for V_{SS}) in A3 for NS-on-NS CFET SRAM bitcell design. (V_{SP}: spacing between the active region and VBPR)

connecting V_{DD} via the BSM layer can be achieved through either BPR or nanoscale through-silicon vias (nTSV) [25, 31], with this work assuming nTSV. Due to space limitations in both the MEOL and frontside MINT layer, V_{SS} is placed on the frontside M1 layer, forming a direct connection to the top PD device through a supervia. This approach deviates from the conventional MEOL to BEOL connection sequence (VINT-MINT-V0-M1), requiring the inclusion of a supervia from M0A to M1 for V_{SS} connection to eliminate the need for extra MINT rails. The formation of the cross-coupled inverter entails a gate-to-drain merge achieved by locally removing the gate spacer, eliminating the need for two MINT tracks [18]. Hence, this approach results in wider BL and BLB, making use of the saved area in the MINT layer. The proposed solution effectively addresses the limited routing space to accommodate each metal line in these ultra-scaled SRAM bitcell layouts.

Takeaways

- **Key Invention—DIW as GC Process Replacement:** Bitcell area can be further scaled beyond A5 by filling the DIW into the narrow gap created by spacer deposition for isolation before forming the gate stack. This DIW method allows for a much narrower gap width than what is achievable with the etching process in GC.
- **Challenges in DIW Approach:** It requires precise process control to fill the dielectrics into narrow holes and to find suitable materials that can withstand the subsequent process flow.

(continued)

- **A3 Seq and Mono CFET SRAM Bitcell Designs:** These designs propose using the DIW process flow instead of GC and self-aligned gate/M0A merges in seq designs, achieving up to 29% bitcell area scaling compared to their A5 counterparts. However, this aggressive area scaling results in a narrower W_{WL} due to the limited space for placing both V_{DD} and V_{SS} at the backside.

4.7 HP/LP Seq and Mono CFET SRAM Bitcell Designs in A5 and A3

In the previous section, the introduction covered standard and alternative CFET SRAM bitcell types derived from HD NS-on-NS CFET SRAM bitcell design. Besides pursuing the smallest area as a HD (1:1:1) option, HP (2:2:1) and LP (2:1:1) bitcell configurations offer faster operating speeds and lower power consumption, respectively, by adjusting the fin or W_{NS} ratio among PU, PG, and PD devices [4, 5]. The subsequent section delves into the challenges associated with the formation of varying fin numbers (#fin) or W_{NS}. In HP or LP seq CFET SRAM bitcell designs, introducing variations in #fin or W_{NS} between each tier is feasible due to the separate fabrication of the top and bottom tiers. However, addressing the intricate procedures of top fin trimming and the removal of redundant top devices becomes particularly crucial in the context of HP and LP mono counterparts, where the top and bottom tiers are tied together. To tackle these challenges, a process flow guideline is proposed based on A3 mono NS-on-NS CFET SRAM bitcell design. Note that the distinctions between seq and mono approaches align with those of HD counterparts, including additional gate merge requirements and the absence of top device cuts (Sect. 4.4). A similar process might be applicable to the remaining channel types (fin and FS) and A5 HP and LP bitcell designs, although these are not elaborated on here to avoid redundancy.

4.7.1 HP A3 Mono NS-on-NS CFET SRAM Bitcell Design

Figures 4.18 and 4.19 showcase the top and cross-sectional views, respectively, of the HP A3 mono NS-on-NS CFET SRAM bitcell layout design. To enhance the performance of the SRAM bitcell in deeply scaled nodes for HP applications, the primary approach involves increasing the PU and PG transistor's channel width. The extremely squeezed cell design poses challenges in the development of an additional process flow for HP bitcell, as discussed in this section and Sect. 4.7.3. In HP bitcell design, broader top and bottom PU, PG, and PD active regions are

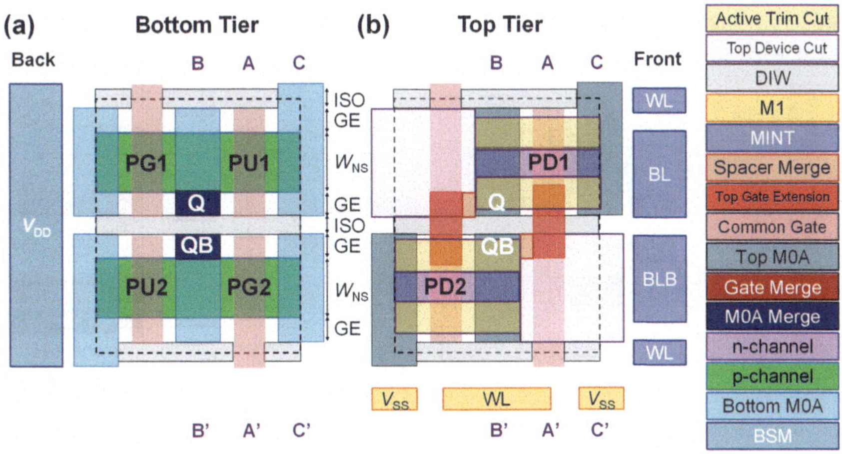

Fig. 4.18 Top views of (**a**) bottom and (**b**) top tiers for HP A3 mono NS-on-NS CFET SRAM bitcell layout design. ($W_{BL} = 32$ nm; $W_{WL} = 41$ nm; $W_{cell} = 78$ nm; $H_{cell} = 94$ nm)

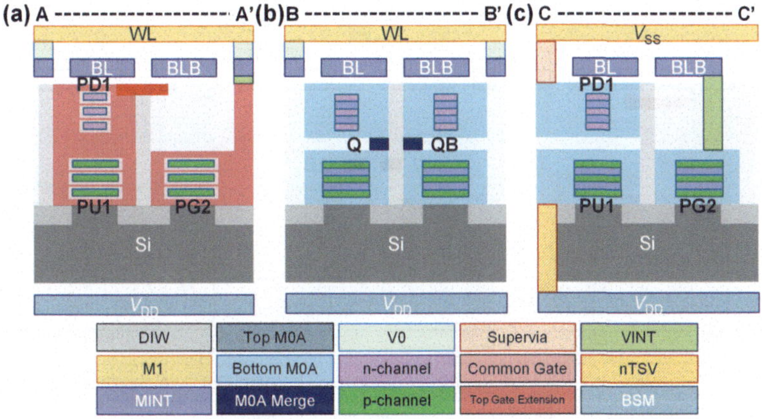

Fig. 4.19 Cross-sectional views along the (**a**) gate (A-A'), (**b**) internal nodes (B-B'; across WL), and (**c**) edge (C-C'; across V_{DD} and V_{SS}) for HP A3 mono NS-on-NS CFET SRAM bitcell layout design

initially fabricated. During the RMG and top device removal process, it's essential to trim the active region of the top PD devices (marked in yellow box) to achieve a W_{NS} ratio of 2:2:1 (PU:PG:PD) as an option for HP bitcell design.

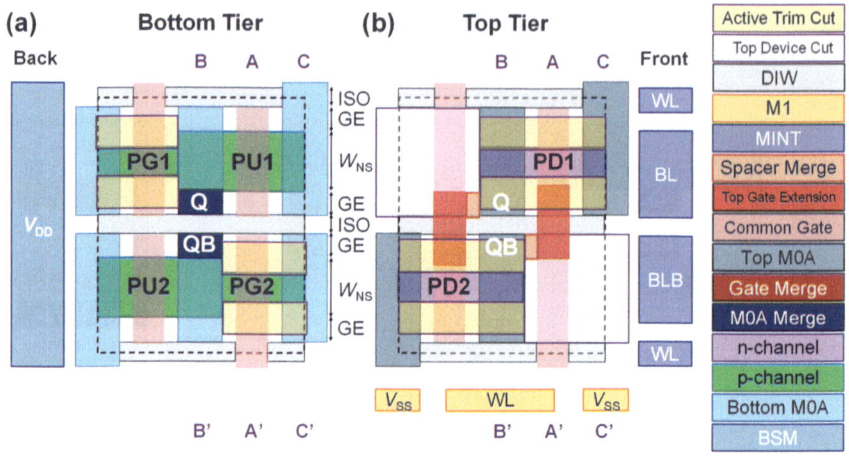

Fig. 4.20 Top views of (**a**) bottom and (**b**) top tiers for HP A3 mono NS-on-NS CFET SRAM bitcell layout design. ($W_{BL} = 32$ nm; $W_{WL} = 41$ nm; $W_{cell} = 78$ nm; $H_{cell} = 94$ nm)

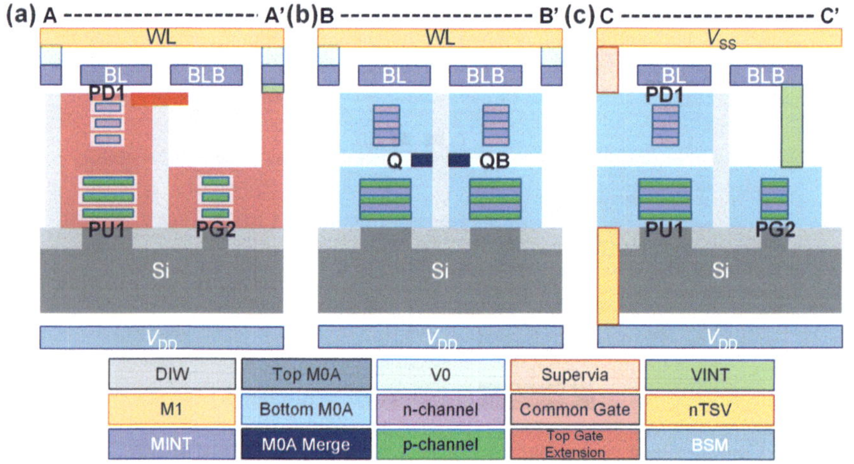

Fig. 4.21 Cross-sectional views along the (**a**) gate (A-A'), (**b**) internal nodes (B-B'; across WL), and (**c**) edge (C-C'; across V_{DD} and V_{SS}) for HP A3 mono NS-on-NS CFET SRAM bitcell layout design

4.7.2 LP A3 Mono NS-on-NS CFET SRAM Bitcell Design

Figures 4.20 and 4.21 present the top and cross-sectional views, respectively, of the LP A3 mono NS-on-NS CFET SRAM bitcell layout design. To improve the read stability characteristics in LP applications, it is a common practice to design the PU device with a greater transistor's channel width compared to the PG and PD devices in SRAM bitcell design. This section, along with Sect. 4.7.3, addresses the challenges presented by the highly scaled bitcell within a dedicated process flow. In LP bitcell design, the active regions of the top and bottom PU, PG, and PD active regions are also first formed with wider width like HP counterparts. During the RMG and top device removal process, the active regions of both top PD and bottom PG devices are trimmed (marked in yellow box) to attain a W_{NS} ratio of 2:1:1 (PU:PG:PD) as an option for LP bitcell design.

4.7.3 *Major Challenges and Highlight Process Features in HP and LP A3 Mono CFET SRAM Bitcell Designs*

4.7.3.1 Active Trim Cut

Apart from the previously discussed active cut procedure for top device removal (Sect. 4.4.3.1), there is a necessity for active trimming of top or bottom channels in both HP and LP mono NS-on-NS CFET SRAM bitcell designs to form different #fin or W_{NS} ratio. Respective process flows for HP and LP mono NS-on-NS CFET SRAM bitcell designs are outlined in Fig. 4.22. The primary distinction between HD and HP or LP cases is the inclusion of the active trim process.

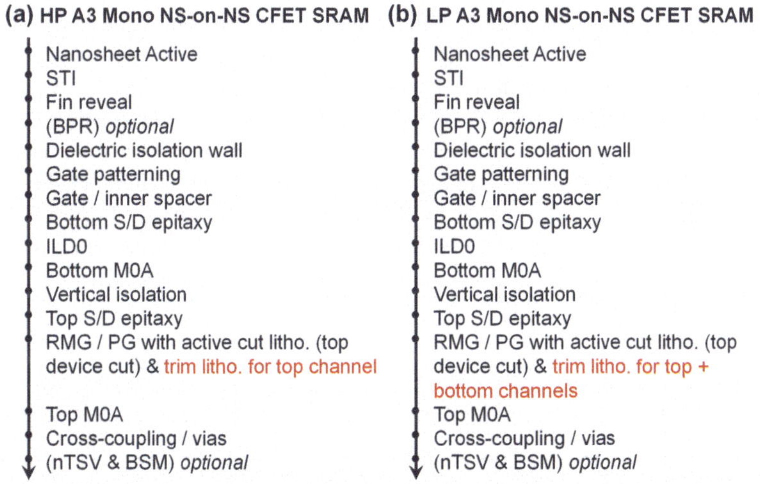

Fig. 4.22 Process flow description of (**a**) HP and (**b**) mono NS-on-NS CFET SRAM bitcell designs

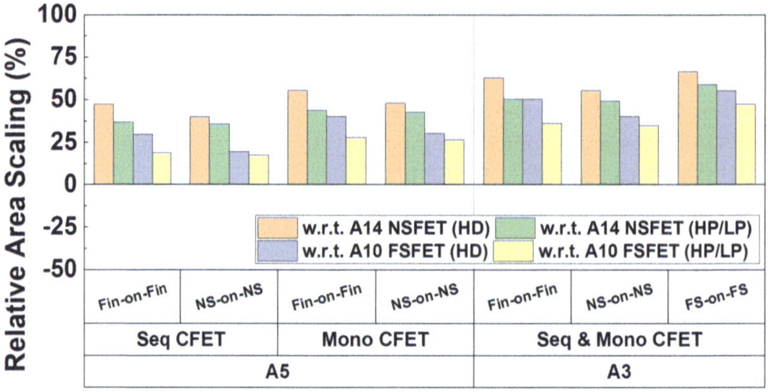

Fig. 4.23 Relative area scaling of HD, HP, and LP A5 and A3 CFET SRAM bitcells vs A14 NSFET and A10 FSFET counterparts

4.7.4 Area Scaling from A5 to A3 in HP and LP Bitcell Designs

The scaling booster from A5 to A3 for both HP and LP options is consistent with HD counterparts, involving the integration concepts such as DIW and self-aligned M0A/gate merges (Sect. 5.4). Substituting litho-defined GCs with DIW decreases tip-to-tip gate spacing from 12/16 nm (seq/mono) to 7 nm. This facilitates a bitcell area scaling of up to 11% for A3 HP and LP CFET SRAM bitcell designs compared to their A5 counterparts. Additionally, the adoption of self-aligned M0A and gate merges in A3 seq CFET SRAM bitcell design eliminates extra spacing (15% of the bitcell area in HP and LP cases) required between the active region and the top-bottom connection of M0A and gate. Implementing both techniques during the migration from A5 to A3 yields a significant bitcell area scaling of up to 21% for both HP and LP bitcell options. Furthermore, an additional 5% area gain is achieved for A3 HP and LP seq NS-on-NS CFET SRAM bitcell compared to its A5 HD counterparts (Fig. 4.23).

Takeaways

- **Key Process Distinction in HP and LP Bitcell Designs:** To achieve different #fin and W_{NS} ratios (PU:PG:PD), an additional process step is needed. This step involves trimming the top or bottom channels with precise selectivity and depth control.

4.8 Conclusions

Assessment of bitcell layouts, critical process flows, and interconnect solutions for seq and mono CFET SRAM bitcell designs has been conducted in both A5 and A3. For a comprehensive understanding, all the top views of HD, HP, and LP stacked Fin, NS, and FS CFET SRAM bitcell layout designs are presented in Appendix (Sects. A.2 and A.3).

HD A5 CFET SRAM bitcell design demonstrates a remarkable bitcell area scaling advantage, offering up to 55% compared to A14 NSFET and 40% compared to A10 FSFET counterparts. The distinctive advantage of A5 mono CFET SRAM bitcell over its seq counterparts is a bitcell area reduction of up to 15%, even with a larger GC requirement. By introducing DIW and self-aligned M0A/gate merge, up to 29% and 21% area scaling is predicted in HD and HP/LP A3 CFET SRAM bitcell design, respectively, as compared to its A5 counterparts.

SRAM bitcell area scaling can continue utilizing CFET technology. However, the intricate and compact nature of CFET SRAM bitcell structures introduces several challenges in process and routing development. These challenges include parasitic top device removal, cross-coupled formation, self-aligned M0A/gate merges, DIW formation, and routing design. Furthermore, the development of an additional active trim cut process is necessitated by the broader transistor's channel width design for both HP and LP applications. Consequently, this chapter explores potential process integration and interconnect designs tailored to seq and mono CFET SRAM bitcell designs. In the following chapter, an alternative CFET structure will be introduced in the A3 SRAM bitcell design. This modification aims to reduce the via connection length from the BL and WL to PG devices, addressing the potential RC issues in both seq and mono counterparts.

References

1. Chang, J., et al. (2017). 12.1 A 7 nm 256 Mb SRAM in high-k metal-gate FinFET technology with write-assist circuitry for low-VMIN applications. In *2017 IEEE International Solid-State Circuits Conference (ISSCC)* (pp. 206–207). https://doi.org/10.1109/ISSCC.2017.7870333
2. Chang, J., et al. (2020). 15.1 A 5 nm 135 Mb SRAM in EUV and high-mobility-channel FinFET technology with metal coupling and charge-sharing write-assist circuitry schemes for high-density and low-VMIN applications. In *2020 IEEE International Solid-State Circuits Conference—(ISSCC)* (pp. 238–240). https://doi.org/10.1109/ISSCC19947.2020.9062967
3. Chehab, B., et al. (2021). Design-technology co-optimization of sequential and monolithic CFET as enabler of technology node beyond 2 nm. In Yuan, C.-M. & R.-H. Kim (Eds.), *Design-process-technology co-optimization XV* (Vol. 11614, p. 116140D). International Society for Optics and Photonics. SPIE. https://doi.org/10.1117/12.2583395
4. Clinton, M., et al. (2017). 12.3 a low-power and high-performance 10nm SRAM architecture for mobile applications. In *2017 IEEE International Solid-State Circuits Conference (ISSCC)* (pp. 210–211). https://doi.org/10.1109/ISSCC.2017.7870335
5. Guo, Z., et al. (2019). A 23.6-Mb/mm^2 SRAM in 10-nm FinFET technology with pulsed-pMOS TVC and stepped-WL for low-voltage applications. *IEEE Journal of Solid-State Circuits, 54*(1), 210–216. https://doi.org/10.1109/JSSC.2018.2861873

6. Guo, Z., et al. (2020). A 10nm SRAM design using gate-modulated self-collapse write assist enabling 175mV VMIN reduction with negligible power overhead. In *2020 IEEE Symposium on VLSI Circuits* (pp. 1–2). https://doi.org/10.1109/VLSICircuits18222.2020.9162782
7. Gupta, A., et al. (2021). Buried power rail metal exploration towards the 1 nm node. In *2021 IEEE International Electron Devices Meeting (IEDM)* (pp. 22.5.1–22.5.4). https://doi.org/10.1109/IEDM19574.2021.9720684
8. Huang, C.-Y., et al. (2020). 3-D self-aligned stacked NMOS-on-PMOS nanoribbon transistors for continued moore's law scaling. In *2020 IEEE International Electron Devices Meeting (IEDM)* (pp. 20.6.1–20.6.4). https://doi.org/10.1109/IEDM13553.2020.9372066
9. Jang, D., et al. (2017). Device exploration of nanosheet transistors for sub-7-nm technology node. *IEEE Transactions on Electron Devices, 64*(6), 2707–2713. https://doi.org/10.1109/TED.2017.2695455
10. Liu, H.-H., Salahuddin, S. M., & Chan, B. (BT). (2023). *BIT Cell for SRAM.*
11. Liu, H.-H., Salahuddin, S. M., & Chan, B. (BT). (2023). *Bit Cell With Isolating Wall.*
12. Liu, H.-H., et al. (2023). CFET SRAM DTCO, interconnect guideline, and benchmark for CMOS scaling. *IEEE Transactions on Electron Devices, 70*(3), 883–890. https://doi.org/10.1109/TED.2023.3235701
13. Liu, H.-H., et al. (2023). CFET SRAM with double-sided interconnect design and DTCO benchmark. *IEEE Transactions on Electron Devices, 70*, 1–8. https://doi.org/10.1109/TED.2023.3305322
14. Liu, H.-H., et al. (2023). DTCO of sequential and monolithic CFET SRAM. In R.-H. Kim & N. V. Lafferty (Eds.), *DTCO and computational patterning II* (Vol. 12495, p. 124950Z). International Society for Optics and Photonics. SPIE. https://doi.org/10.1117/12.2657524
15. Loubet, N., et al. (2017). Stacked nanosheet gate-all-around transistor to enable scaling beyond FinFET. In *2017 Symposium on VLSI Technology* (pp. T230–T231). https://doi.org/10.23919/VLSIT.2017.7998183
16. Natarajan, S., et al. (2014). A 14nm logic technology featuring 2nd-generation FinFET, air-gapped interconnects, self-aligned double patterning and a 0.0588 μm2 SRAM cell size. In *2014 IEEE International Electron Devices Meeting* (pp. 3.7.1–3.7.3). https://doi.org/10.1109/IEDM.2014.7046976
17. Rachmady, W., et al. (2019). 300mm heterogeneous 3D integration of record performance layer transfer germanium PMOS with silicon NMOS for low power high performance logic applications. In *2019 IEEE International Electron Devices Meeting (IEDM)* (pp. 29.7.1–29.7.4). https://doi.org/10.1109/IEDM19573.2019.8993626
18. Ryckaert, J., et al. (2018). The complementary FET (CFET) for CMOS scaling beyond N3. In *2018 IEEE Symposium on VLSI Technology* (pp. 141–142). https://doi.org/10.1109/VLSIT.2018.8510618
19. Salahuddin, S., et al. (2020). Buried power SRAM DTCO and system-level bench-marking in N3. In *2020 IEEE Symposium on VLSI Technology* (pp. 1–2). https://doi.org/10.1109/VLSITechnology18217.2020.9265076
20. Schuddinck, P., et al. (2019), Device-, circuit-block-level evaluation of CFET in a 4 track library. In *2019 Symposium on VLSI Technology* (pp. T204–T205). https://doi.org/10.23919/VLSIT.2019.8776513
21. Schuddinck, P., et al. (2022). PPAC of sheet-based CFET configurations for 4 track design with 16nm metal pitch. In *2022 IEEE Symposium on VLSI Technology and Circuits (VLSI Technology and Circuits)* (pp. 365–366). https://doi.org/10.1109/VLSITechnologyandCir46769.2022.9830492
22. Singh, T., et al. (2017). 3.2 Zen: A next-generation high-performance ×86 core. In *2017 IEEE International Solid-State Circuits Conference (ISSCC)* (pp. 52–53). https://doi.org/10.1109/ISSCC.2017.7870256
23. Singh, T. et al. (2018). Zen: An energy-efficient high-performance 86 core. *IEEE Journal of Solid-State Circuits, 53*(1), 102–114. https://doi.org/10.1109/JSSC.2017.2752839

24. Singh, T., et al. (2020). 2.1 Zen 2: The AMD 7nm energy-efficient high-performance x86-64 microprocessor core. In *2020 IEEE International Solid-State Circuits Conference—(ISSCC)* (pp. 42–44). https://doi.org/10.1109/ISSCC19947.2020.9063113
25. Sisto, G., et al. (2021). IR-drop analysis of hybrid bonded 3D-ICs with backside power delivery and μ-n-TSVs. In *2021 IEEE International Interconnect Technology Conference (IITC)* (pp. 1–3). https://doi.org/10.1109/IITC51362.2021.9537541
26. Song, T. et al. (2014). 13.2 A 14nm FinFET 128Mb 6T SRAM with VMIN-enhancement techniques for low-power applications. In *2014 IEEE International Solid-State Circuits Conference Digest of Technical Papers (ISSCC)* (pp. 232–233). https://doi.org/10.1109/ISSCC.2014.6757413
27. Song, T. et al. (2017). A 10 nm FinFET 128 Mb SRAM with assist adjustment system for power, performance, and area optimization. *IEEE Journal of Solid-State Circuits, 52*(1), 240–249. https://doi.org/10.1109/JSSC.2016.2609386
28. Song, T. et al. (2018). A 7nm FinFET SRAM using EUV lithography with dual write-driver-assist circuitry for low-voltage applications. In *2018 IEEE International Solid - State Circuits Conference—(ISSCC)* (pp. 198–200). https://doi.org/10.1109/ISSCC.2018.8310252
29. Subramanian, S., et al. (2020). First monolithic integration of 3D complementary FET (CFET) on 300mm wafers. In *2020 IEEE Symposium on VLSI Technology* (pp. 1–2). https://doi.org/10.1109/VLSITechnology18217.2020.9265073
30. Veloso, A., et al. (2014). (Invited) replacement metal gate/high-k last technology for aggressively scaled planar and FinFET-based devices. *ECS Transactions, 61*(2), 225. https://doi.org/10.1149/06102.0225ecst
31. Veloso, A., et al. (2021). Enabling logic with backside connectivity via n-TSVs and its potential as a scaling booster. In *2021 Symposium on VLSI Technology* (pp. 1–2).
32. Weckx, P., et al. (2019). Novel forksheet device architecture as ultimate logic scaling device towards 2nm. In *2019 IEEE International Electron Devices Meeting (IEDM)* (pp. 36.5.1–36.5.4). https://doi.org/10.1109/IEDM19573.2019.8993635

Chapter 5
Hybrid CFET SRAM Bitcell Designs with Stacked NS and FS Channels Towards A3

Static random-access memory (SRAM) plays a vital role in CMOS integrated circuits (ICs) and significantly impacts power, performance, and area (PPA) in system-on-chip (SoC) designs [21–23]. To achieve higher chip performance and energy efficiency, SRAM bitcell scaling has been a common approach. Several device options have been explored to continue SRAM bitcell scaling beyond 3-nm (N3) technology nodes, such as nanosheet (NS) [13], forksheet (FS) [29], and complementary field-effect transistor (CFET) [10–12]. Among different transistor technologies, CFET stands out for its superior scalability [4, 14, 17, 19, 20, 27] due to stacked NMOS and PMOS devices on top of each other. However, the aggressive CFET SRAM scaling comes with challenges such as complex interconnect routing design and increased parasitic RC [10, 11, 13, 18]. Hence, this work primarily focuses on addressing these issues.

In Chap. 4, the introduction of sequential (seq) and monolithic (mono) CFETs successfully boosts SRAM bitcell area scaling towards 3-Å-compatible technology (A3) [10, 12]. This involves detailed background information and exploration of challenging process flow. Notably, A3 seq and mono CFET SRAM bitcell designs achieve remarkable bitcell area scaling, providing up to 63% and 50% reductions compared to 14-Å-compatible technology (A14) and 10-Å-compatible technology (A10) counterparts, respectively. However, we discover that this stacked architecture introduces taller via connections from the frontside back-end-of-line (BEOL) signals to the bottom pass-gate (PG) devices (Figs. 5.1 and 5.2), resulting in increased bitline (BL) and wordline (WL) capacitances compared to non-stacked SRAM bitcells (Sect. 5.1) [10, 11]. Further research supporting the fact that the higher parasitic capacitance hinders power and performance (PP) gains between nodes is detailed in Chap. 6 [10].

In this chapter, we introduce the hybrid CFET as an alternative device option for SRAM bitcell design in A3, alongside the seq and mono CFETs. The hybrid CFET combines process features from both seq and mono CFETs [11]. The process flow of hybrid CFET involves simultaneous active fabrication for both

H.-H. Liu, F. Catthoor, *Circuit-Technology Co-Optimization of SRAM Design in Advanced CMOS Nodes*, https://doi.org/10.1007/978-3-031-76109-6_5

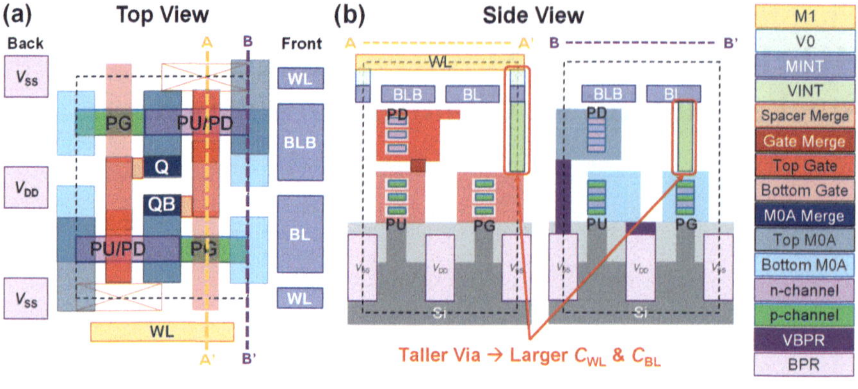

Fig. 5.1 (**a**) Top and (**b**) side views of HD A5 seq NS-on-NS CFET SRAM bitcell layout design (W_{BL}/W_{WL} = 33.5/65 nm). The vias connected from the frontside BL and WL to the bottom PG devices are relatively taller than non-stacked SRAM bitcell (e.g., Fin, NS, and FSFET SRAM bitcells). Both seq and mono CFET SRAM bitcells employ the same interconnect routing design at each node

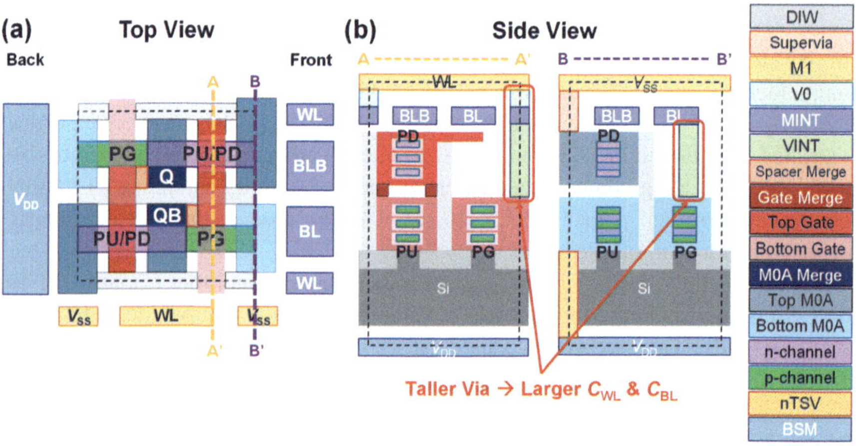

Fig. 5.2 (**a**) Top and (**b**) side views of HD A3 seq NS-on-NS CFET SRAM bitcell layout design (W_{BL}/W_{WL} = 21/41 nm). The vias connected from the frontside BL and WL to the bottom PG devices are relatively taller than non-stacked SRAM bitcell (e.g., Fin, NS, and FSFET SRAM bitcells). Both seq and mono CFET SRAM bitcells employ the same interconnect routing design at each node

top and bottom tiers, akin to mono CFETs, while the top and bottom gates are formed separately with vertical via connection, similar to seq CFETs. Afterwards, the fabrication entails initially creating the devices and respective interconnect on one side, followed by wafer flipping, bonding, and substrate removal to extend the fabrication to the other side. This approach allows us to apply identical BEOL design rules for both sides, aligning with the rules used for frontside metal layers (e.g., MINT, M1, M2) in previous designs. This concept, termed "double-sided interconnect features" [11], eliminates the need for distinct backside delivery networks with larger metal pitch (MP) and dimension requirements. Consequently, signals and power rails can be flexibly placed on any side of the hybrid CFET SRAM bitcell due to tighter MP patterning capabilities on both sides. Notably, in the seq and mono counterparts, the placement of signal lines (WL and BL) at the backside was not suggested. This is because it could lead to higher capacitance induced by longer via connections through the thick substrate, and it requires excessively smaller MP, which is not feasible at the backside [11]. To evaluate its potential in future SRAM bitcell development, we conducted a comprehensive design technology co-optimization (DTCO) in Chaps. 5 and 6. The hybrid CFET offers several advantages over its previous seq and mono counterparts for A3 bitcell designs: (1) relaxation of WL metal width by removing the necessity of placing one of the power rails on the same metal layer as WL (Sect. 4.6.5 in Chap. 4); (2) shallower via connections from signal lines to bottom PG devices achieved through substrate removal; and (3) flexibility of placing PG devices on top or bottom tiers as n-type or p-type devices due to double-sided interconnect features. These enhancements result in significant reductions in parasitic resistance and capacitance (RC), leading to improvements in power, performance, and area (PPA), as detailed in Chap. 6.

Section 5.1 discusses the challenges encountered in seq and mono CFET SRAM bitcell designs in 5-Å-compatible technology (A5) and 3-Å-compatible technology (A3). It emphasizes the significant role played by vias in facilitating signal connections and their impact on the overall capacitance. Section 5.2 examines design strategies and their primary challenges for minimizing via height, aiming to reduce via capacitance. Section 5.3 introduces the fundamental concepts, process flow, and interconnect routing capabilities specific to hybrid CFET designs. In Sect. 5.4, a comprehensive outline of SRAM bitcell specifications is provided, drawing a comparison between NMOS- and PMOS PG-based SRAM bitcell configurations. In hybrid CFET SRAM bitcells, the double-sided interconnect features provide flexibility in the placement of PG devices, allowing them to be positioned on either the top or bottom tier as either n-type or p-type, unlike the assumptions made in seq and mono counterparts where PG devices are assumed to be p-type. Section 5.5 delves into innovative ideas for A3 hybrid CFET SRAM bitcell design. These layout and process design concepts are universally applied to all flavours including high-density (HD), high-performance (HP), and low-power (LP) bitcells. Section 5.6 presents an overview of design rules, technology labels, BEOL assumptions, and bitcell area benchmarks from A14 to A3. Finally, Sect. 5.7 concludes this chapter, summarizing the key findings and insights. The proposed bitcell layout designs are presented in Appendix A.4.

5.1 Previous Challenges in Seq and Mono CFET SRAM Bitcell Designs

Figures 5.1 and 5.2 illustrate the schematic views of previous A5 and A3 seq CFET SRAM bitcell designs, respectively. The interconnect routing designs in these

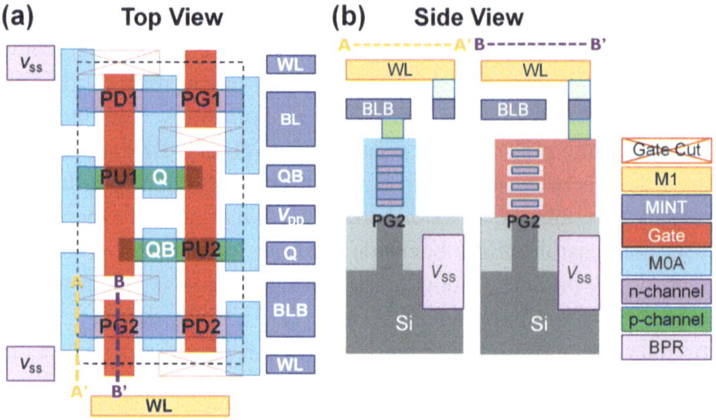

Fig. 5.3 (**a**) Top and (**b**) side views of HD A14 NSFET SRAM bitcell layout design (W_{BL}/W_{WL} = 27/70 nm)

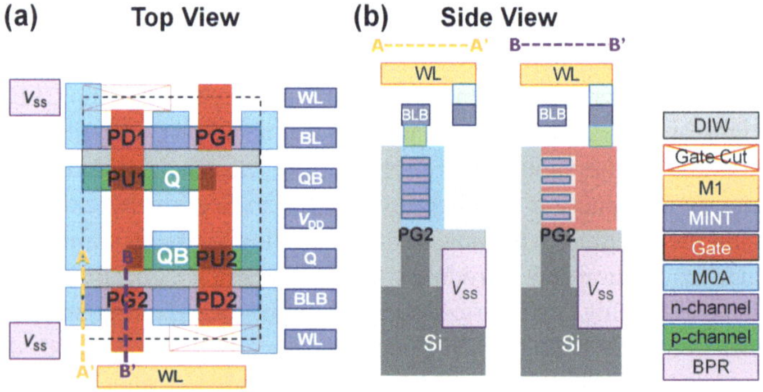

Fig. 5.4 (**a**) Top and (**b**) side views of HD A10 FSFET SRAM bitcell layout design (W_{BL}/W_{WL} = 11/70 nm)

figures are identical to their mono counterparts [10]. Although A5 CFET SRAM bitcell brings up to 55% area scaling compared to A14 NSFET counterpart, further area scaling beyond CFET structure becomes limited due to the requirement for large gate-cuts (GC) [10]. To achieve additional area scaling (up to 29%) from A5 to A3 CFET SRAMs, the self-aligned M0A/gate merges and dielectric isolation wall (DIW) as GC replacement were introduced in the previous chapter [10]. However, the stacked structure in CFET SRAM bitcells still leads to relatively taller via connections from the frontside BL and WL signals to the bottom PG devices (Figs. 5.1 and 5.2) compared to non-stacked (NS and FS FETs) SRAM bitcells (Figs. 5.3 and 5.4) [10–12].

As depicted in Fig. 5.5, these taller vias contribute significantly to the parasitic BL and WL capacitances (C_{BL} and C_{WL}) per cell in HD A3 seq and mono CFET SRAM bitcell designs [11]. These increased BL and WL via capacitances (C_{via}) impede the node-to-node PP improvements when transitioning from non-stacked to

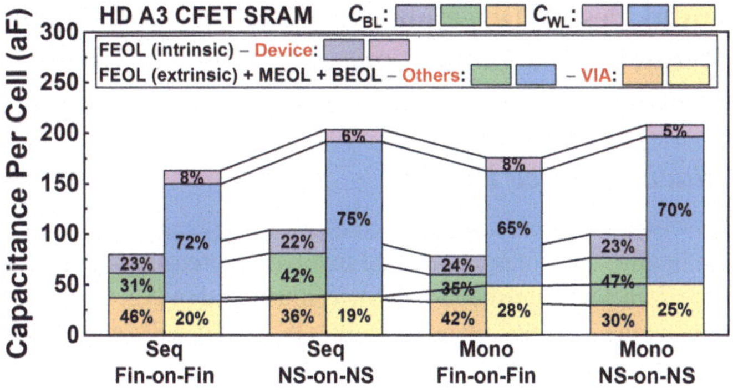

Fig. 5.5 Breakdown of parasitic capacitance per cell of BL and WL includes FEOL (intrinsic and extrinsic), MEOL, and BEOL capacitances for HD A3 seq and mono CFET SRAM bitcells with stacked Fin and FS channels [10]. Note that the purple and pink bars represent only the intrinsic device capacitances, the extrinsic device capacitances are included in the green and blue bars

stacked SRAM bitcells [10]. When examining the total WL capacitance, a larger portion of capacitance arises from the part comprising extrinsic gate capacitance and BEOL capacitance. Note that in the Quickcap simulation environment, distinguishing the impact solely from the changed WL width (W_{WL}) in metal capacitance extraction is challenging. The BEOL capacitance of WL is greater than BL due to the wider and longer metal rail. As a result, only a minor fraction (around 5–8% of the total breakdown of capacitance for WL) is attributed to the intrinsic capacitance, making it a relatively insignificant factor when compared to the BL. More detailed information about extracting front-end-of-line (FEOL), middle-end-of-line (MEOL), and BEOL parasitic capacitance contribution is described in Sect. 2.1.2.2 of Chap. 2.

As SRAM bitcell designs continue to aggressively scale beyond A14, the routing congestion becomes a significant bottleneck for HD option [12]. In ultra-scaled nodes, backside routing is one of the solutions to address the limited space at the frontside BEOL [10, 12, 13, 29]. Previous HD A5 seq and mono CFET SRAM bitcell designs have successfully reduced bitline (R_{BL}) and wordline (R_{WL}) widths by relocating both supply (V_{DD}) and ground (V_{SS}) voltages to the backside buried power rail (BPR), taking advantage of wider widths (Fig. 5.1) [12, 17]. However, the spacing between the backside power via and the active region becomes insufficient when migrating to A3 designs [10]. Thus, only the V_{DD} can be placed at the backside, while V_{SS} needs to be moved back to the frontside. This configuration makes the W_{WL} narrow again and enlarges R_{WL}, as both V_{SS} and WL are on the same frontside layer (M1) (Fig. 5.2). Note that despite sharing the same metal layer for both WL and V_{SS}, the W_{WL} (41 nm) remains wider compared to the BL width (W_{BL}) (21 nm) in A3 design.

5.2 Potential Design Directions and Their Major Challenges for Reducing Via Capacitance Through Via Height Shortening

5.2.1 Reducing Device Height

To reduce via height, one potential method involves decreasing the fin height (H_{fin}) or the number of stacked nanosheets (*NNS*), as illustrated in Fig. 5.6a. However, this approach reduces the effective channel width (W_{ch}) and will severely degrade the overall PP and margins, as discussed in Sect. 6.5.1. Therefore, decreasing device height is not recommended for reducing via capacitance due to its detrimental impact on device performance.

5.2.2 Positioning PG Devices on the Top Tiers

Another potential strategy is to place PG devices on the top tiers, thereby shortening the via height, as shown in Fig. 5.6b. However, this requires too complex fabrication processes on the bottom tier, such as cross-coupled connections and redundant PG device removal, which must be carried out under the shadow of the top devices [10, 12]. These processes pose significant challenges and necessitate additional process development efforts, making this approach impractical for reducing via capacitance.

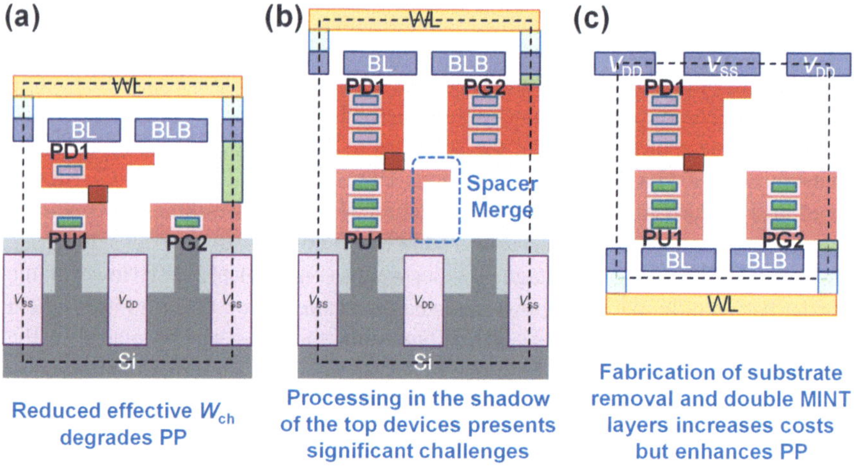

Fig. 5.6 Side views of via height reduction approaches: (**a**) reducing device height, (**b**) positioning PG devices on the top tiers, and (**c**) removing substrate and fabricating double MINT layers

5.2.3 Removing Substrate and Fabricating Double MINT Layers

Figure 5.6c demonstrates that via height can be reduced by removing the substrate and fabricating an additional MINT layer beneath the devices. Although this approach increases process complexity and cost, it has the potential to improve PP and margins. It is important to emphasize that the backside reveal process is less challenging compared to placing PG devices on the top tier. The process flow for selective wet etching has been published and is well-documented [9]. Furthermore, while the fabrication of a double MINT layer will increase costs, it can be justified if it leads to significant PP enhancement over time. Therefore, it is worth exploring a novel transistor architecture and process flow to achieve this design, referred to as the "hybrid CFET" architecture introduced in the next section.

Takeaways

- **Drawback in Seq and Mono CFET SRAM Bitcell Designs:** The stacked architecture results in relatively tall via connections from the frontside BL and WL signals to the bottom PG devices. These tall vias significantly increase the C_{BL} and C_{WL}, potentially hindering improvements in PPA from node to node.
- **Motivated Design Strategy:** Reduce capacitance overhead by shortening via height with an alternative transistor structure.
- **Non-Recommended Approach:** Reducing NNS or H_{fin} will severely degrade the PP. Additionally, placing PG devices on the top tier would overly complicate the process due to the presence of top devices.
- **Recommended Approach:** Implement substrate removal and fabricate signal lines close to the bottom PG devices.

5.3 Hybrid Complementary Transistor

5.3.1 Coventor Process Flow

Figure 5.7 depicts the key process flow of hybrid CFET using Coventor process emulator,[1] offering a brief overview of the integration flow. The active formation of hybrid CFET is similar to mono CFET, where the top and bottom tiers are tied and

[1] Acknowledgment is given to Hans Mertens for the Coventor process flow advancements.

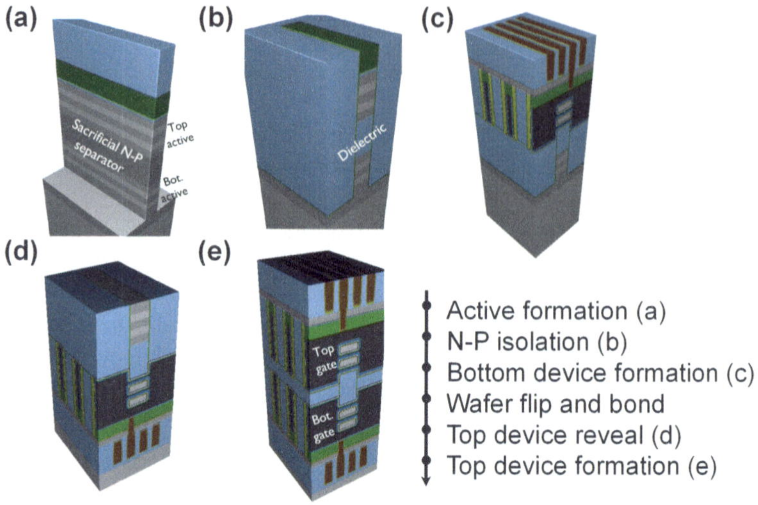

Fig. 5.7 Hybrid CFET process flow description with Coventor process emulator. (**a**) Top and bottom active formation. (**b**) NMOS and PMOS tiers isolation. (**c**) Bottom devices formation. (**d**) Top device reveal. (**e**) Top devices formation

formed together (Fig. 5.7a). This alignment of the active regions between the top and bottom tiers in hybrid CFET eliminates the risk of non-alignment seen in seq CFET designs. The NMOS and PMOS tiers are initially separated by the sacrificial epitaxy layer and then isolated by filling the dielectric after sacrificial layer removal (Fig. 5.7b). Next, the bottom devices and their metal stack are fabricated on the bottom side of the CFET stack (Fig. 5.7c) followed by wafer flipping and bonding. After the substrate thinning and removal (Fig. 5.7d), the top device and topside interconnect are formed (Fig. 5.7e). The selective wet etching during substrate reveal [9] ensures that the substrate surface remains even, avoiding variations in via heights. The process of separate top and bottom tier fabrication acts like seq CFET, thereby causing the top-bottom gates to be non-aligned. However, hybrid CFET enables the double-sided routing capability using the traditional frontside BEOL design rules on both sides. Moreover, the contacts on both sides are exposed to the surrounding oxide without noticeable capacitance impacts due to the assumption of substrate removal. Note that for simplicity, the components of top and bottom stacks are fixed in this work, but in fact they can be reversible to align with logic's BEOL assumptions.

5.3.2 *Comparison Between Seq, Mono, and Hybrid CFETs*

A thorough process comparison between seq and mono CFET was presented in [4, 11, 14, 17, 19, 20, 27]. Seq CFET involves separate top-bottom tier fabrication,

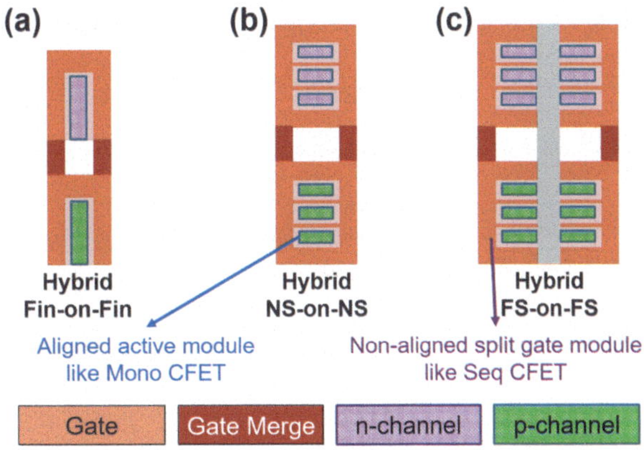

Fig. 5.8 Cross-sectional views of the gate for hybrid (**a**) Fin-on-Fin, (**b**) NS-on-NS, and (**c**) FS-on-FS CFETs

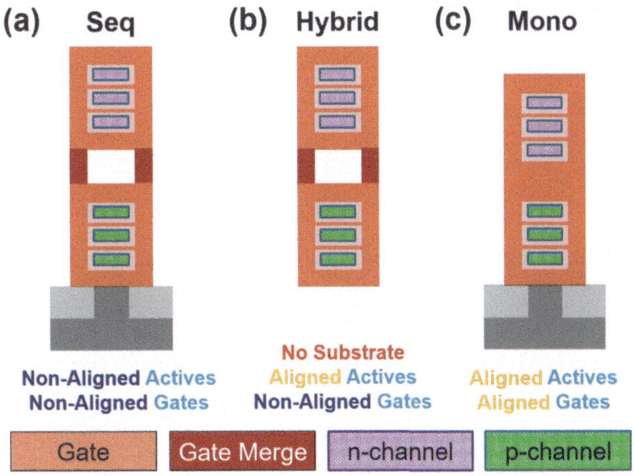

Fig. 5.9 Comparison of the cross-sectional views of the gate for (**a**) seq, (**b**) hybrid, and mono (**c**) NS-on-NS CFETs

non-aligned active region and gate (split gate), a higher number of masks, layer transfer, and bonding processes. On the other hand, mono CFET is characterized by inherent top-bottom fabrication, self-aligned active region and gate (common gate), fewer masks, and high aspect ratio (AR) processes. Hybrid CFET, as the name indicates, combines the process flow elements of both seq and mono CFET schemes, as discussed in Sect. 5.3.1. The key distinction of hybrid CFET lies in the wafer flipping, layer bonding, and substrate removal processes. This flow facilitates the implementation of double-sided interconnect features, wherein the top and bottom devices are created separately using the same design rules employed for frontside BEOL in previous designs. Figure 5.8 presents the schematic views of hybrid CFET with stacked Fin, NS, and FS channels. Figure 5.9 and Table 5.1 provide a comparison of the highlighted process features of seq, mono, and hybrid CFET.

Apart from the standard fabrication cost, seq CFET incurs additional expenses due to layer transfer and repeated modules in both tiers (e.g., replacement metal gate, gate cut, nanosheet patterning, etc.). In contrast, both mono and hybrid CFETs benefit from cost reduction by minimizing these repetitive modules. However, the cost increases again in Hybrid CFET due to the inclusion of extra processes like

Table 5.1 Comparison of process features in seq, mono, and hybrid CFETs (NA: not aligned; A: aligned; AR: aspect ratio)

	Seq CFET	Mono CFET	Hybrid CFET
Active	NA	A	A
Gate	NA	A	NA
	Split gate	Common gate	Split gate
CFET Formation	Separate top-bottom tier fabrication followed by layer transfer and bonding process	Tied top-bottom tier fabrication	Active modules like mono CFET, rest of the flows like seq CFET with wafer flipping and substrate reveal
Process	More masks	High AR	Reduced number of high-AR steps
Additional Gate Merge	Yes	No	Yes
Cost	Higher	Lower	Similar to seq CFET at most and higher than mono CFET
BEOL Categories (this work)	Frontside (above FEOL) Backside (below substrate)		Topside Bottom side (No substrate) (Reversible sides to align with logic's BEOL)

substrate removal and double MINT layer fabrication, which go beyond mono CFET. Consequently, the qualitative estimation suggests that the cost of hybrid CFET is expected to be similar to seq CFET at most and higher than mono CFET.

Takeaways

- **Seq CFET:** Fabricates the top and bottom tiers separately, featuring layer transfer and bonding processes.
- **Mono CFET:** Fabricates the top and bottom tiers together, featuring higher AR and self-aligned processes.
- **Hybrid CFET:** Forms the active layers together but fabricates the top and bottom devices and interconnects separately. This involves wafer flipping, layer bonding, and substrate removal processes.

5.4 SRAM Bitcell Specification for Hybrid CFET Technology

In seq and mono CFET SRAM bitcell designs (Chap. 4), it is consistently assumed that the PG devices are positioned on the p-type bottom tier, exclusively as unipolar PMOS devices. This decision is made to avoid the necessity of fabricating certain processes on the bottom tier (e.g., cross-coupled connection, redundant PG device removal, etc. [10, 12]), which would be in the shadow of the top devices. Otherwise, it may require additional efforts in process flow development beyond the existing steps. However, in the hybrid CFET SRAM bitcell designs, it is possible to place either the PMOS pull-up (PU) devices or NMOS pull-down (PD) devices on the bottom tier due to the double-sided and stack-flipping features. As a result, the 6-transistor (6T) hybrid CFET SRAM bitcell is capable of being designed with n- or p-type PG (N-PG or P-PG) (Fig. 5.10). To explore a comprehensive PPA comparison among HD, HP, and LP applications in N-PG or P-PG-based configurations, please refer to Chaps. 6 and 7.

5.5 HD, HP, and LP Hybrid CFET SRAM Bitcell Designs in A3

5.5.1 HD A3 Hybrid NS-on-NS CFET SRAM Bitcell Design

Figures 5.11/5.12 and 5.13/5.14 depict the top and cross-sectional views, respectively, of the HD A3 hybrid NS-on-NS CFET SRAM bitcell layout design with P-/N-PG. Note that the interconnect solutions remain the same for both P- and N-PG cases, with the only difference being the swapping of the n- and p-channel components, including their respective power rails. It is evident that the W_{WL} of HD A3 hybrid CFET SRAM bitcells become wider, resulting in less R_{WL} compared to A3 seq and mono counterparts (Figs. 4.12 and 4.14 in Chap. 4). This

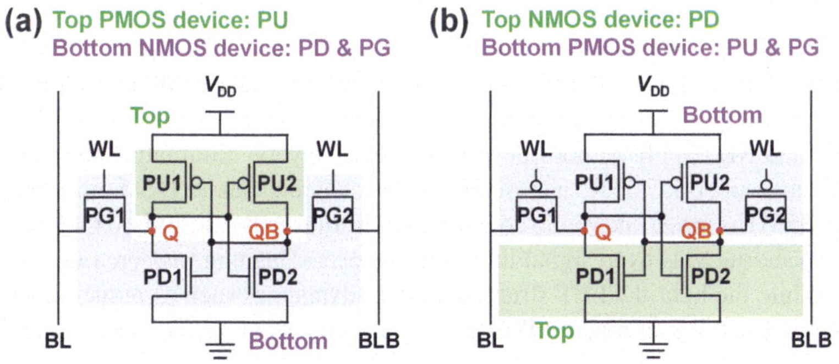

Fig. 5.10 Schematic views of HD 6T CFET SRAM bitcell design with (**a**) N-PG (**b**) P-PG

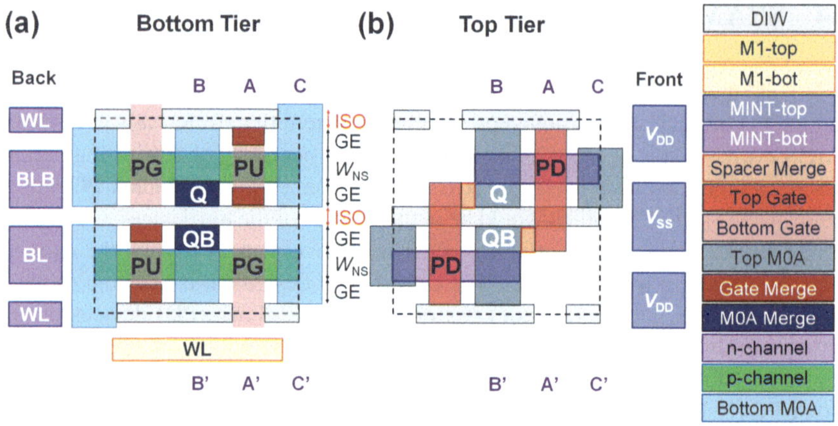

Fig. 5.11 Top views of (**a**) bottom and (**b**) top tiers for HD A3 hybrid NS-on-NS CFET SRAM with P-PG bitcell layout design. ($W_{BL} = 21$ nm; $W_{WL} = 65$ nm; $W_{cell} = 78$ nm; $H_{cell} = 72$ nm)

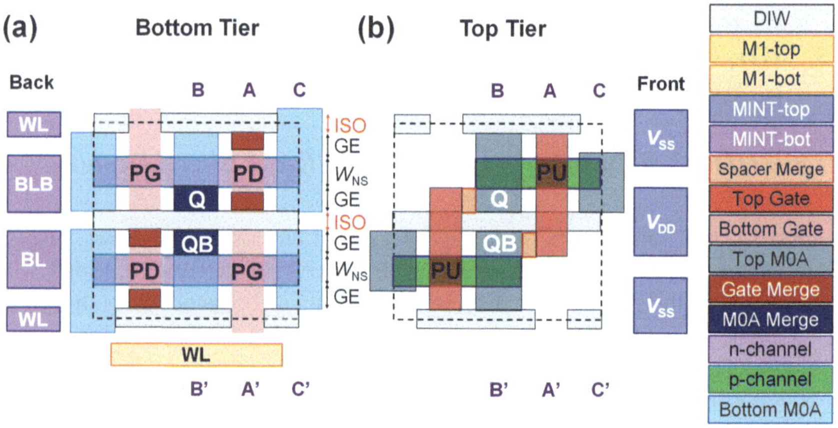

Fig. 5.12 Top views of (**a**) bottom and (**b**) top tiers for HD A3 hybrid NS-on-NS CFET SRAM with N-PG bitcell layout design. ($W_{BL} = 21$ nm; $W_{WL} = 65$ nm; $W_{cell} = 78$ nm; $H_{cell} = 72$ nm)

widening occurs due to the separate placements of power rails and signal lines at their respective metal layers. Furthermore, the design rules for frontside metal (MINT and M1) apply to both power and signal layers, eliminating the necessity of utilizing supervia, BPR, nanoscale through-silicon vias (nTSV), and backside metal (BSM) as in seq and mono counterparts. Additionally, the hybrid CFET design achieves shorter vias from signal lines to PG devices, leading to decreased C_{BL} and C_{WL}. While the hybrid CFET structures offer advantages such as reducing RC and providing flexibility in placing PG devices, it's important to note that the total cell

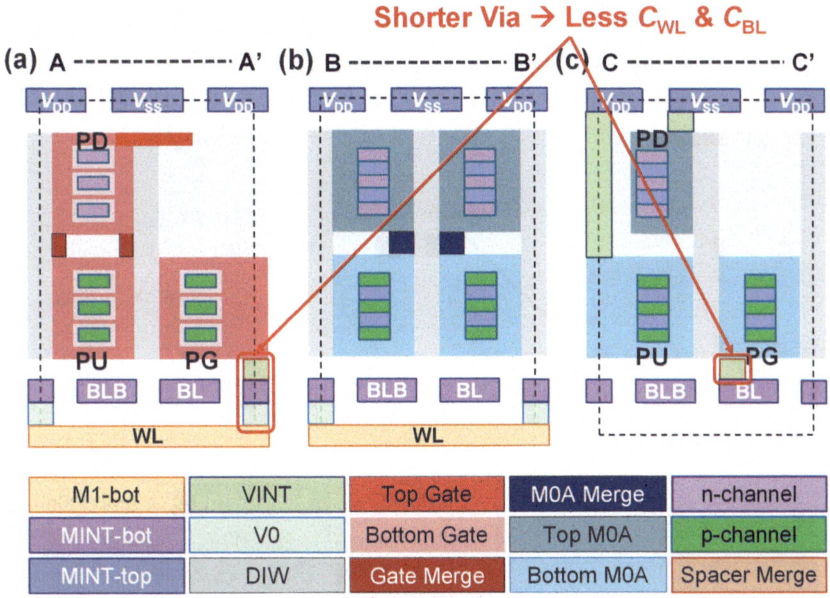

Fig. 5.13 Cross-sectional views along the (**a**) gate (A-A'), (**b**) internal nodes (B-B'; across WL), and (**c**) edge (C-C'; across V_{DD} and V_{SS}) for HD A3 hybrid NS-on-NS CFET SRAM with P-PG bitcell layout design

height (H_{cell}) and cell width (W_{cell}) remain unchanged compared to previous A3 seq and mono CFET bitcell designs.

5.5.2 HP/LP A3 Hybrid NS-on-NS CFET SRAM Bitcell Designs

In HP/LP A3 hybrid CFET SRAM bitcell designs, the fundamental approach involves configuring the distinct number of fin (#fin) or fin height (H_{fin})/NS width (W_{NS}) of PU, PD, and PG devices. These process aspects are aligned with seq counterparts, facilitated by the separate fabrication of the top and bottom tiers as outlined in Sect. 4.7 of Chap. 4. Additionally, the strategic placement of power rails and signal lines on opposing sides mirrors that of HD counterparts, leading to broader WL dimensions compared to A3 seq and mono counterparts. To avoid the redundancy of presenting similar layout design figures, the corresponding layout designs for all bitcell variations are presented in Appendix A.4.

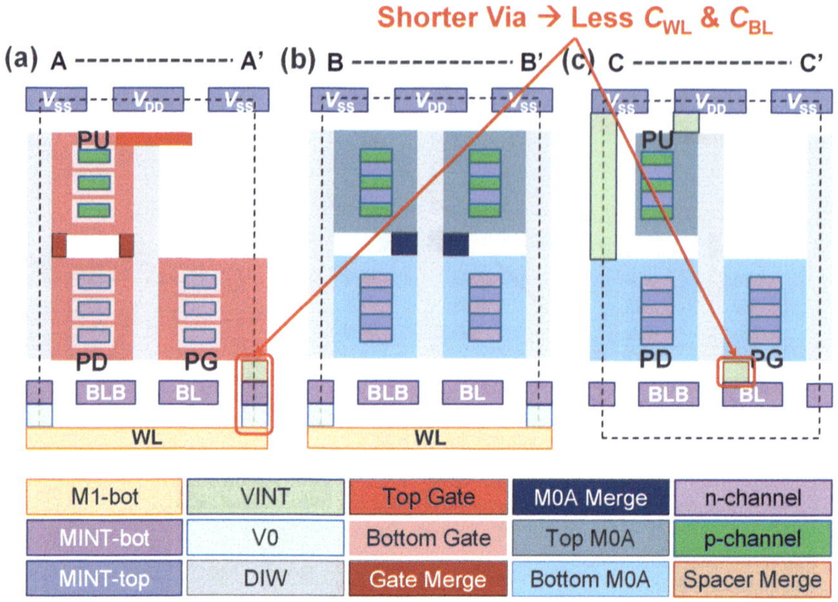

Fig. 5.14 Cross-sectional views along the (**a**) gate (A-A'), (**b**) internal nodes (B-B'; across WL), and (**c**) edge (C-C'; across V_{DD} and V_{SS}) for HD A3 hybrid NS-on-NS CFET SRAM with N-PG bitcell layout design

5.5.3 *Major Challenges and Highlight Process Features in A3 Hybrid CFET SRAM Bitcell Designs*

5.5.3.1 Common Active Region Formation

The top and bottom active regions are created simultaneously in hybrid CFET SRAM bitcell design like mono counterparts. This allows for the interchangeability of PMOS and NMOS epitaxy layers. Notably, the PG device's channel type is determined by the bottom epitaxy, resulting in either P- or N-PG-based hybrid CFET SRAM bitcell.

5.5.3.2 Separate Top and Bottom Device Fabrication

In hybrid CFET SRAM bitcell design, the top and bottom gate stack and interconnect layer are fabricated separately. This process includes wafer flipping, bonding, and substrate reveal in the middle. Consequently, it necessitates additional gate merging, akin to seq CFET counterpart. Moreover, this approach unlocks the potential for double-sided routing by eliminating the thick substrate constraint on the bottom and introducing wafer flipping.

5.5.4 MEOL and Backside Routing Challenge and Solution

In Chap. 4, the BEOL routing for A3 seq and mono CFET SRAM bitcell designs has been categorized as frontside (above active) and backside (below substrate) placements (Table 5.1) [10]. BL and WL signals are always designed at the frontside due to the tight pitch requirement and limited capacitance benefits of using backside signals (longer backside vias cause larger capacitance than frontside vias). As a result, the via height of stacked CFET SRAM bitcell connecting from frontside signals to the bottom PG devices is taller than non-stacked counterparts resulting in larger C_{BL} and C_{WL}. Furthermore, high AR supervia connections between V_{SS} and PD devices, as well as narrower WL, were assumed due to the insufficient space at backside [10]. The previous design approach introduced more difficulties with the via process and the drawback of larger R_{WL}, which could pose high risk when pursuing such a challenging process.

In this chapter, the problems of increased R_{WL}, C_{BL}, and C_{WL}, and high AR supervia are effectively addressed through the novel design of hybrid CFET SRAM bitcell. Hybrid CFET incorporates 3D processing on both sides of the wafer with substrate thinning and removal assumptions, thereby providing double-sided routing capability. This design flexibility allows for the placement of signal lines either on the topside or bottom-side BEOL under the same metal stack design rules in SRAM bitcell. Consequently, in hybrid CFET SRAM bitcell design, the via connections for signal lines become shallower by placing BL and WL at the side that is closer to PG devices compared to seq and mono counterparts. Furthermore, V_{DD} and V_{SS} power rails can be placed at the topside without requiring extra metal rail at the same metal layer as the signal lines.

5.5.5 Parasitics Comparison for HD A3 NS-on-NS CFET SRAM Bitcell

Figure 5.15 demonstrates the parasitic RC results of HD A3 seq, mono, and hybrid CFET SRAM bitcells. R_{BL} remains the same across the seq, mono, and hybrid NS-on-NS CFET cases due to the unchanged W_{BL}. In A3 seq and mono CFET SRAM bitcell designs, WL and V_{SS} were positioned at the same frontside metal layer (M1) due to the limited space at the backside resulting in narrower W_{WL} [10]. However, in hybrid CFET SRAM bitcell, 41% of R_{WL} reduction is achieved by having wider WL since the power rails (V_{DD} and V_{SS}) are placed on the opposite side of BEOL. Although wider WL may contribute to the increase in C_{WL}, it only represents a part of the total C_{WL} in hybrid CFET SRAM bitcell design. Fortunately, the overall C_{WL} is still decreased due to the stronger impact of shorter via height. Thus, hybrid NS-on-NS CFET SRAM bitcell exhibits significant reductions of 25% in C_{BL} and 35% in C_{WL} primarily by shortening the via height as compared to seq and mono counterparts.

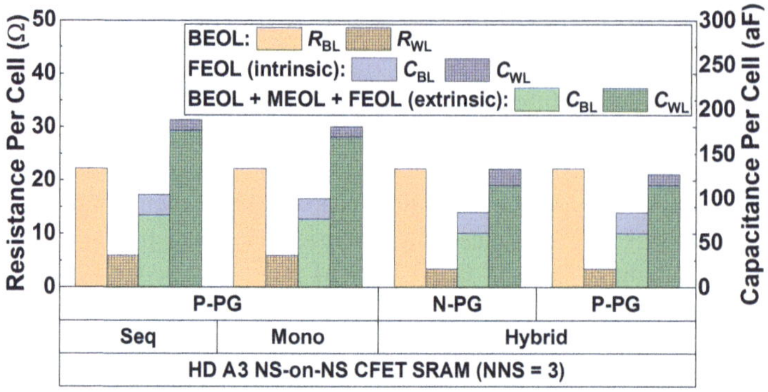

Fig. 5.15 Parasitic resistance and capacitance per cell of BL and WL for HD A3 seq, mono, and hybrid NS-on-NS CFET SRAM bitcell designs

Takeaways

- **Benefits in Proposed Hybrid CFET SRAM Bitcell Designs Compared to Seq and Mono Counterparts:**
  - **Reduction of C_{BL} and C_{WL}:** Achieved by shortening the via connections to signal lines.
  - **Wider WL:** Accomplished by placing signal lines and power lines on opposite sides of the interconnect layers.

5.6 Summary of 6T SRAM Bicell Design and Area Benchmark

Chapters 4 and 5 comprehensively explore the design aspects of seq, mono, and hybrid CFET SRAM bitcells in A5 and A3, including HD, HP, and LP options. For more descriptions of SRAM bitcell setup, please refer to Sect. 2.1.2 in Chap. 2. Notably, each bitcell layout design is summarized in Appendix A.2, A.3, and A.4. This section provides a lucid overview of design rules, technology labels, BEOL assumptions, and bitcell area benchmarks. The provided information will play a crucial role in the ensuing subarray-level and macro-level PPA analysis discussed in Chaps. 6 and 7, respectively. It is important to note that the design assumptions in this work are relatively aggressive compared to those in the recently published work [3]. Future design roadmaps should be continuously updated to align with the latest design rules.

5.6.1 Design Rules and Technology Label

The SRAM bitcell layout designs presented in Appendix A.2, A.3, and A.4 strictly adhere to their respective design criteria, as outlined in Table 5.2 for precise specifications. Given the intricacies involved in the design and technology assumptions for various SRAM bitcell configurations, Tables 5.3 and 5.4 serve as a comprehensive reference for the following subarray- and macro-level PPA analysis in Chaps. 6 and 7. It succinctly summarizes the technology labels and BEOL assumptions with distinct characteristics for each SRAM bitcell design, ranging from A14 to A3. It is important to highlight that the 2-nm (N2) NSFET and 7-Å-compatible technology (A7) FSFET SRAM bitcell designs have been excluded from the PPA comparison. The reverse scaling from N3 FinFET to N2 NSFET, attributed to wider active width, could potentially degrade both subarray- and macro-level PPA due to increased area and total RC of WL and BL (Sect. 5.6.2). Therefore, A14 NSFET is chosen to represent NS-based SRAM bitcell designs, offering bitcell area scaling with a relatively larger potential in achieving node-to-node gain. Conversely, the A7 FSFET SRAM bitcell design provides only a 9% bitcell area scaling, albeit at the expense of a narrower BL compared to A10 counterparts. This increase in R_{BL} may result in a worse write margin, leading to a higher write failure risk. Consequently, A10 FSFET is selected to represent FS-based SRAM bitcell designs.

As SRAM technology undergoes scaling, design parameters such as gate length (L_g), metal pitch (MP), contacted poly pitch (CPP), critical dimension (CD), and spacing of each metal layer should be proportionally scaled to enhance the bitcell area density, as outlined in Table 5.2 [10]. However, the scalability of GC is hindered by limitations in metal patterning via lithography. Additionally, there is an observed increase in GC for A5 mono CFET SRAM bitcell designs due to high AR process requirements. Consequently, a solution involving DIW with a 7 nm width has been proposed for A3 CFET SRAM bitcell designs. Furthermore, the concept of self-aligned gate/M0A merge is introduced in A3 seq CFET SRAM bitcell designs, obviating the need for additional specifications present in A5 counterparts, such as gate to internal node contact (GI), gate to internal node extension (GIE), and gate to internal node peak (GIP) shown in Sect. 4.4.1 of Chap. 4. On the other hand, the gate extension (GE) increases from 7.5 to 9 nm when transitioning to A3, attributed to the double work function in DIW process described in Sect. 4.6.4.2 of Chap. 4. Stress values for each technology are derived from the compact model [8, 20] under nominal V_{DD} (0.7 V) and temperature (27 °C) conditions.

Table 5.3 presents 18 bitcell design options ranging from A14 to A3, along with their corresponding nomenclature delineating the differences across various design parameters. These parameters include transistor type, channel type, PG device type, fin height (H_{fin}), number of stacked nanosheets (*NNS*), fin width (W_{fin}), nanosheet width (W_{NS}), BL width (W_{BL}), BL length (L_{BL}), cell width (W_{cell}), WL width (W_{WL}), WL length (L_{WL}), and cell height (H_{cell}). The labels x, s, m, and h denote non-stacked FET, seq CFET, mono CFET, and hybrid CFET, respectively, while 1, 2, and 3 signify Fin-, NS-, and FS-based channels, respectively. Additionally, n and

Table 5.2 6T SRAM bitcell design rules and other parameters from A14 to A3. (S: Seq CFET; M: Mono CFET; H: Hybrid CFET)

Parameters	A14 & A10		A5 & A3			
Transistor Type	NS (A14) FS (A10)		Seq/Mono (A5) Seq/Mono/Hybrid (A3)			
Gate Length (L_g)	15 nm		12 nm			
Metal Pitch (MP)	18 nm		16 nm			
Contacted Poly Pitch (CPP)	42 nm		39 nm			
M0A/MINT Spacing	9 nm		7 nm			
M1 Spacing	14 nm		10 nm			
Gate-Cut (GC)	12 nm		12 nm (Seq) 16 nm (Mono)			
Gate Extension (GE)	7.5 nm		7.5 nm (A5) 9 nm (A3)			
Gate to Internal Node Contact (GI) Gate to Internal Node Extension (GIE) Gate to Internal Node Peak (GIP)	–		9 nm (A5 Seq)			
Isolation Wall Width (ISO)	–		7 nm			
Nominal Supply Voltage (V_{DD})	0.7 V					
Nominal Temperature (TEM)	27 °C					
	NS	FS	S/M Fin	M NS	S/H NS	S/M/H FS
N-channel Stress (S_N) (GPa)	0.694		0.962	0.923	0.8936	0.946
P-channel Stress (S_P) (GPa)	1.7		1.850	1.7412	1.820	1.77

p designate N-PG- and P-PG-based SRAM bitcell, while a and b represent wider W_{NS} of 15 and 20 nm, respectively. The listed H_{fin}, NNS, and W_{fin} values are in line with optimized logic dimensions [8, 20]. It is noteworthy that the scaling of bitcell area in this book is predicated on a fixed W_{NS} of 11 nm as technology progresses, with exceptions found in A3.h3pa and A3.h3pb. W_{BL} is determined by H_{cell}, the number of metal tracks, and MP at the MINT layer. L_{BL} is equivalent to W_{cell}, twice the CPP. L_{WL} equals H_{cell}, predominantly decided by W_{fin}, W_{NS}, GC, and GE.

In nano-scaled technology nodes, conventional copper (Cu) interconnects encounter several challenges, including significantly increased resistivity, reduced reliability, and limitations in liner scaling to sustain interconnect RC scaling beyond 20 nm in MP [28]. Ruthenium (Ru) has been proposed as an alternative metal material for nano-scale interconnects due to its barrierless nature, resistance to electromigration and oxidation, and high melting point [16, 28, 30]. Consequently, Ru is assumed for MINT to M3 metal layers in A14 to A3 technologies (Table 5.4). Additionally, for less-scaled MP in M4 to M12 metal layers, Cu interconnects with a 3.5 nm TaNCo barrier-liner (CuTaNCo) are assumed due to their superior resistivity compared to other barrier materials [5] (Table 5.4).

Table 5.3 Representative names of A14, A10, A5, and A3 technology indicating its transistor, channel, PG type, H_{fin}, NNS, W_{fin}, W_{NS}, W_{BL}, L_{BL}, W_{cell}, W_{WL}, L_{WL}, and H_{cell}

Tech	Transistor	Channel	PG Type	H_{fin}(nm)/ NNS	$W_{\text{fin}} W_{\text{NS}}$(nm)	Name	W_{BL}(nm)	$L_{\text{BL}} = W_{\text{cell}}$(nm)	W_{WL}(nm)	$L_{\text{WL}} = H_{\text{cell}}$(nm)
A14	NSFET	NS	NMOS	4	11	A14.x2n	27	84	70	150
A10	FSFET	FS		3		A10.x3n	11			112
A5	Seq CFET	Fin	PMOS	35	5	A5.s1p	27.5	78	65	85
		NS		2	11	A5.s2p	33.5			97
	Mono CFET	Fin		35	5	A5.m1p	21			72
		NS		2	11	A5.m2p	27			84
A3	Seq CFET	Fin		35	5	A3.s1p	15		41	60
		NS		3	11	A3.s2p	21			72
		FS				A3.s3p	12			54
	Mono CFET	Fin		35	5	A3.m1p	15			60
		NS		3	11	A3.m2p	21			72
		FS				A3.m3p	12			54
	Hybrid CFET	NS	NMOS	3	11	A3.h2n	21		68	72
		FS				A3.h3n	12			54
		NS	PMOS			A3.h2p	21			72
		FS				A3.h3p	12			54
					15	A3.h3pa	16			62
					20	A3.h3pb	21			72

Table 5.4 BEOL assumptions for A14, A10, A5, and A3 nodes at imec. (P: pitch; M: material; CD: critical dimension; S: spacing; T: thickness)

	A14 & A10					A5 & A3				
Metal Layer	P	M	CD	S	T	P	M	CD	S	T
MINT (M0A)	18	Ru	10	8	27	16	Ru	9	7	24
MINT (Poly)	18	Ru	10	8	27	16	Ru	9	7	24
M1	28	Ru	16	12	49	26	Ru	16	10	39
M2	18	Ru	10	8	27	16	Ru	9	7	24
M3	28	Ru	16	12	49	26	Ru	16	10	39
M4	28	CuTaNCo	16	12	28	28	CuTaNCo	16	12	28
M5	28	CuTaNCo	16	12	28	28	CuTaNCo	16	12	28
M6	80	CuTaNCo	40	40	80	48	CuTaNCo	24	24	48
M7	80	CuTaNCo	40	40	80	48	CuTaNCo	24	24	48
M8	80	CuTaNCo	40	40	80	80	CuTaNCo	40	40	80
M9	80	CuTaNCo	40	40	80	80	CuTaNCo	40	40	80
M10	80	CuTaNCo	40	40	80	80	CuTaNCo	40	40	80
M11	80	CuTaNCo	40	40	80	80	CuTaNCo	40	40	80
M12	80	CuTaNCo	40	40	80	80	CuTaNCo	40	40	80

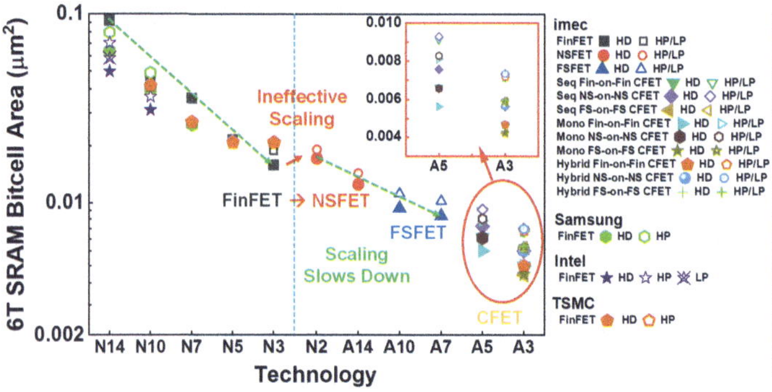

imec Design Rules		
Tech	CPP	MP
N14	90	64
N10	64	44
N7	56	40
N5	48	28
N3	45	22
N2	45	21
A14	42	18
A10	42	18
A7	39	16
A5	39	16
A3	39	16

Fig. 5.16 HD/HP/LP 6T SRAM bitcell area scaling benchmark across different companies, namely imec [10–12, 18, 29], Samsung Electronics [24–26], Intel [7, 15], and TSMC [1, 2, 6], and corresponding imec's key design rules from N14 to A3

5.6.2 HD, HP, and LP 6T SRAM Bitcell Area Benchmark

Figure 5.16 compiles data from imec [10–12, 18, 29], Samsung Electronics [24–26], Intel [7, 15], and TSMC [1, 2, 6], providing an extensive analysis of the 6T SRAM bitcell areas across HD, HP, and LP bitcell design categories ranging from 14-nm (N14) to A3. Figure 4.23 (Chap. 4) presents the relative area scaling of HD, HP, and LP CFET SRAM bitcells from A14 to A3. A3 CFET SRAM bitcells offer up to 63% and 50% of bitcell area scaling compared to A14 NSFET and A10 FSFET counterparts, respectively. Additionally, Fig. 5.17 offers a clear overview

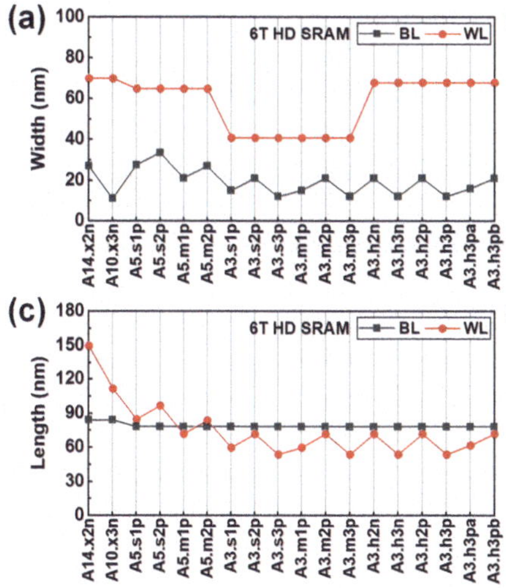

Fig. 5.17 BL and WL width and length for (**a**)(**c**) HD and (**b**)(**d**) HP/LP SRAM bitcell designs from A14 to A3. (BL length = W_{cell}; WL length = H_{cell})

of the evolution in BL and WL sizes within 6T SRAM bitcells. It's worth noting that BL length is identical to the W_{cell}, which is precisely twice the contacted poly pitch (CPP). In parallel, WL length corresponds to the H_{cell}. Considering these insights, the gradual narrowing of BL may pose challenges, leading to increased R_{BL} which in turn impacts node-to-node PP enhancement in ultra-scaled nodes. Further exploration of RC implications is provided in Sect. 6.2 of Chap. 6.

Takeaways

- The N2 NSFET and A7 FSFET SRAM bitcell designs are excluded from the research focus, as they are likely to degrade the PPA results compared to the A14 and A10 counterparts, respectively.
- The PPA evaluation will be conducted on the A14 NSFET, A10 FSFET, A5 CFET, and A3 CFET SRAM bitcell designs.

5.7 Conclusions

This chapter conducts an evaluation of bitcell layouts, critical process flows, and interconnect solutions for hybrid CFET SRAM bitcell designs in A3. For a comprehensive understanding, the top views of HD, HP, and LP stacked NS and FS

CFET SRAM bitcell layout designs are systematically presented in the Appendix (Sect. A.4).

In both A5 and A3 seq and mono CFET SRAM bitcell designs, the increased height of vias between signals and bottom PG devices introduces a higher RC issue compared to the non-stacked architecture. Additionally, the H_{cell} reduction from A5 to A3 results in a narrower W_{WL}, leading to a larger R_{WL}. This chapter introduces the hybrid CFET into A3 SRAM bitcell design to address the aforementioned issues in seq and mono counterparts.

Benefiting from its double-sided interconnect capability, the A3 hybrid CFET SRAM bitcell design allows for shorter via height and a wider WL width. This innovative bitcell design demonstrates a remarkable reduction of up to 41%, 25%, and 35% in R_{WL}, C_{BL}, and C_{WL}, respectively, when compared to seq and mono counterparts in HD option. Consequently, it emerges as a promising candidate among other CFET alternatives, an assertion substantiated by the forthcoming PPA analysis in Chaps. 6 and 7. It is crucial to highlight that the incorporation of the double-sided process may introduce heightened complexity, necessitating thorough development and validation with experimental data in future work.

References

1. Chang, J., et al. (2017). 12.1 A 7 nm 256 Mb SRAM in high-k metal-gate FinFET technology with write-assist circuitry for low-VMIN applications. In *2017 IEEE International Solid-State Circuits Conference (ISSCC)* (pp. 206–207). https://doi.org/10.1109/ISSCC.2017.7870333
2. Chang, J., et al. (2020). 15.1 A 5 nm 135 Mb SRAM in EUV and high-mobility-channel FinFET technology with metal coupling and charge-sharing write-assistCircuitry schemes for high-density and low-VMIN applications. In *2020 IEEE International Solid-State Circuits Conference (ISSCC)* (pp. 238–240). https://doi.org/10.1109/ISSCC19947.2020.9062967
3. Chang, J., et al. (2023). A 3 nm 256 Mb SRAM in FinFET technology with new array banking architecture and write-assist circuitry scheme for high- density and low-VMIN applications. In *2023 IEEE Symposium on VLSI Technology and Circuits (VLSI Technology and Circuits)* (pp. 1–2). https://doi.org/10.23919/VLSITechnologyandCir57934.2023.10185287
4. Chehab, B., et al. (2021). Design-technology co-optimization of sequential and monolithic CFET as enabler of technology node beyond 2 nm. In C.-M. Yuan & R.-H. Kim (Eds.), *Design-process-technology co-optimization XV* (Vol. 11614, p. 116140D). International Society for Optics and Photonics, SPIE. https://doi.org/10.1117/12.2583395
5. Ciofi, I., et al. (2019). RC benefits of advanced metallization options. *IEEE Transactions on Electron Devices, 66*(5), 2339–2345. https://doi.org/10.1109/TED.2019.2902031
6. Clinton, M., et al. (2017). 12.3 A low-power and high-performance 10 nm SRAM architecture for mobile applications. In *2017 IEEE International Solid-State Circuits Conference (ISSCC)* (pp. 210–211). https://doi.org/10.1109/ISSCC.2017.7870335
7. Guo, Z., et al. (2020). A 10 nm SRAM design using gate-modulated self-collapse write assist enabling 175 mV VMIN reduction with negligible power overhead. In *2020 IEEE Symposium on VLSI Circuits* (pp. 1–2). https://doi.org/10.1109/VLSICircuits18222.2020.9162782
8. Jang, D., et al. (2017). Device exploration of NanoSheet transistors for sub-7-nm technology node. *IEEE Transactions on Electron Devices, 64*(6), 2707–2713. https://doi.org/10.1109/TED.2017.2695455

9. Jourdain, A. et al. (2020). Extreme wafer thinning and nano-TSV processing for 3D heterogeneous integration. In *2020 IEEE 70th Electronic Components and Technology Conference (ECTC)* (pp. 42–48). https://doi.org/10.1109/ECTC32862.2020.00020
10. Liu, H.-H., et al. (2023). CFET SRAM DTCO, interconnect guideline, and benchmark for CMOS scaling. *IEEE Transactions on Electron Devices, 70*(3), 883–890. https://doi.org/10.1109/TED.2023.3235701
11. Liu, H.-H., et al. (2023). CFET SRAM with double-sided interconnect design and DTCO benchmark. *IEEE Transactions on Electron Devices* 1–8. https://doi.org/10.1109/TED.2023.3305322
12. Liu, H.-H., et al. (2023). DTCO of sequential and monolithic CFET SRAM. In R.-H. Kim & N.V. Lafferty (Eds.), *DTCO and computational patterning II* (Vol. 12495, p. 124950Z). International Society for Optics and Photonics. SPIE. https://doi.org/10.1117/12.2657524
13. Liu, H.-H., et al. (2022). Extended methodology to determine SRAM write margin in resistance-dominated technology node. *IEEE Transactions on Electron Devices, 69*(6), 3113–3117. https://doi.org/10.1109/TED.2022.3165738
14. Mirabelli, G., et al. (2023). Design-technology co-optimization overview of CFET architecture. In R.-H. Kim & N. V. Lafferty (Eds.), *DTCO and computational patterning II* (Vol. 12495, p. 124950M). International Society for Optics and Photonics. SPIE. https://doi.org/10.1117/12.2657726
15. Natarajan, S., et al. (2014). A 14 nm logic technology featuring 2nd-generation FinFET, air-gapped interconnects, self-aligned double patterning and a 0.0588 μm2 SRAM cell size. In *2014 IEEE International Electron Devices Meeting* (pp. 3.7.1–3.7.3). https://doi.org/10.1109/IEDM.2014.7046976
16. Pokhrel, A., et al. (2022). MP18–26 Ru direct-etch integration development with leakage improvement and increased aspect ratio. In *2022 IEEE International Interconnect Technology Conference (IITC)* (pp. 61–63). https://doi.org/10.1109/IITC52079.2022.9881278
17. Ryckaert, J., et al. (2018). The Complementary FET (CFET) for CMOS scaling beyond N3. In *2018 IEEE Symposium on VLSI Technology* (pp. 141–142). https://doi.org/10.1109/VLSIT.2018.8510618
18. Salahuddin, S., et al. (2020). Buried power SRAM DTCO and system-level bench-marking in N3. In *2020 IEEE Symposium on VLSI Technology* (pp. 1–2). https://doi.org/10.1109/VLSITechnology18217.2020.9265076
19. Schuddinck, P., et al. (2019). Device-, Circuit-Block-level evaluation of CFET in a 4 track library. In *2019 Symposium on VLSI Technology*. T204–T205. https://doi.org/10.23919/VLSIT.2019.8776513
20. Schuddinck, P., et al. (2022). PPAC of sheet-based CFET configurations for 4 track design with 16 nm metal pitch. In *2022 IEEE Symposium on VLSI Technology and Circuits (VLSI Technology and Circuits)* (pp. 365–366). https://doi.org/10.1109/VLSITechnologyandCir46769.2022.9830492
21. Singh, T., et al. (2020). 2.1 Zen 2: the AMD 7 nm energy-efficient high-performance x86-64 microprocessor core. In *2020 IEEE International Solid-State Circuits Conference (ISSCC)* (pp. 42–44). https://doi.org/10.1109/ISSCC19947.2020.9063113
22. Singh, T., et al. (2017). 3.2 Zen: a next-generation high-performance ×86 core. In *2017 IEEE International Solid-State Circuits Conference (ISSCC)* (pp. 52–53). https://doi.org/10.1109/ISSCC.2017.7870256
23. Singh, T., et al. (2018). Zen: an energy-efficient high-performance 86 core. *IEEE Journal of Solid-State Circuits, 53*(1), 102–114. https://doi.org/10.1109/JSSC.2017.2752839
24. Song, T., et al. (2014). 13.2 A 14 nm FinFET 128Mb 6T SRAM with VMIN-enhancement techniques for low-power applications. In *2014 IEEE International Solid-State Circuits Conference Digest of Technical Papers (ISSCC)* (pp. 232–233). https://doi.org/10.1109/ISSCC.2014.6757413
25. Song, T., et al. (2017). A 10 nm FinFET 128 Mb SRAM with assist adjustment system for power, performance, and area optimization. *IEEE Journal of Solid-State Circuits, 52*(1), 240–249. https://doi.org/10.1109/JSSC.2016.2609386

26. Song, T., et al. (2018). A 7 nm FinFET SRAM using EUV lithography with dual write-driver-assist circuitry for low-voltage applications. In *2018 IEEE International Solid – State Circuits Conference (ISSCC)* (pp. 198–200). https://doi.org/10.1109/ISSCC.2018.8310252
27. Subramanian, S., et al. (2020). First monolithic integration of 3D complementary FET (CFET) on 300 mm wafers. In *2020 IEEE Symposium on VLSI Technology* (pp. 1–2). https://doi.org/10.1109/VLSITechnology18217.2020.9265073
28. Tőkei, Z., et al. (2020). Inflection points in interconnect research and trends for 2 nm and beyond in order to solve the RC bottleneck. In *2020 IEEE International Electron Devices Meeting (IEDM)* (pp. 32.2.1–32.2.4). https://doi.org/10.1109/IEDM13553.2020.9371903
29. Weckx, P., et al. (2019). Novel forksheet device architecture as ultimate logic scaling device towards 2 nm. In *2019 IEEE International Electron Devices Meeting (IEDM)* (pp. 36.5.1–36.5.4). https://doi.org/10.1109/IEDM19573.2019.8993635
30. Wen, L. G., et al. (2016). Ruthenium metallization for advanced interconnects. In *2016 IEEE International Interconnect Technology Conference/Advanced Metallization Conference (IITC/AMC)* (pp. 34–36). https://doi.org/10.1109/IITC-AMC.2016.7507651

Part III
SRAM Subarray- and Macro-level DTCO PPA Analysis

Chapter 6
Subarray-Level Performance-Power-Area Benchmark of CFET SRAM Subarray

Given the significant impact of the static random-access memory (SRAM) bitcell area on both subarray-level and macro-level power, performance, and area (PPA) [7, 8], the SRAM bitcell scaling becomes increasingly critical in advanced technology nodes. Chapters 4 and 5 explore innovative strategies to further reduce the footprint of the 6-transistor (6T) SRAM bitcell within sequential (seq), monolithic (mono), and hybrid complementary field-effect transistors (CFET) architectures in 5-Å-compatible technology (A5) and 3-Å-compatible technology (A3) [7–9].

In Chap. 4, the substitution of gate-cuts (GC) with dielectric isolation walls (DIW) leads to up to 17% reduction in bitcell area for high-density (HD) A3 seq and mono CFET SRAM bitcell designs compared to A5 counterparts [7]. However, this innovative CFET SRAM bitcell approach comes with certain trade-offs due to its aggressive bitcell area scaling. These trade-offs include narrower bitlines (BL) and wordlines (WL), resulting in increased BL resistance (R_{BL}) and WL resistance (R_{WL}) by factors ranging from 1.3 to 3 and from 4% to 70%, respectively, compared to A5 counterparts. Furthermore, CFET SRAM bitcells exhibit larger total capacitance compared to nanosheet (NS) and forksheet (FS) FET counterparts due to the tall via connections from BL and WL to the bottom pass-gate (PG) devices. These significant increases in resistance and capacitance (RC) act as a bottleneck for node-to-node power and performance (PP) enhancements. It highlights the importance of back-end-of-line (BEOL) optimization for advanced SRAM bitcell scaling. Therefore, this chapter provides interconnect design guidelines to address these challenges and enable further improvements in subarray-level PPA.

Chapter 5 introduces the benefits of hybrid CFET in SRAM bitcell design, with a specific focus on decreasing WL width and via height [8]. In the case of HD A3 hybrid CFET SRAM bitcell, it offers substantial improvements, including a 41% reduction in R_{WL}, a 25% decrease in C_{BL}, and a 35% decrease in C_{WL} when compared to seq and mono counterparts. This chapter demonstrates that the reduction in RC significantly improves write margin (WM) and read delay (RD) by 75 mV and 26%, respectively, in HD scenarios (Sect. 6.6). Moreover, the HD

H.-H. Liu, F. Catthoor, *Circuit-Technology Co-Optimization of SRAM Design in Advanced CMOS Nodes*, https://doi.org/10.1007/978-3-031-76109-6_6

A3 hybrid CFET design achieves an 80% enhancement in the macro-level energy-delay-area product (EDAP) compared to the 14-Å-compatible technology (A14) as discussed in Chap. 7. These compelling results motivate further research and development of the process flow for hybrid CFET device in SRAM technology in the future.

This chapter focuses on the subarray-level PPA analysis results of all the device options, ranging from A14 to A3, across high-density (HD), high-performance (HP), and low-power (LP) subarray designs. For the technology labels assigned to each bitcell and the assumed BEOL conditions, please refer to Sect. 5.6 in Chap. 5. Section 6.1 provides the experimental setup of SRAM subarray design. Section 6.2 presents the overall RC benchmark. Section 6.3 displays the results for hold and read stability under worst-case temperature conditions and a Pelgrom coefficient (A_{VT}) of 1 mV·μm. Section 6.4 explores the effects of channel stress on both RSNM and WM. The stress engineering associated with the top device in CFET architecture is anticipated to pose challenges [12, 13], potentially impacting the RSNM and WM outcomes, which are the crucial metrics for assessing SRAM functionality. Consequently, this section exclusively scrutinizes RSNM and WM to ensure functionality prior to conducting additional PPA analyses. Section 6.5 delves into the influence of channel geometry on each respective metric. Section 6.6 offers the comprehensive subarray-level PPA benchmark, along with guidelines for adjusting subarray size and optimizing interconnects. Finally, Sect. 6.7 concludes this chapter.

6.1 Experimental Setup of SRAM Subarray Design

The simulation circuits include address/timing control, precharge circuits, row decoder, column multiplexer (4–1; M4 or 8–1; M8), sense amplifiers, write drivers, and the SRAM bitcell array. The subarray circuit designs across different technology nodes are almost identical. The only differences lie in the WL and precharge signals, which are designed as active high/low and low/high, respectively, for N-PG/P-PG-based SRAM subarrays. The simulation methodologies follow the description in Chap. 2. Table 6.1 presents subarray sizes for HD (HD-1 and HD-2), HP, and LP cases. The reason for having two types of HD designs with the same capacity is explained in Sect. 6.2. To ensure a fair comparison, all benchmark simulations are executed at the nominal temperature (27 °C), nominal power supply (V_{DD}) (0.7 V), and typical–typical (TT) corner. This chapter employs compact models as reported in A14 [4] and A5/A3 [13], with consistent assumptions regarding effective off current (~0.02 nA) for each technology.

Table 6.1 SRAM subarray size for each node. (N_{row}: number of row; N_{col}: number of column)

Design	PU:PG:PD (N-/P-PG)	N_{row} (R)	N_{col} (C)	Subarray Size
HD-1	1:1:1	256	288	$1024 \times 72M4$
HD-2	1:1:1	128	576	$1024 \times 72M8$
HP	1:2:2/2:2:1	256	144	$1024 \times 36M4$
LP	1:1:2/2:1:1	256	144	$1024 \times 36M4$

6.2 Parasitic RC Benchmark of SRAM Bitcell Designs

Figure 6.1 illustrates the parasitic resistance (R_{BL} and R_{WL}) and capacitance (C_{BL} and C_{WL}) associated with BL and WL benchmarking across HD, HP, and LP SRAM bitcells. The extraction of these RC values followed the methodology outlined in Sect. 2.1.2.2 of Chap. 2, using Quickcap. These parameters significantly influence subarray performance and energy consumption. The nominal results consider the specified logic dimensions, including fin height (H_{fin}), number of stacked NS (NNS), and NS width (W_{NS}), as outlined in Table 5.3 of Chap. 5.

The cell height scaling has a direct impact on BL width resulting in increased R_{BL} and the demand for R_{BL} reduction techniques. Figure 6.1 demonstrates a consistent and more substantial reduction in R_{WL} between nodes compared to the reduction in R_{BL}. This disparity primarily arises from the aggressive scaling of cell height, in contrast to the more constrained scaling of cell width [7, 8]. Thus, WL width exceeds BL width in each technology, and this trend is further influenced by thickness assumptions based on design rules (Table 5.4), resulting in less R_{WL} compared to R_{BL}. Hence, it is advisable to reconsider the SRAM subarray's design, favouring a longer WL and a shorter BL configuration (HD-2) instead of a typical square-like (HD-1) subarray to mitigate the risk of write failure [10].

Despite the implementation of the buried power rail (BPR) technique in both A14 [9] and A10 [15] SRAM bitcell designs, there is still a significant increase in R_{BL} as the technology scales down from A14 to A10. To address this challenge, the spacer merge technique is introduced in A5 (already with BPR assumption) to reduce R_{BL} by decreasing the number of MINT metal tracks. However, R_{BL} once again increases from A5 to A3, highlighting the need for back-end-of-line (BEOL) optimization for SRAM bitcell scaling. Furthermore, seq and mono CFET SRAM bitcells exhibit higher BEOL capacitance than their NSFET and FSFET counterparts due to the tall via (VINT) connections from BL/WL to the bottom PG devices. These increased RC values in A3 CFET SRAM bitcells create a bottleneck for node-to-node PP gain. Therefore, a hybrid CFET SRAM bitcell design is proposed in Chap. 5 to reduce R_{WL}, C_{BL}, and C_{WL} without any R_{BL} degradation compared to other A3 counterparts.

Figure 6.2 displays relative changes in RC parameters for HD cases, referencing A14.x2n and A5.s2p. Transitioning from A14.x2n to A5.s2p reveals significant RC improvements, including a 16% gain in R_{BL}, a 15% reduction in R_{WL}, and an impressive 31% decrease in C_{WL}. These enhancements result from bitcell area scaling and the application of the spacer merge technique. However, as the

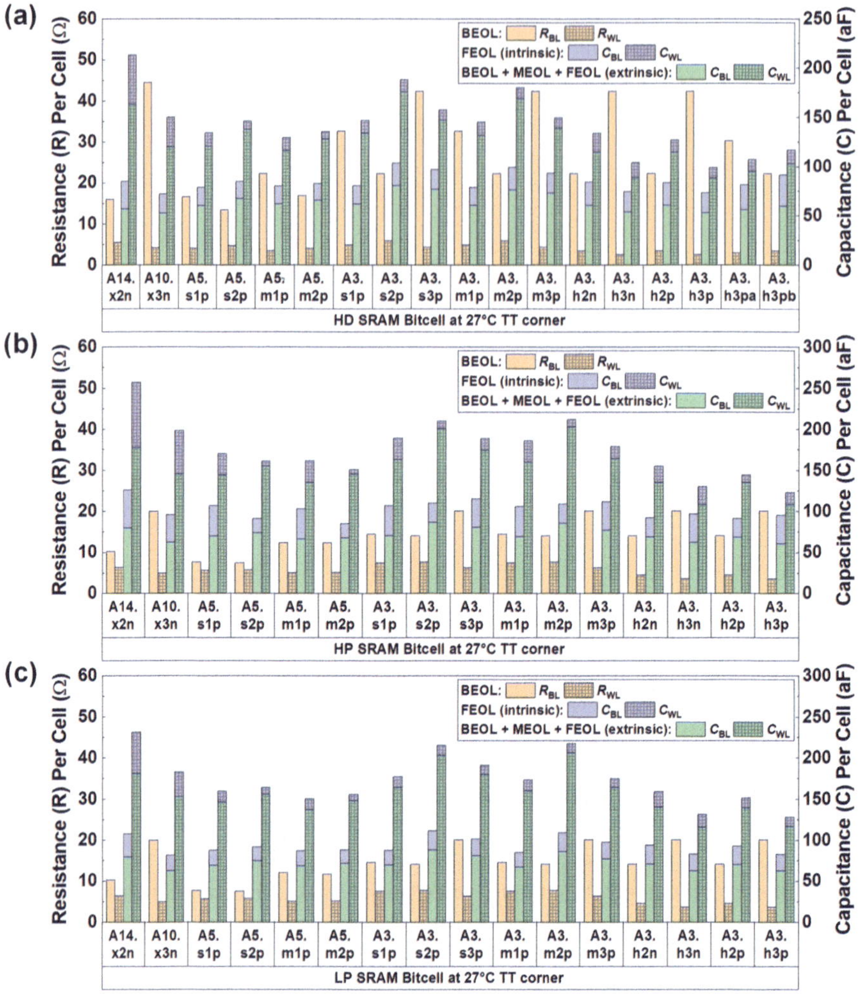

Fig. 6.1 Parasitic RC per cell of BL and WL for (**a**) HD, (**b**) HP, and (**c**) LP SRAM bitcells

analysis progresses from A5 to A3, the RC gains become less pronounced, primarily due to an increase in R_{BL}. Moving from A5.s2p to A3.h3p, R_{BL} experiences a notable 216% increase, while R_{WL}, C_{BL}, and C_{WL} register respective decreases of 44%, 13%, and 32%. Consequently, A3.h3pa and A3.h3pb have been devised with wider BL widths to mitigate R_{BL} increments to 126% and 66%, respectively. This adjustment comes at the cost of increased bitcell area and slightly reduced reductions in R_{WL}, C_{BL}, and C_{WL}. It is intriguing to compare the PPA outcomes based on these three distinct BL width designs (A3.h3p, A3.h3pa, and A3.h3pb). For HP and LP cases, the parasitic RC trends align with those observed in HD cases without high R_{BL} problem due to the larger cell height.

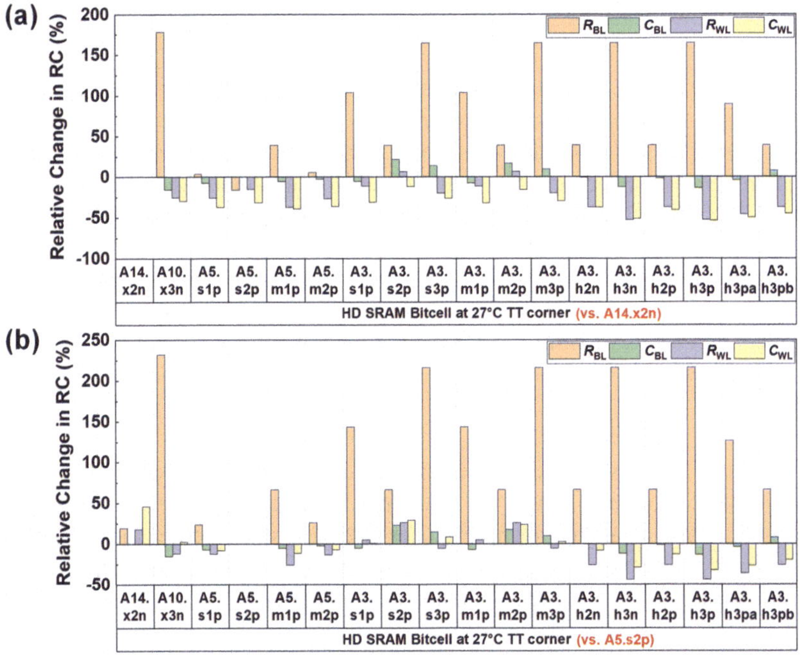

Fig. 6.2 Relative changes in parasitic RC per cell of BL and WL for HD cases w.r.t. (**a**) A14.x2n and (**b**) A5.s2p SRAM bitcells

Takeaways

- **Challenges in Deeply Scaled Nodes:**
  - Aggressive bitcell area scaling significantly increases R_{BL}, creating a severe bottleneck for ensuring write functionality and PP improvements beyond A14.
  - Continuing with traditional bitcell area scaling may offer diminishing returns for subarray- and macro-level PP improvements.
- **Motivated Design Strategy:** To identify the most promising SRAM bitcell designs for advanced nodes, and evaluate the PPA results for all designs from A14 to A3. This assessment will help determine whether it is worthwhile to continue pursuing the bitcell area scaling in interconnect-dominant nodes.

6.3 Hold and Read Stability

Both hold and read static noise margin (HSNM and RSNM) values are derived from voltage transfer characteristics (VTCs), quantified by the side length of the largest square that can be accommodated within the butterfly curves [14]. For a comprehensive understanding of the simulation methodologies, please refer to Chap. 2. Stability tests are conducted under the worst-case temperature of 80 °C and identical A_{VT} (1 mV·μm), employing 6-sigma (6σ) Monte Carlo (MC) simulation with threshold voltage (V_T) sweeping. It's important to note that A_{VT} may decrease with technology scaling [11], necessitating a more precise determination via an extended process, variability, and temperature-dependent analysis. As an initial approximation, A_{VT} set at 1 aligns with the transistor size range, as documented in [3, 6].

Figure 6.3 presents the HSNM and RSNM results across all HD, HP, and LP SRAM cases. It's evident that HSNM remains comfortably within the bounds ensuring stable data retention at 0.7 V and may extend to even lower V_{DD} levels in hold mode. This LP application necessitates an extensive co-optimization at both the bitcell and circuit levels, incorporating variability models. However, it goes beyond the current scope of this study and should be explored in future research. On the other hand, the worst-case scenarios ($\mu - 6\sigma$; μ: mean value; σ: standard deviation) for RSNM in each case approach the thermal voltage limits, approximately ~30 mV at 80 °C. Thus, RSNM stands as the critical criterion to satisfy and analyse, serving as the central focus in deeply scaled nodes. To maintain stability in the standard subarray design, it is recommended to keep the V_{DD} at the nominal value of 0.7 V. However, using a dual rail V_{DD} technique could help address stability issues [5]. This method involves keeping the V_{DD} of unselected bitcells high enough to ensure stability while reducing the V_{DD} of selected bitcells to save energy. Although this approach would result in an area penalty, it is a promising avenue for future research in deeply scaled nodes.

In CFET SRAM bitcells, the increased V_{DD} resistance (R_{VDD}) in the Fin-on-Fin configurations results in less favourable SNM compared to the NS-on-NS counterparts, primarily due to the longer VBPR connection (Fig. 6.4). Among the HD, HP, and LP cases, the latter two exhibit significantly improved SNM values due to stronger PU devices. For a more comprehensive discussion, the following subsection will provide detailed insights.

Takeaways

- **Challenges in Read Stability:** In deeply scaled nodes, the worst-case read stability approaches the thermal limit. This suggests that reducing V_{DD} is not advisable if we want to maintain SRAM functionality.

(continued)

- **Motivated Design Strategy:**
  - Maintain V_{DD} at its nominal value of 0.7 V for the standard subarray designs used in this work.
  - Explore alternative circuit techniques to address this issue in future work.

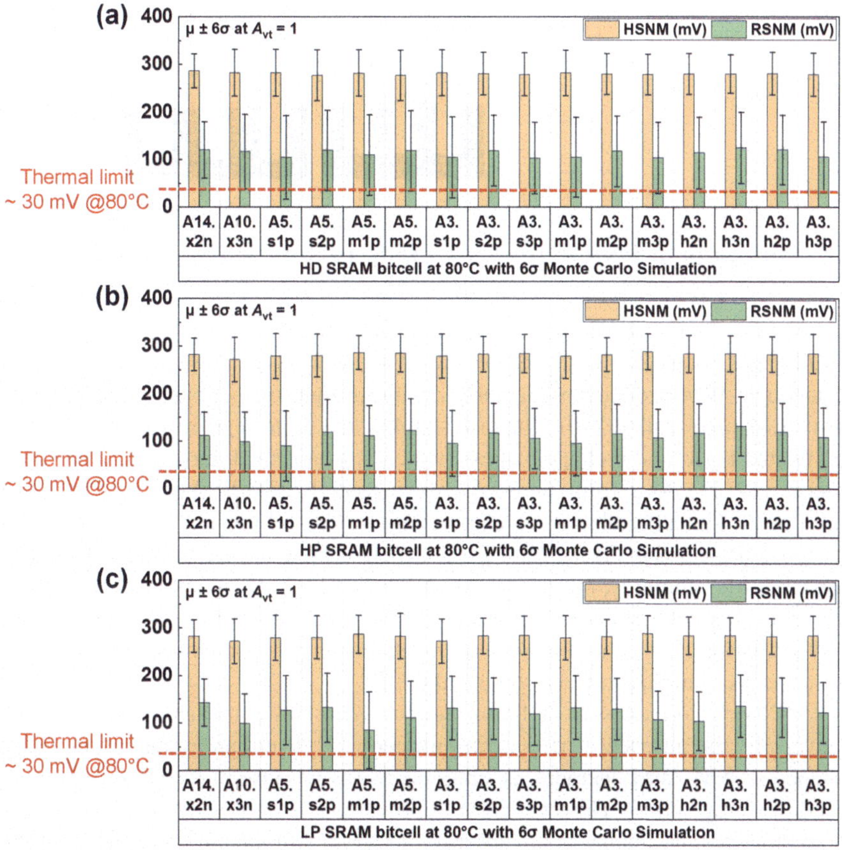

Fig. 6.3 μ ± 6σ HSNM and RSNM based on MC simulation at 80 °C for (**a**) HD, (**b**) HP, and (**c**) LP SRAM bitcells

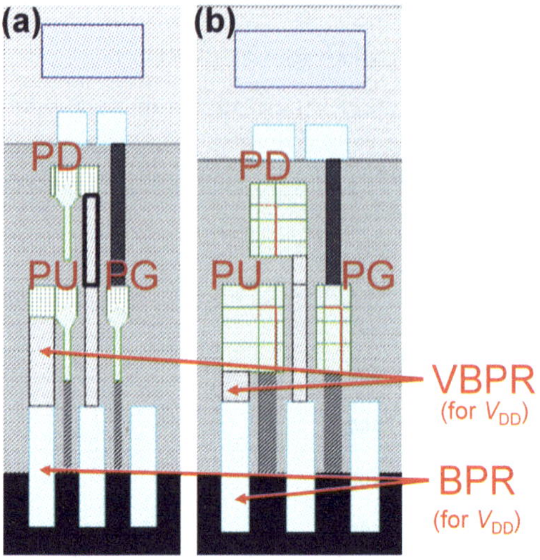

Fig. 6.4 Cross-sectional view of HD A5 mono (**a**) Fin-on-Fin and (**b**) NS-on-NS CFET SRAM bitcells

6.4 Impacts of PMOS and NMOS Stress on RSNM and WM

For the functional read and write operations in an SRAM subarray, it is imperative to fulfill the criteria for both RSNM and WM. Moreover, a thorough verification of adequate WM becomes particularly crucial in light of its potential degradation with bitcell area scaling [10]. The read stability and write capability of SRAM may be affected by n- and p-channel stress, which is established during the fabrication process [12, 13]. As a result, the impact of stress on RSNM/WM on HD A5 and A3 CFET SRAM bitcells/subarrays is explored by varying NMOS and PMOS stress (S_N and S_P), as depicted in Fig. 6.5/6.6. These experiments are specifically conducted on HD cases for simplicity, but the trends in HP and LP cases should exhibit similar behaviour. It's noteworthy that N-PG or P-PG-based SRAM bitcells with opposite read and write paths share comparable mechanisms of stress impact, as discussed further and summarized in Figs. 6.5d and 6.6d.

RSNM deteriorates as stress diminishes due to the slower switching time of VTC. Conversely, the absence of stress can mitigate read disturbance and consequently enhance RSNM. WM degrades as stress decreases because of the slower charging speed of storage nodes. However, WM improvement is observed when there is less contention current, thanks to the absence of stress. Stress analysis reveals that the S_N impact is less pronounced than S_P. It's essential to note that stress in CFET devices is primarily determined by the source/drain (S/D) rather than substrate strain [13]. Even if the substrate is removed during the hybrid CFET process or devices are positioned on the top tier, the impact on channel stress should be relatively minimal. As a result, the stress impact does not increase in hybrid CFET SRAM bitcells in comparison to seq and mono counterparts. RSNM is determined by the drive strength of transistors between PG and PD/PU devices and the resistance

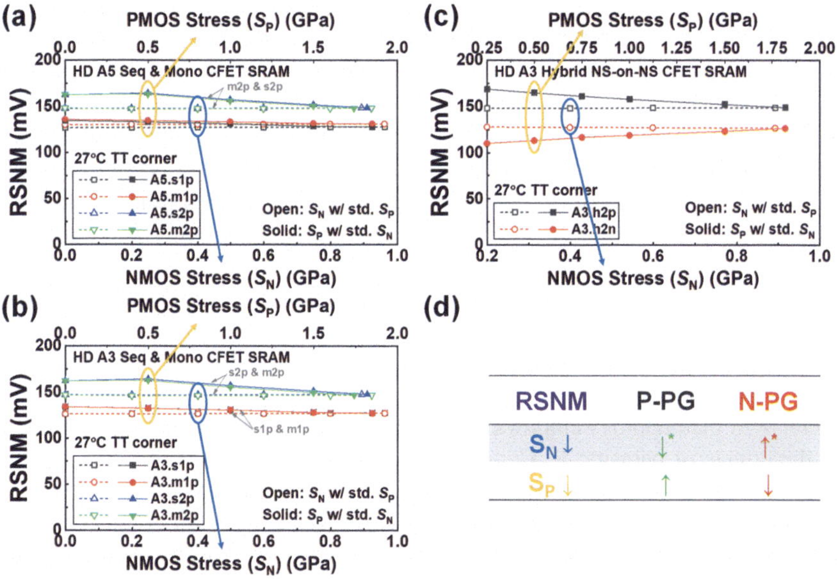

RSNM	P-PG	N-PG
S_N ↓	↓*	↑*
S_P ↓	↑	↓

Fig. 6.5 (**a**)(**b**)(**c**)(**d**) RSNM with varying NMOS and PMOS stress (S_N and S_P) of A5 and A3 CFET SRAM bitcells/subarrays. *S_N impact is weaker than S_P

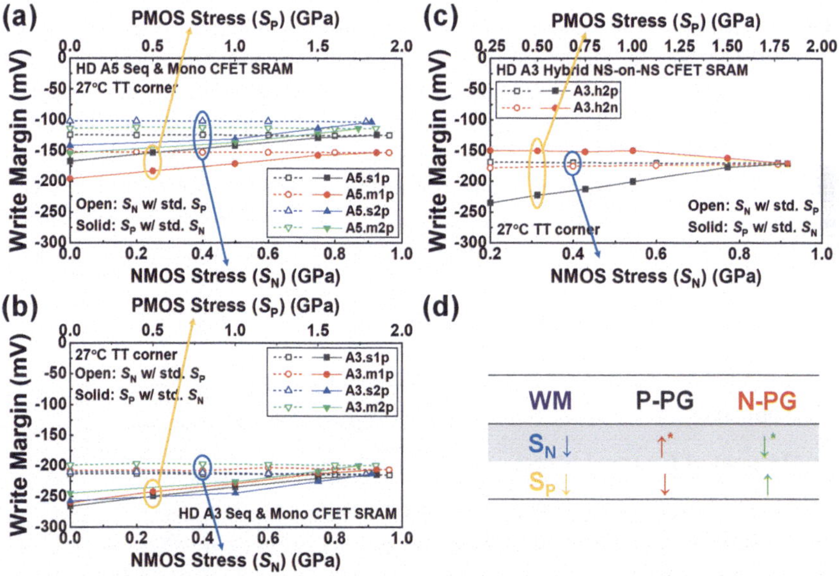

WM	P-PG	N-PG
S_N ↓	↑*	↓*
S_P ↓	↓	↑

Fig. 6.6 (**a**)(**b**)(**c**)(**d**) WM with varying NMOS and PMOS stress (S_N and S_P) of A5 and A3 CFET SRAM bitcells/subarrays. *S_N impact is weaker than S_P

of V_{SS}/V_{DD} resistance (R_{VSS}/R_{VDD}), which arises from wires and connected vias in the read path for N/P-PG-based SRAM bitcells. In this context, the discussion primarily focuses on the impact of R_{power} on RSNM since the ratio between each transistor remains constant due to the use of the same NMOS and PMOS models. Larger R_{power} values have an adverse effect on RSNM. In both A5 and A3 seq and mono CFET SRAM bitcells, the greater R_{VDD} in the Fin-on-Fin (s1p and m1p) configurations leads to inferior RSNM compared to the NS-on-NS (s2p and m2p) counterparts. This can be attributed to the longer VBPR/nTSV connection shown in Fig. 6.4. In the case of A5, the mono Fin-on-Fin (A5.m1p) configuration exhibits slightly lower R_{power} than the seq counterpart, resulting in improved RSNM, primarily due to wider VBPR assumptions at the edge. In A3, similar RSNM values are achieved between seq and mono cases because they share identical V_{DD} and nTSV widths.

Our WM methodology employs transient simulation to comprehensively account for the effects of parasitic R, C, and PG transistors, as reported in Chap. 3 [10]. Consequently, WM is influenced by R_{BL}, C_{BL}, the drive strength of transistors between PG and PU/PD devices, and R_{VSS}/R_{VDD} arising from wires and connected vias in the write path for N/P-PG-based SRAM subarrays. For this discussion, we will primarily focus on the impacts of R_{BL} and R_{power} since the strengths of NMOS and PMOS devices are comparable, and C_{BL} exhibits no significant differences compared to R_{BL}. Lower R_{BL} and higher R_{power} contribute to an improved WM. Reduced R_{BL} can be achieved through a wider BL, which naturally accompanies a larger cell height (as illustrated in Fig. 6.1). In both A5 and A3 seq and mono CFET SRAM bitcells, the Fin-on-Fin (s1p and m1p) configurations exhibit both higher R_{BL} and R_{VSS} compared to the NS-on-NS (s2p and m2p) counterparts. This is attributed to the smaller cell height and the longer VBPR/supervia connection, respectively. Thus, differences in WM between each type of SRAM bitcells result from the interplay of these two factors. For instance, the mono Fin-on-Fin case with lower R_{VSS} still exhibits the poorest WM among A5 counterparts due to its larger R_{BL}. In contrast, the A5 seq NS-on-NS case boasts the best WM because it has the smallest R_{BL} even with less R_{VSS}. In general, the overall WM of A5 CFET SRAM subarrays surpasses those of A3 counterparts due to the lower R_{BL}.

Takeaways

- **Impact of Channel Stress on RSNM and WM:**
  - The impact of S_N is less pronounced compared to S_P.
  - **RSNM:** Generally, RSNM deteriorates as stress diminishes due to the slower switching time of VTC. However, beyond a certain point, the absence of stress can reduce read disturbance, consequently enhancing RSNM.

(continued)

- **WM:** Typically, WM degrades as stress decreases due to the slower charging speed of storage nodes. Nevertheless, WM improves when there is less contention current due to the absence of stress.

6.5 Impacts of Channel Geometry

The device's margin and PPA outcomes are significantly influenced by the dimensions of the channel stack, including H_{fin}, *NNS*, and W_{NS}. To enhance the performance of the transistor, it is common practice to increase H_{fin}, *NNS*, and W_{NS}. However, augmenting W_{NS} leads to a larger cell height, resulting in potential degradation at the macro level. Additionally, it is imperative to initially validate the compatibility of H_{fin} and *NNS* assumptions between SRAM and logic cells to avoid extra fabrication steps. As a result, Sect. 6.5.1 aims to explore the impact of varying H_{fin} and *NNS* on each metric while keeping the bitcell area constant with a fixed W_{NS}. After verifying the optimized H_{fin} and *NNS* for each SRAM bitcell option in alignment with logic assumptions, Sect. 6.5.2 will provide insights into the impacts of W_{NS}.

6.5.1 *Impacts of H_{fin} and NNS*

In this section's figures, the channel stacked dimensions of H_{fin} and *NNS* for logic are highlighted with the green box. Specifically, H_{fin} is set at 35 nm and *NNS* is defined as 4/3/2/3 in A14/A10/A5/A3, as detailed in Table 5.3 of Chap. 5 [8, 12, 13, 15]. To substantiate the suitability of the logic assumptions regarding H_{fin} and *NNS* for the CFET SRAM bitcell at the same cell size, an exploration into the impacts of varying H_{fin} or *NNS* on HSNM, RSNM, WM, RD, read energy (RE), write delay (WD), and write energy (WE) is conducted across HD (HD-1 and HD-2), HP, and LP cases. Before delving into the data, the following paragraphs introduce the two primary phenomena that govern the determination of optimized points for each metric.

For HSNM/RSNM (Figs. 6.7, 6.8, 6.9, and 6.10), two/three competing phenomena occur simultaneously with the increase in H_{fin} or *NNS*: (1) improvement due to increased transistor's drive strength, (2) degradation due to increased R_{power} from taller via (e.g., R_{VSS} and R_{VDD} for N- and P-PG-based SRAM bitcells, respectively), and (3) degradation due to more read disturbance (only for RSNM). Generally, the enhancement is more significant than the degradation. Hence, both HSNM and RSNM first increase and then saturate at a certain point with increasing H_{fin} and *NNS*.

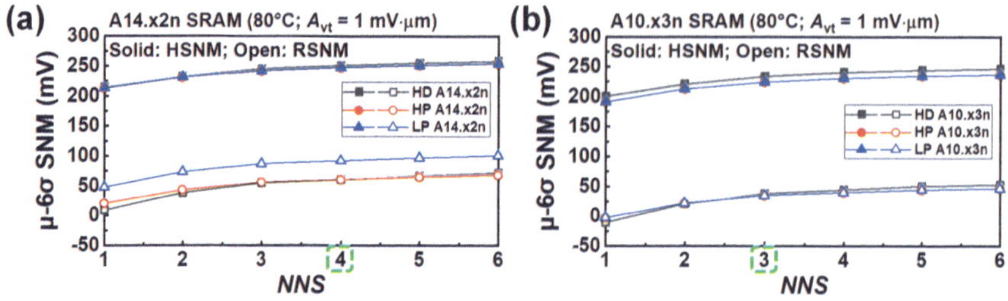

Fig. 6.7 SNM with varying *NNS* of HD, HP, and LP (**a**) A14.x2n and (**b**) A10.x3n bitcells

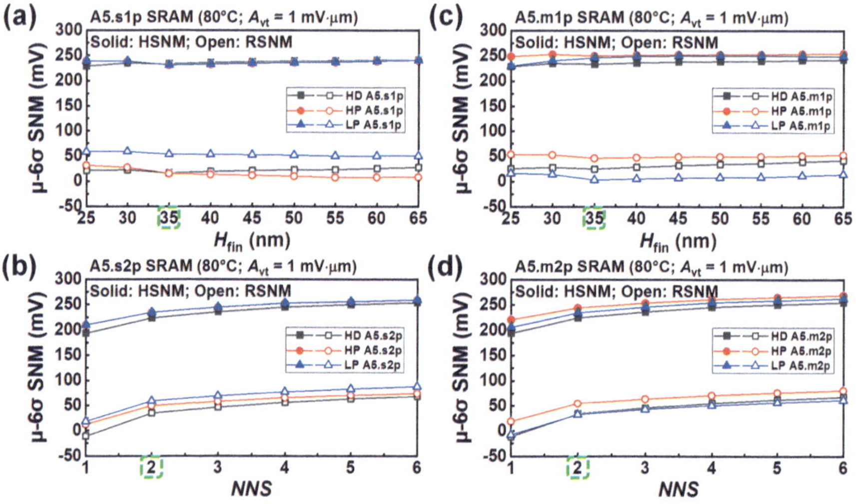

Fig. 6.8 SNM with varying H_{fin} and *NNS* of HD, HP, and LP (**a**) A5.s1p, (**b**) A5.s2p, (**c**) A5.m1p, and (**d**) A5.m2p bitcells

For the remaining metrics including WM, speeds, and energies (Figs. 6.11, 6.12, 6.13, 6.14, 6.15, 6.16, 6.17, 6.18, 6.19, 6.20, 6.21, and 6.22), two competing phenomena occur simultaneously as H_{fin} or *NNS* increase: (1) improvement due to increased transistor's drive strength and (2) degradation because of a larger voltage drop caused by increased device capacitance. Note that the current flow to the bitcells also passes through parasitic capacitors, resulting in a voltage drop on BL. Consequently, these metrics initially improve and then start degrading or saturating at certain points with increasing H_{fin} and *NNS*.

The subsequent subsections will elaborate on the influence of channel stack dimensions on individual metrics. To conclude, the distinctions among CFET SRAM subarrays are mostly determined by the overall impacts of R_{WL}, R_{BL}, and C_{BL}. The simulation results confirm that the optimized H_{fin} and *NNS* of SRAM closely align with or are compatible with logic dimensions. It's important to highlight that comparative assessments of overall subarray-level outcomes hold

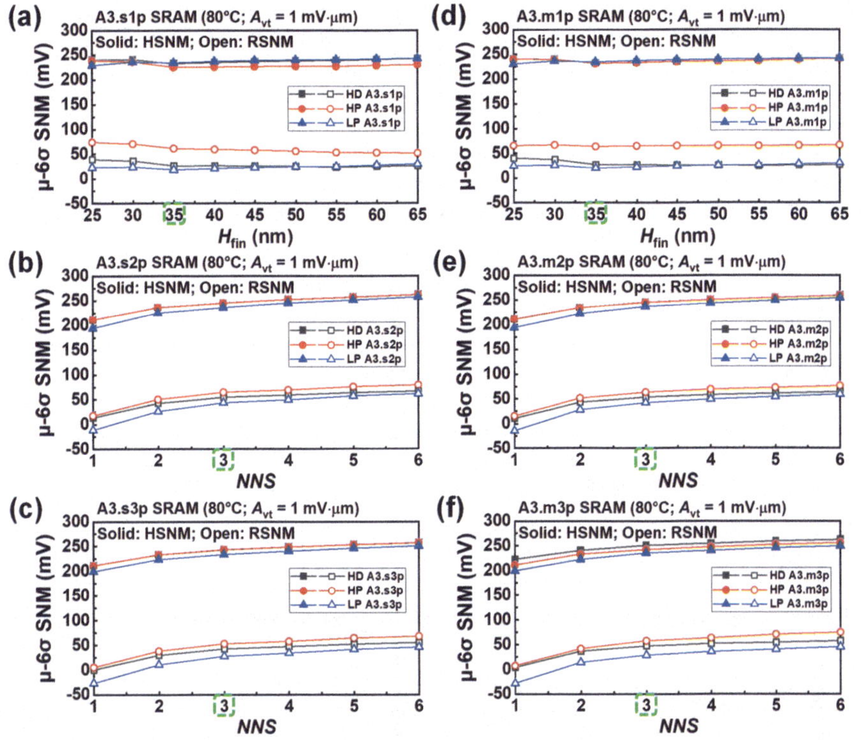

Fig. 6.9 SNM with varying H_{fin} and *NNS* of HD, HP, and LP (**a**) A3.s1p, (**b**) A3.s2p, (**c**) A3.s3p, (**d**) A3.m1p, (**e**) A3.m2p, and (**f**) A3.m3p bitcells

validity solely within identical subarray sizes (HD-1 and HD-2 or HP and LP). Moreover, any determination of superior design should conscientiously incorporate the macro-level simulation detailed in Chap. 7.

6.5.1.1 SNM

In this subsection, the provided data encompass HSNM and RSNM with diverse H_{fin} and *NNS* values. The focus lies particularly on RSNM to guarantee read functionality under stricter thermal limit criterion when compared to HSNM. This paragraph provides a summary of the SNM results ranging from A14 to A3. Detailed explanations for each individual result will be presented in the subsequent paragraphs. In most SRAM technologies, HP configurations consistently outperform HD and LP counterparts. Conversely, LP emerges as the optimized design for A14.x2n, A5.s1p, and A5.s2p (Figs. 6.7a and 6.8a, b). Building on the preceding discussion in Sect. 6.4, it is evident that Fin-on-Fin CFET SRAM bitcells yield inferior SNM results compared to NS-on-NS configurations due to larger R_{power} (Figs. 6.8

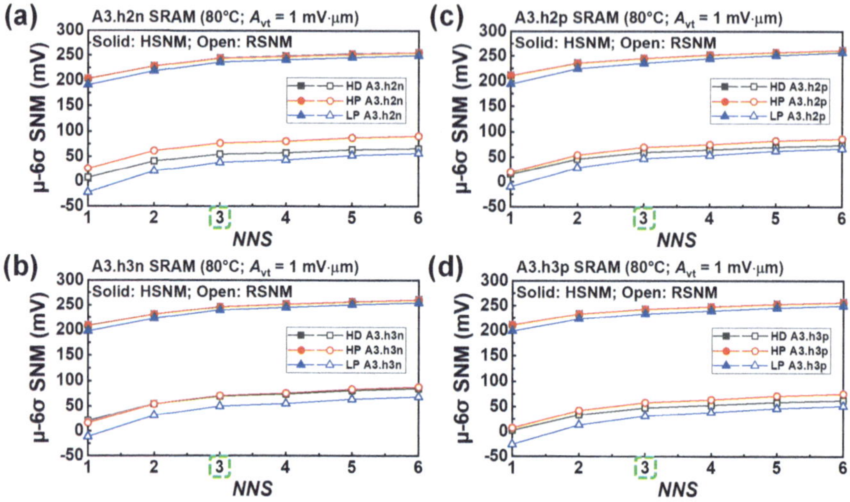

Fig. 6.10 SNM with varying H_{fin} and *NNS* of HD, HP, and LP (**a**) A3.h2n, (**b**) A3.h3n, (**c**) A3.h2p, and (**d**) A3.h3p bitcells

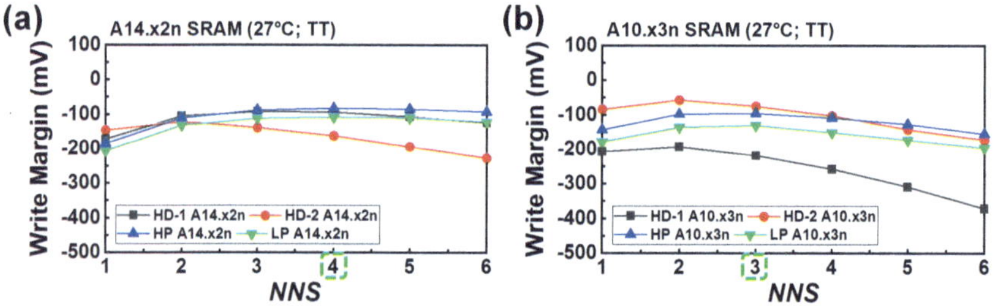

Fig. 6.11 WM with varying *NNS* of HD, HP, and LP (**a**) A14.x2n and (**b**) A10.x3n subarrays

and 6.9). Notably, the HD and LP configurations for A5.s1p (Fig. 6.8a), A5.m1p (Fig. 6.8c), A3.s1p (Fig. 6.9a), and A3.m1p (Fig. 6.9a) raise significant concerns regarding the assurance of read functionality under the worst-case conditions, as the RSNM approaches or even falls below thermal limits. In such scenarios, only the HP configurations prove resilient under these stringent contexts. The saturation of SNM is observed as H_{fin} or *NNS* reaches the logic assumptions. Hence, the optimized values of H_{fin} and *NNS* in SRAM subarray are deemed compatible with logic.

For A14.x2n and A10.x3n SRAM bitcells (Fig. 6.7), the SNM demonstrates an increasing trend with higher *NNS*, attributed to enhanced transistors' drive strength. This trend saturates around $NNS = 4$ for A14.x2n and 3 for A10.x3n due to increased R_{power}. The optimized *NNS* values for A14.x2n and A10.x3n SRAM bitcells are found to be correlated with logic dimension. Variations across the HD, HP, and LP configurations remain minimal, which are less than 3 mV for HSNM in A14.x2n and 10 mV for both HSNM and RSNM in A10.x3n. However, in the case of LP A14.x2n

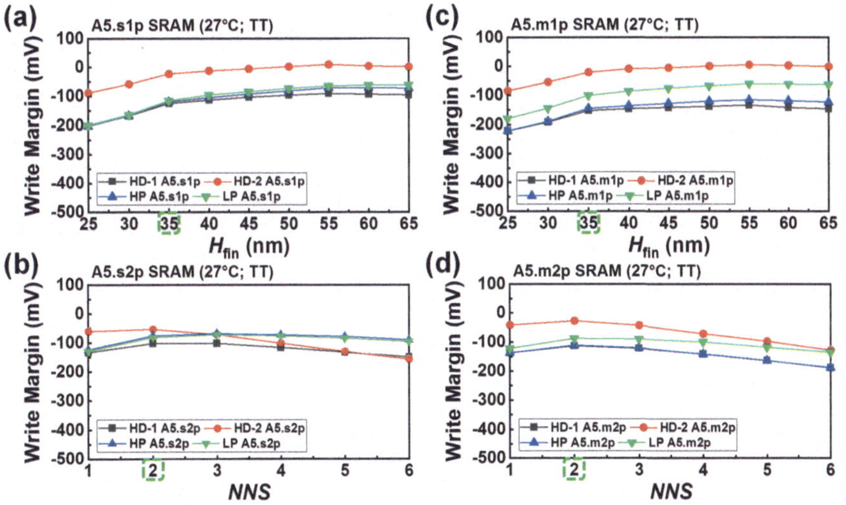

Fig. 6.12 WM with varying H_{fin} and *NNS* of HD, HP, and LP (**a**) A5.s1p, (**b**) A5.s2p, (**c**) A5.m1p, and (**d**) A5.m2p subarrays

with $NNS = 4$, the RSNM notably exceeds that of its HD and HP counterparts by 41 and 32 mV, respectively. The assurance of read stability is maintained for HD, HP, and LP configurations in both A14.x2n and A10.x3n, as RSNM surpasses the thermal limit.

For A5 SRAM bitcells (Fig. 6.8), a discernible upward trend in SNM is solely evident in A5.s2p and A5.m2p, plateauing when *NNS* reaches 2 aligned with logic dimension. Conversely, the impact of increased R_{power} is akin to heightened transistors' drive strength for A5.s1p and A5.m1p, resulting in negligible alterations in SNM. The disparities in HSNM are as follows: 0, 17, 13, and 21 mV between HD, HP, and LP A5.s1p, A5.m1p, A5.s2p, and A5.m2p, respectively, when $H_{\text{fin}} = 35$ nm. In A5.s1p, only the LP configuration surpasses the thermal threshold, yielding up to 38 mV of RSNM enhancements compared to counterparts. In A5.m1p, the only and optimized choice to ensure read stability is the HP option, offering up to 44 mV of RSNM improvements compared to counterparts. LP A5.s2p and HP A5.m2p options demonstrate superior RSNM, with up to 25 and 23 mV enhancements, respectively, compared to their counterparts. Notably, the HD, HP, and LP configurations persist beyond the thermal limit in both A5.s2p and A5.m2p bitcells.

For A3 CFET SRAM bitcells (Figs. 6.9 and 6.10), similar outcomes have been observed when compared to the A5 variants. These include: (1) balanced trade-offs between transistors' drive strength and R_{power} as H_{fin} increases for Fin-based CFET scenarios; and (2) initially increasing then stabilizing curves as *NNS* increases for NS- and FS-based CFET scenarios. The HP configurations emerge as the preferred option, exhibiting enhanced RSNM by up to 43, 21, 25, 45, 22, 29, 39, 21, 23, and 27 mV in comparison to their A3.s2p, A3.s3p, A3.m2p, A3.m3p, A3.h2n, A3.h3n, A3.h2p, and A3.h3p counterparts, respectively, when $H_{\text{fin}} = 35$ nm and *NNS*

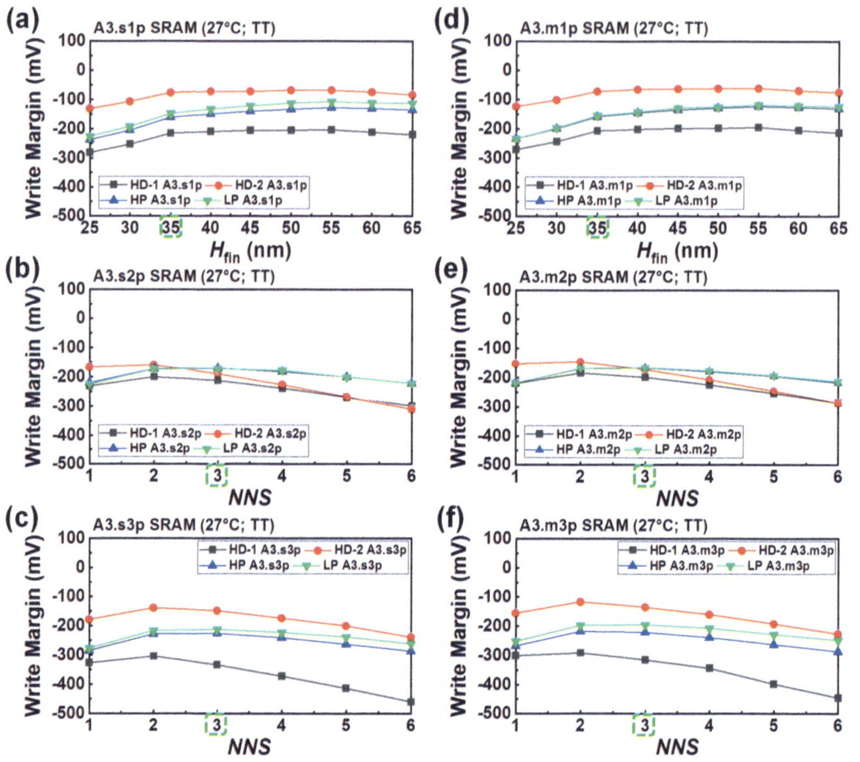

Fig. 6.13 WM with varying H_{fin} and *NNS* of HD, HP, and LP (**a**) A3.s1p, (**b**) A3.s2p, (**c**) A3.s3p, (**d**) A3.m1p, (**e**) A3.m2p, and (**f**) A3.m3p subarrays

= 3. It is noteworthy that the HD, HP, and LP configurations consistently exceed the thermal criteria for nearly all A3 CFET SRAM bitcells, except for A3.s1p, A3.m1p, A3.s3p, and A3.m3p, where only the HP configurations demonstrate viability.

6.5.1.2 Write Margin (WM)

In this subsection, the WM results across various Hfin and NNS for each SRAM technology are presented. This paragraph summarizes the results of the WM analysis spanning from A14 to A3. Detailed explanations for each result will follow in subsequent paragraphs. WM is mostly determined by the total R_{BL} and C_{BL} in an SRAM subarray, as discussed in Sect. 6.4 [7]. In general, a more positive WM indicates a lower requirement for write-assist techniques. A stronger write-assist circuit could speed up the write operation at the cost of higher power consumption. Thus, it is recommended to attain a more positive WM for SRAM subarray design to mitigate the write failure risk and significant area and power overhead associated with the write-assist techniques. HD-2 design generally provides the best WM

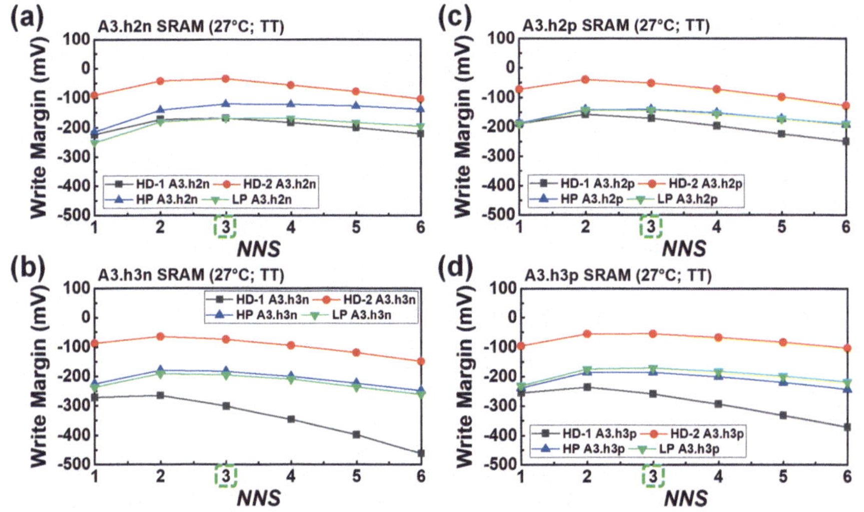

Fig. 6.14 WM with varying H_{fin} and *NNS* of HD, HP, and LP (**a**) A3.h2n, (**b**) A3.h3n, (**c**) A3.h2p, and (**d**) A3.h3p subarrays

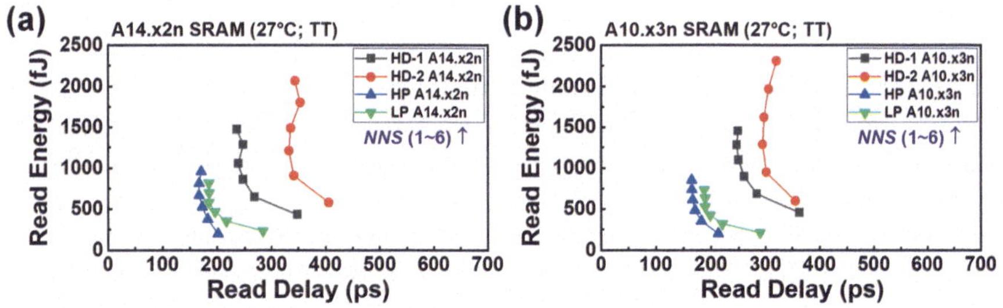

Fig. 6.15 Pareto distribution of RD and RE with varying *NNS* of HD, HP, and LP (**a**) A14.x2n and (**b**) A10.x3n subarrays

results over HD-1, HP, and LP counterparts in each design type due to less total R_{BL} and C_{BL} of a shorter total BL length except for A14.x2n (Fig. 6.11a) and A3.s2p (Fig. 6.13b). HP and LP designs exhibit better WM than HD-1 mainly because of the less total R_{WL} coming from a shorter total WL length. In most cases, the optimized H_{fin} and *NNS* for WM aligns with logic assumptions.

For A14.x2n SRAM subarrays (Fig. 6.11a), the HP bitcell option demonstrates superior WM compared to alternatives, attributed to its expedited write path and reduced R_{BL}. Discrepancies in WM remain under 25 mV across HD-1, HP, and LP subarray choices, with WM initially rising before plateauing at *NNS* = 4. However, HD-2 yields comparable WM only at *NNS* = 2, thereafter exhibiting a substantial degradation as *NNS* increases, with 79 mV degradation compared to the HP counterpart at *NNS* = 4. This underscores that the RC of BL per cell in HD-1 A14.x2n remains sufficiently low to ensure write ability, obviating the necessity

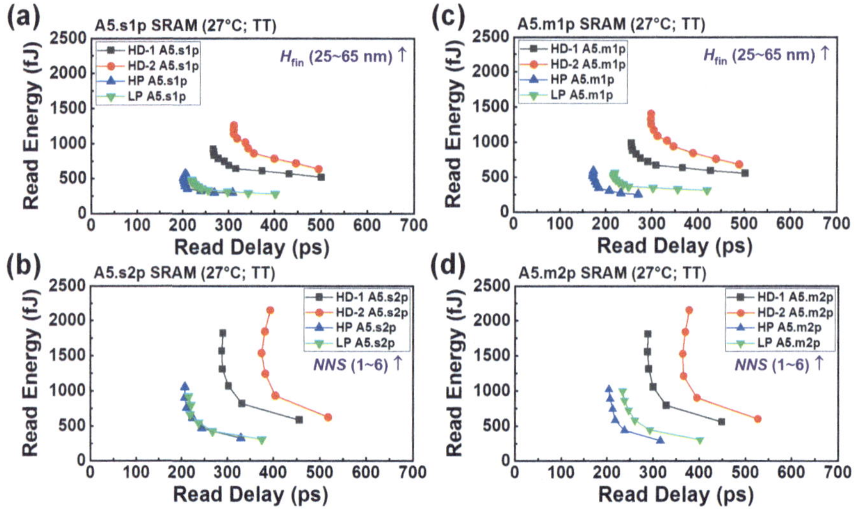

Fig. 6.16 Pareto distribution of RD and RE with varying H_{fin} and *NNS* of HD, HP, and LP (**a**) A5.s1p, (**b**) A5.s2p, (**c**) A5.m1p, and (**d**) A5.m2p subarrays

for subarray size rearrangement to HD-2. Conversely, for A10.x3n SRAM subarrays (Fig. 6.11b), the WM of the HD-2 configuration surpasses its counterpart, displaying 144 mV improvement compared to the HD-1 scenario. This underscores the need for HD-2 design when the RC of BL per cell exceeds tolerable limits (Fig. 6.1), leading to diminished WM.

In the A5 CFET SRAM subarrays (Fig. 6.12), notable WM enhancements are observed in the HD-2 subarrays. Compared to their HD-1 counterparts in A5.s1p, A5.m1p, and A5.m2p, HD-2 subarrays offer significant increases in WM of 94, 133, and 85 mV, respectively, at *NNS* = 2. These enhancements are attributed to the increased R_{BL} as the cell height scales to A5 (Fig. 6.1). However, in the A5.s2p scenario, the WM enhancement diminishes to 50 mV at *NNS* = 2 when comparing HD-2 to HD-1 subarrays, owing to decreased R_{BL} with relatively larger cell height in A5. Furthermore, as *NNS* reaches 3 and beyond, the WM of HD-2 A5.s2p subarray falls below that of HP counterparts. Notably, optimized WM values are achieved when H_{fin} and *NNS* dimensions align with those of the logic.

For A3 CFET SRAM subarrays (Figs. 6.13 and 6.14), the relationship between WM trends and increasing values of H_{fin} and *NNS* reveals a balance between two factors: (1) enhanced WM stemming from the heightened drive strength of transistors; and (2) saturated or declined WM due to elevated device capacitance. Among the A3.s1p, A3.m1p, A3.s3p, A3.m3p, A3.h2n, A3.h3n, A3.h2p, and A3.h3p subarrays, HD-2 configurations demonstrate superiority over other variants, exhibiting WM enhancements of 141, 137, 188, 183, 135, 228, 121, and 206 mV, respectively, compared to HD counterparts. Conversely, the optimized configuration for A3.s2p and A3.m2p is HP subarray design, resulting in WM improvements of 44

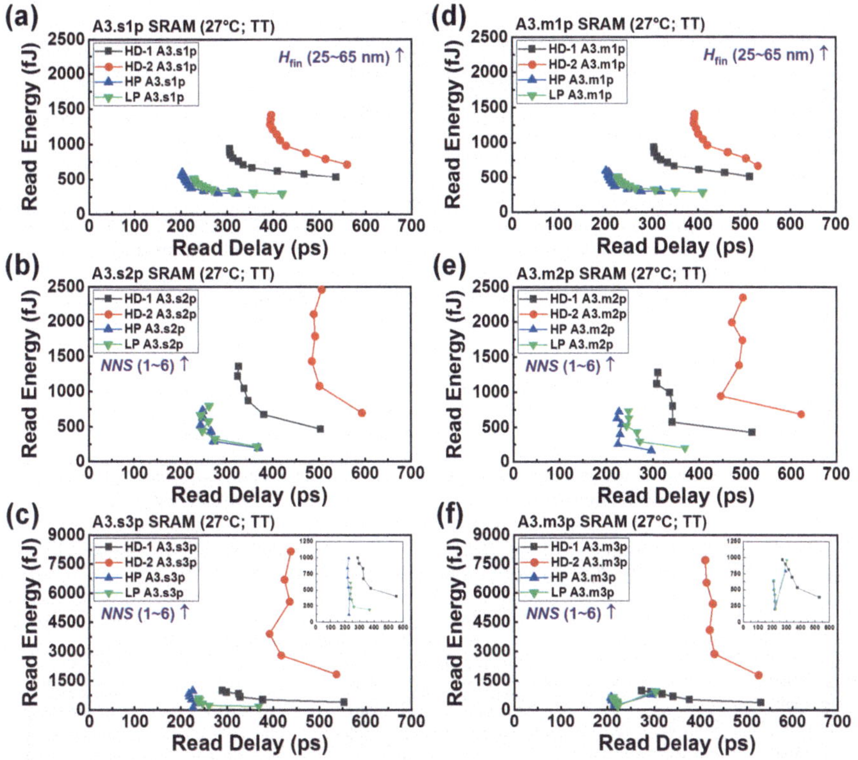

Fig. 6.17 Pareto distribution of RD and RE with varying H_{fin} and *NNS* of HD, HP, and LP (**a**) A3.s1p, (**b**) A3.s2p, (**c**) A3.s3p, (**d**) A3.m1p, (**e**) A3.m2p, and (**f**) A3.m3p subarrays

and 34 mV, respectively, relative to HD counterparts. Notably, the maximum WM for optimized subarray designs does not align with $NNS = 3$ deviating from the logic dimension, observed in A3.s3p, A3.m3p, A3.h3n, and A3.h2p subarrays. This discrepancy is attributed to the amplified impact of increased capacitance, leading to WM degradation as *NNS* surpasses 2.

6.5.1.3 Read Delay (RD) and Read Energy (RE)

In this subsection, the Pareto distribution of RD and RE across diverse H_{fin} and *NNS* is presented. This paragraph provides a summary of the findings obtained from RD and RE, spanning from A14 to A3. Subsequent paragraphs will offer detailed elucidations for each of these results. RD is mainly affected by the total R_{WL} and C_{BL} in an SRAM subarray, and less RC results in better performance [7]. RE, i.e., the active read energy of the subarray, is primarily determined by integration time (RD) and total C_{BL}. As H_{fin} and *NNS* increase, RD initially improves before reaching saturation or degradation. This behaviour is influenced by the competing

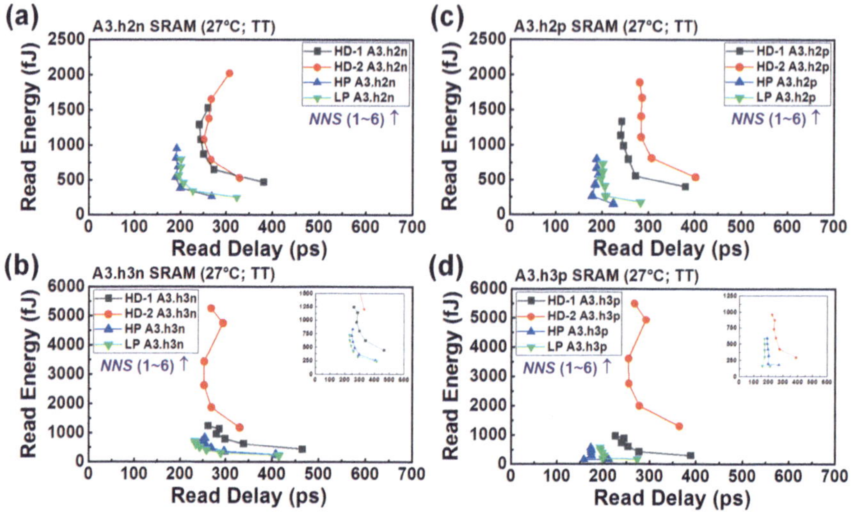

Fig. 6.18 Pareto distribution of RD and RE with varying H_{fin} and *NNS* of HD, HP, and LP (**a**) A3.h2n, (**b**) A3.h3n, (**c**) A3.h2p, and (**d**) A3.h3p subarrays

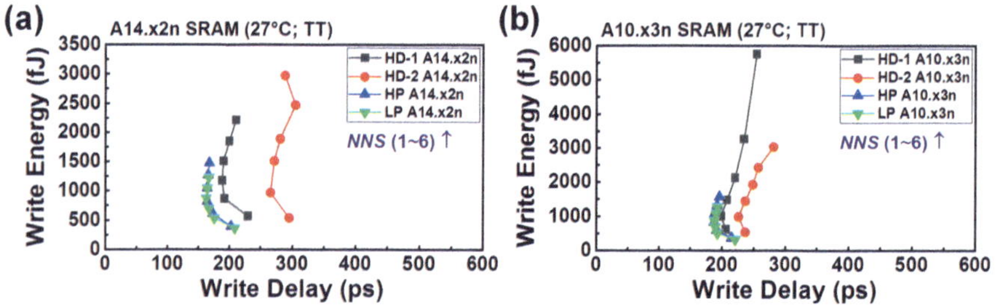

Fig. 6.19 Pareto distribution of WD and WE with varying *NNS* of HD, HP, and LP (**a**) A14.x2n and (**b**) A10.x3n subarrays

phenomena introduced at the beginning of this section. Meanwhile, RE continues to rise as the channel stack dimensions expand, driven by increased total C_{BL} and C_{WL}. This observation implies that the optimized values for H_{fin} and *NNS* should be established when RD reaches saturation. Otherwise, elevating H_{fin} and *NNS* may lead to additional RE consumption without substantial RD improvement. The LP and HP cases exhibit minimal RD and RE compared to their HD-1 and HD-2 counterparts due to the smaller subarray size. Additionally, in most cases, LP demonstrates smaller RD than HP under the same Hfin or NNS, owing to a stronger PD:PG ratio (for N-PG) or PU:PG ratio (for P-PG). Notably, RE of HP and LP exhibits similarity due to the comparable RC values for both BL and WL in the SRAM subarray (Sect. 6.2). On the other hand, the HD-2 displays higher RD and RE compared to HD-1, stemming from the increased total WL length of SRAM subarray. However, HD-1 may suffer from higher write failure risk due to the poorer

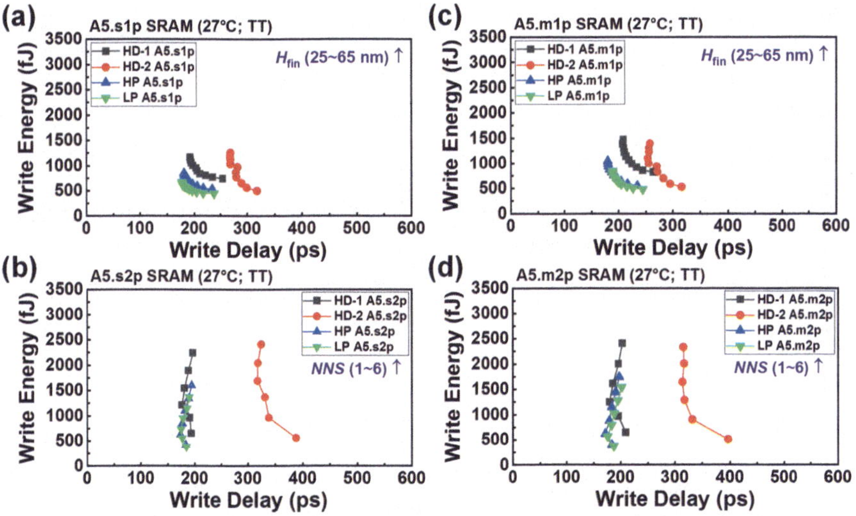

Fig. 6.20 Pareto distribution of WD and WE with varying H_{fin} and *NNS* of HD, HP, and LP (**a**) A5.s1p, (**b**) A5.s2p, (**c**) A5.m1p, and (**d**) A5.m2p subarrays

WM compared to HD-2. This trade-off leads to the conclusion that sacrificing delay and energy may be necessary to ensure reliable write functionality.

For A14.x2n and A10.x3n subarrays (Fig. 6.15), HP demonstrates superior RD and RE over LP due to its faster read path. Additionally, the RD and RE in HP also outperform that of HD-1 and HD-2 due to its smaller subarray size. When comparing HD designs of equal capacity, HD-1 exhibits RD/RE enhancements of 29%/30% and 12%/31% for A14.x2n (*NNS* = 4) and A10.x3n (*NNS* = 3), respectively, compared to their HD-2 counterparts. Referring to the WM discussion for A14.x2n (Fig. 6.11a), it is evident that the HD-2 subarray design need not be considered due to its degradation in both WM and read metrics. However, for A10.x3n, the HD-2 option is recommended to achieve a higher WM value at the expense of increased RD and RE overhead. Conversely, HP not only offers compromised WM between that of HD-1 and HD-2 but also demonstrates the most significant RD/RE enhancements of 30%/36% and 34%/45% for A14.x2n (*NNS* = 4) and A10.x3n (*NNS* = 3), respectively, compared to HD-1 counterparts. Thus, HP emerges as the optimized subarray design for these two SRAM technologies. Moreover, the optimized *NNS* values of 4 and 3 for A14.x2n and A10.x3n, respectively, align with logic dimensions.

In A5 CFET SRAM subarrays (Fig. 6.16), the read metrics of HP or LP excel among other configurations due to a smaller subarray size. HD-1 also exhibits superior RD and RE results compared to HD-1, attributable to a shorter WL length at the same subarray size. HP A5.s1p, A5.s2p, A5.m1p, and A5.m2p subarrays manifest RD/RE enhancements of 36%/40%, 26%/58%, 43%/52%, and 27%/44%, respectively, at H_{fin} = 35 nm and *NNS* = 2, in contrast to HD-1 subarrays. HD-1

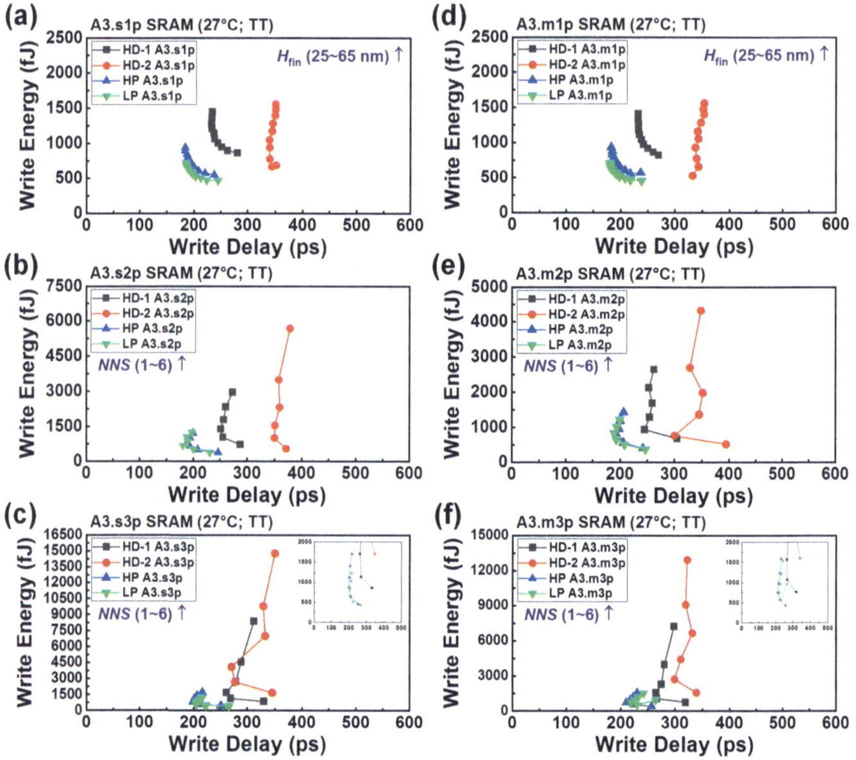

Fig. 6.21 Pareto distribution of WD and WE with varying H_{fin} and *NNS* of HD, HP, and LP (**a**) A3.s1p, (**b**) A3.s2p, (**c**) A3.s3p, (**d**) A3.m1p, (**e**) A3.m2p, and (**f**) A3.m3p subarrays

A5.s1p, A5.s2p, A5.m1p, and A5.m2p subarrays also yield RD/RE enhancements of 7%/23%, 18%/13%, 6%/25%, and 17%/12%, respectively, at $H_{fin} = 35$ nm and $NNS = 2$, relative to HD-2 subarrays. Furthermore, the optimized values of H_{fin} and *NNS* align with the logical dimension. Concerning the advantages of WM for HD-1 subarrays, the design must either sacrifice RD and RE to achieve the most favourable WM or opt for compromised HP alternatives to achieve superior SNM, RD, and RE, albeit at the risk of write failure and additional area and energy overhead from the write-assist technique.

In A3 CFET SRAM subarrays (Figs. 6.17 and 6.18), the RD and RE characteristics of HP or LP configurations outshine HD-1, HD-2, and LP counterparts. Specifically, the HD-1 configuration outperforms HD-2 in all cases except for the RD in A3.h3n. Furthermore, the results for HP and LP configurations are nearly indistinguishable except for A3.m2p, A3.h2p, and A3.h3p. For HP A3.s1p, A3.s2p, A3.s3p, A3.m1p, A3.m2p, A3.m3p, A3.h2n, A3.h3n, A3.h2p, and A3.h3p subarrays at $H_{fin} = 35$ nm and $NNS = 3$, RD/RE enhancements of 39%/45%, 23%/50%, 32%/19%, 38%/45%, 33%/50%, 35%/53%, 24%/38%, 14%/45%, 28%/45%, and 31%/49%, respectively, are demonstrated compared to

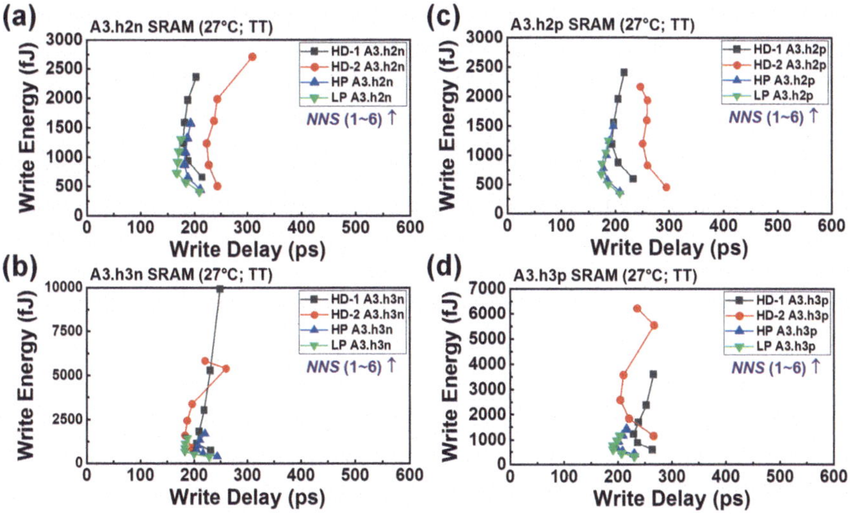

Fig. 6.22 Pareto distribution of WD and WE with varying H_{fin} and *NNS* of HD, HP, and LP (**a**) A3.h2n, (**b**) A3.h3n, (**c**) A3.h2p, and (**d**) A3.h3p subarrays

HD-1 counterparts. Conversely, HD-1 A3.s1p, A3.s2p, A3.s3p, A3.m1p, A3.m2p, A3.m3p, A3.h2n, A3.h3n, A3.h2p, and A3.h3p subarrays at H_{fin} = 35 nm and *NNS* = 3 exhibit RD/RE improvements of 13%/30%, 28%/40%, 16%/83%, 13%/30%, 29%/42%, 19%/83%, 0%/20%, −18%/70%, 10%/29%, and 0%/78%, respectively, compared to HD-2 counterparts. Notably, the negative sign indicates that the RD of HD-2 A3.h3n surpasses that of HD-1 counterpart by 18%. The trade-off between optimizing WM and RD/RE is also evident in A3, with optimized H_{fin} and *NNS* values aligning with logic dimensions except for HP A3.m2p, HP A3.h2p, and HP A3.h3p subarrays.

6.5.1.4 Write Delay (WD) and Write Energy (WE)

This subsection demonstrates the Pareto distribution of WD and WE with varying H_{fin} and *NNS*. This paragraph presents a summary of the findings derived from WD and WE, covering the range from A14 to A3. The following sections will provide comprehensive explanations for each of these outcomes. WD is influenced not only by the total R_{WL} and C_{BL} in an SRAM subarray but also by the WM results. As detailed in Sect. 2.4 of Chap. 2, the write driver voltage is set according to the WM value to represent diverse requirements of write-assist techniques across distinct SRAM technologies. A poorer WM necessitates a larger boost capacitor for a higher write-assist voltage level, incurring additional energy overhead. This elevated voltage level or reduced RC for BL and WL may result in faster data switching times, thereby enhancing WD. WE, i.e., the active write energy of the subarray, is mainly determined by integration time (WD), total C_{BL}, and the applied write-assist techniques. The trends observed in WD and WE closely mirror those of RD and RE, as outlined in the preceding subsection (Sect. 6.5.2). The HP and LP cases exhibit superior WD and WE outcomes owing to their smaller subarray size compared to HD-1 and HD-2 counterparts. Although HD-2 has effectively improved WM compared to HD-1, it comes with a dramatic degradation in WD and WE due to the larger total R_{WL} in an SRAM subarray with extended total WL length. This trade-off between write failure risk and PP has been observed in the previous subsection (Sect. 6.5.1.3), prompting a quest for an alternative design approach beyond utilizing the HD-2 subarray size for co-optimization between WM and PP. Consequently, the WNS is optimized for A3 SRAM subarrays to enhance WM at the cost of increased bitcell area scaling, as discussed in the subsequent section (Sect. 6.5.2). Notably, even the HD-2 A3.h3p subarray shows 11% improvement in WD, but it still suffers from 2.1× degradation in WE compared to HD-1 counterpart. Most of the optimized channel stack dimensions appear compatible with logic dimensions. However, for A3 SRAM bitcells, it is advisable to set the optimized result of *NNS* to 2 instead of 3. Future exploration is recommended to consider the optimized channel stack in both SRAM and logic applications. For consistency, the logic assumptions for Hfin and NNS will be maintained for SRAM bitcell designs in this chapter and Chap. 7.

For A14.x2n and A10.x3n subarrays (Fig. 6.19), the optimized configuration, whether HP or LP, demonstrates significant improvements in WD/WE of up to 15%/41% at $NNS = 4$ and 9%/52% at $NNS = 3$, respectively, when compared to their HD-1 counterparts. This improvement is attributed to superior WM and smaller subarray size. Furthermore, the HD-1 A14.x2n and A10.x3n subarrays exhibit better WD/WE ratios by 33%/21% at $NNS = 4$ and 12%/0% at $NNS = 3$, respectively, compared to HD-2 counterparts. Analysis of the RSNM (Fig. 6.7a) and WM (Fig. 6.11a) results suggests that LP configuration is optimized for achieving superior RSNM, WD, WE, and second-best RD and RE for A14.x2n. Conversely, while HD-2 offers the best WM for A10.x3n, it incurs a significant energy overhead compared to its HP or LP counterparts. Therefore, HP configuration is recommended for A10.x3n due to its second-best WM and best RSNM, RD, RE, WD, and WE.

For A5 CFET SRAM subarrays (Fig. 6.20), it is observed that HP and LP configurations yield superior WD and WE performance compared to other subarray arrangements. Notably, the WD characteristics of HD-1 approach those of HP and LP subarrays in A5.s2p and A5.m2p due to reduced RC of BL per cell (Fig. 6.1). Specifically, HP and LP configurations at A5.s1p, A5.s2p, A5.m1p, and A5.m2p with $H_{fin} = 35$ nm and $NNS = 2$ demonstrate WD/WE enhancements of 8%/40%, 0%/39%, 8%/41%, and 0%/29% over HD-1 counterparts, respectively. Conversely, HD-1 exhibits superior RD compared to HD-2 across all scenarios, with HD-1 surpassing HD-2 in RE solely within A5.s2p and A5.m2p. Furthermore, HD-1 configurations in A5.s1p and A5.m1p at $H_{fin} = 35$ nm enhance WD by 23% and 18%, respectively, at the expense of degraded WE by 21% and 30%, respectively. Additionally, HD-1 configurations in A5.s2p and A5.m2p with $NNS = 2$ show WD/WE improvements of 47%/12% and 44%/4%, respectively, relative to HD-2 counterparts. For A5.s2p, it is advised not to opt for the HD-2 configuration due to the minimal WM improvement of less than 30 mV compared to HP and LP counterparts (Fig. 6.12b), despite significant degradation in WD and WE by 47% and 46%, respectively. LP is suggested as the optimized configuration for A5.s2p subarrays to uphold superior read stability (Fig. 6.8b). Conversely, the optimized configuration choice between HP/LP or HD-2 should be a compromise between RSNM/speed/energy and WM considerations for A5.s1p, A5.m1p, and A5.m2p subarrays.

For A3 CFET SRAM subarrays (Figs. 6.21 and 6.22), the trends regarding WD and WE align with previous SRAM technologies, where both HP and LP configurations emerge as the optimized choices. Moreover, improvements in WM for HD-2 designs come at the expense of significant degradation in WD and WE compared to their counterparts. Across HP and LP A3.s1p, A3.s2p, A3.s3p, A3.m1p, A3.m2p, A3.m3p, A3.h2n, A3.h3n, A3.h2p, and A3.h3p subarrays at Hfin = 35 nm and NNS = 3, enhancements in WD/WE reach up to 15%/47%, 28%/52%, 22%/61%, 17%/46%, 24%/49%, 21%/52%, 6%/41%, 13%/62%, 10%/43%, and 17%/50%, respectively, compared to HD-1 counterparts. The WD/WE improvement of HD-1 A3.s1p, A3.s2p, A3.s3p, A3.m1p, A3.m2p, A3.m3p, A3.h2n, A3.h3n, A3.h2p, and A3.h3p subarrays at Hfin = 35 nm and NNS = 3 compared to HD-2 counterparts are 26%/−20%, 28%/13%, 4%/59%, 26%/−17%, 26%/6%, 15%/65%, 20%/0%, −12%/26%, 23%/1%, and −12%/53%, respectively. Notably, negative values denote superior write metrics for HD-2 compared to HD-1. In all cases, HP configurations yield the most favourable results for RSNM, RD, RE, WD, and WE, albeit with compromised WM.

6.5.1.5 Brief Summary

Table 6.2 presents a summary of the relationship between HD (HD-1 and HD-2), HP, and LP subarray configurations across various metrics. Two crucial design criteria for maintaining SRAM functionality are RSNM and WM. Each choice of subarray configuration entails a trade-off among margin, speed, and energy. For instance, selecting LP A5.s1p and HP for A5.m1p, A3.s1p, and A3.m1p is essential to avoid the read stability concerns. However, these selections result in degraded WM in all subarrays or RD and RE in A5.m1p subarray. Prioritizing RSNM as the primary criterion for SRAM functionality suggests that the optimized configurations for A3 subarrays would be HP, yielding the best RSNM, speed, and energy results, while LP configuration would be preferable for A14.x2n, A5.s1p, and A5.s2p. Conversely, prioritizing WM as the main criterion for SRAM subarray design indicates that HP configuration would be favourable for A14.x2n and A3.s2p, while HD-2 configuration would be optimized for the remaining SRAM technologies at the expense of speed and energy penalties. Furthermore, the HD-2 design emerges as the optimized choice for A10.x3n, A3.h3n, and A3.h2p, achieving superior WM and RSNM simultaneously, albeit with speed and energy overheads. It is important

Table 6.2 Optimization of SRAM subarray configurations based on distinct metrics for each SRAM technology within the scope of logic dimension. Underlined text signifies potential concerns regarding read stability

Tech	HSNM	RSNM	WM	RD/RE	WD/WE
A14.x2n	HD≈HP≈LP	LP>HP≈HD	HP≈HD-1>LP>HD-2	HP<LP<HD-1<HD-2	HP≈LP <HD-1 <HD-2
A10.x3n	HD>HP≈LP	HD≈HP≈LP	HD-2>HP>LP>HD-1		
A5.s1p	HD≈HP≈LP	LP>HD≈HP	HD-2>HD-1≈HP≈LP	HP≈LP<HD-1<HD-2	
A5.s2p	HP≈LP>HD	LP>HP>HD	HD-2>HP≈LP>HD-1		
A5.m1p		HP>HD>LP	HD-2>LP>HP≈HD-1	HP<LP<HD-1<HD-2	
A5.m2p	HP>LP>HD	HP>HD≈LP			
A3.s1p	HD≈LP>HP	HP>HD≈LP	HD-2>LP>HP>HD-1	HP≈LP>HD-1<HD-2	
A3.s2p		HP>HD>LP	HP≈LP>HD-2>HD-1		
A3.s3p		HP>HD>LP	HD-2>LP>HP>HD-1		
A3.m1p	HD≈HP≈LP	HP>HD≈HP	HD-2>HP≈LP>HD-1		
A3.m2p	HD≈HP>LP	HP>HD>LP	HP≈LP≈HD-2>HD-1	HP<LP<HD-1<HD-2	
A3.m3p	HD>HP>LP	HP>HD>LP	HD-2>LP>HP>HD-1	HP≈LP<HD-1<HD-2	
A3.h2n	HD≈HP>LP	HP>HD>LP	HD-2>HP>LP≈HD-1		
A3.h3n		HD≈HP>LP	HD-2>HP>LP>HD-1		
A3.h2p			HD-2>HP≈LP>HD-1	HP<LP<HD-1<HD-2	
A3.h3p		HP>HD>LP	HD-2>LP>HP>HD-1		

to emphasize that the ultimate selection of the optimized design choice should be validated through macro-level PPA simulation in Chap. 7.

Takeaways

- **Impact of Channel Geometry (as H_{fin} or *NNS* increases):**
  - **HSNM and RSNM:**
    - Improvement: Increased transistor drive strength.
    - Degradation: Increased R_{power} from taller via.
    - Degradation: More read disturbance (only for RSNM).
  - **WM, Speeds, and Energies:**
    - Improvement: Increased transistor drive strength.
    - Degradation: Larger voltage drops due to increased device capacitance.
- **Trade-offs:**
  - Optimization among margin, speed, and energy involves trade-offs.
  - Example: The HD-2 design emerges as the best option for A10.x3n, A3.h3n, and A3.h2p, attaining superior WM and RSNM while incurring speed and energy overheads.

6.5.2 *Impacts of* W_{NS} *on HD SRAM Bitcells*

The previous section suggests searching for an alternative design approach to improve both WM and PP, instead of altering the subarray size from HD-1 to HD-2. As a result, this section provides the optimization of W_{NS} on HD-1 SRAM bitcells. As the HD SRAM bitcell area scales to A3 technology, the increased R_{BL} poses a challenge to PP improvement, especially for the WM. The exploration focuses on determining whether significant changes in WM and RD can be achieved by relaxing the nanosheet width (W_{NS}) and BL width (W_{BL}) through an increase in cell height. HD-1 A3.h2p and HD-1 A3.h3p are selected for testing this concept, as the issue of insufficient W_{BL} only arises in HD-1 cases.

Expanding the bitcell area of A3 NS-on-NS and FS-on-FS CFET SRAM bitcells to match A5.m2p and A3 stacked Fin/NS counterparts allows for potential PP benefit due to wider W_{NS} and W_{BL}, as depicted in Figs. 6.23 and 6.24. While a broader W_{NS} and W_{BL} can be attained by increasing the bitcell area, it incurs an R_{WL} penalty. Consequently, limited RD improvements are observed in both A3.h2p (4%) and A3.h3p (10%) cases. On the flip side, there are 14 and 92 mV of WM

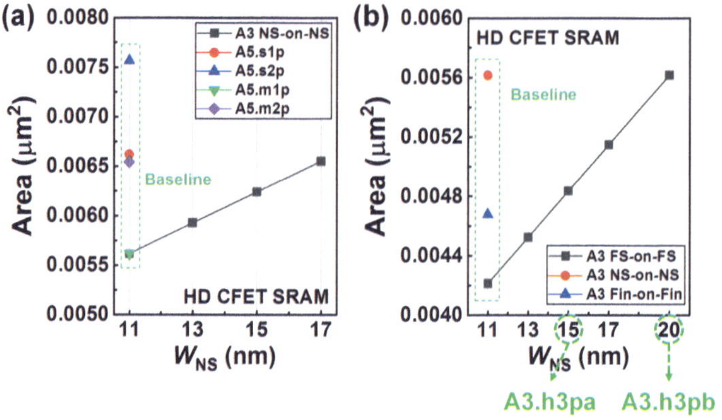

Fig. 6.23 6T SRAM bitcell area comparison of increased W_{NS} in HD A3 (**a**) NS-on-NS and (**b**) FS-on-FS CFET SRAM bitcells with larger SRAM bitcells as baseline

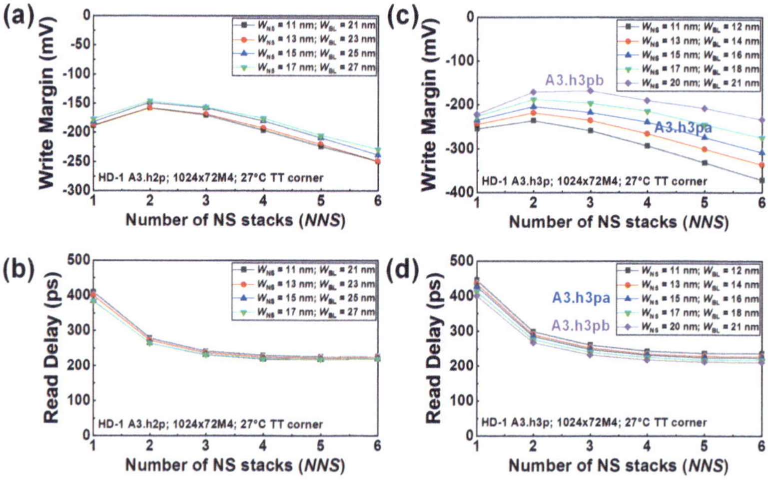

Fig. 6.24 WM and RD for HD-1 (**a**)(**b**) A3.h2p and (**c**)(**d**) A3.h3p subarrays with varying W_{NS} and W_{BL}

enhancements in HD-1 A3.h2p and HD-1 A3.h3p, respectively, attributed to R_{BL} reduction. These results suggest that the current W_{NS} of 11 nm and W_{BL} of 21 nm appear sufficiently wide for A3.h2p. For A3.h3p, it is worth considering the HD-1 subarray designs with increased W_{NS} and W_{BL}, namely A3.h3pa and A3.h3pb (Table 5.3 in Chap. 4), for further testing.

Takeaways

- **Impact of Channel Geometry (as W_{NS} increases):**
  - Increasing bitcell area leads to wider W_{NS} and W_{BL} but comes with R_{WL} penalty.
  - **A3 NS-on-NS CFET SRAM Bitcell Designs:** It is not suggested to widen the W_{NS} in these bitcell designs since the improvement is marginal.
  - **A3 FS-on-FS CFET SRAM Bitcell Designs:** Widening the W_{NS} in these bitcell designs is advised because of the notable improvement, particularly in WM for mitigating the write failure risk.

6.6 Subarray-Level PP Benchmark

Figures 6.25 and 6.26 present the margin, performance, and active energy across A14 to A3 technologies for HD (HD-1 and HD-2), HP, and LP subarrays. These results are acquired at nominal temperature (27 °C) and typical–typical (TT) corners to ensure a fair comparison based on the same device dimensions as logic. The key factors and relationships influencing each metric for double-stack (CFET) SRAM subarrays have been discussed in preceding sections. These concepts are equally relevant to single-stack (NSFET and FSFET) SRAM subarrays with opposing read and write paths, given their NMOS-PG-based structure. Thus, this section explores the distinctions between single- and double-stack SRAM subarrays. It's important to note that the discussion here is based on HD-1 for simplicity, with other cases following a similar pattern.

RSNM of each case is similar due to comparable R_{power} and transistors' drive strengths. A10.x3n has a worse WM than A14.x2n due to a much larger R_{BL} coming from area scaling. WMs of A5 seq and mono CFET SRAM bitcells are improved compared to A10.x3n due to less R_{BL} thanks to the spacer merge technique. However, limited R_{BL} and C_{BL} reduction in A3 CFET SRAM subarrays still cause poor (negative value) WMs. This implies a high risk of write failure [10], which should be solved by further WM enhancement strategies such as BEOL optimization and supplementary DTCO in the future. In the optimization of WM for HD-2 subarray design versus A3.h3pa/A3.h3pb with increased W_{NS}, it is observed that HD-2 A3.h3p subarray demonstrates superior WM results by 113 mV compared to HD-1 A3.h3pb counterpart. Therefore, redesigning the subarray size is considered as the most efficacious method for enhancing WM. The RD between A14.x2n and A10.x3n SRAM subarrays is similar since the latter one has less R_{WL} and C_{BL} but a larger R_{BL}. This compensates for the improved or degraded effects on RD. Although the R_{WL} decreases significantly due to cell height scaling in A5 and A3

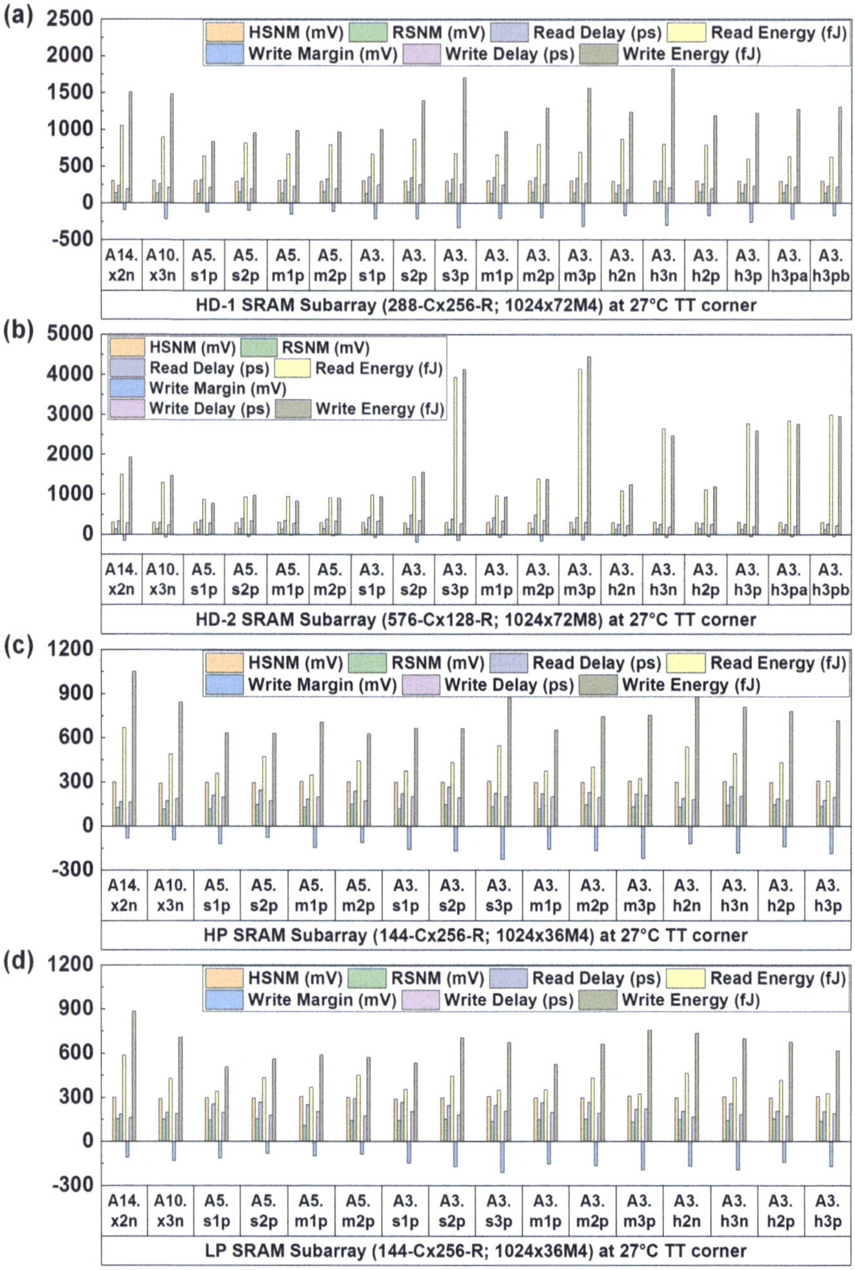

Fig. 6.25 SRAM subarray benchmark of margin, delay, and energy for read and write operations from A14 to A3 with the subarray size of (**a**) HD-1 (288-C×256-R; 1024×72M4), (**b**) HD-2 (576-C×128-R; 1024×72M8), (**c**) HP (144-C×256-R; 1024×36M4), and (**d**) LP (144-C×256-R; 1024×36M4) designs

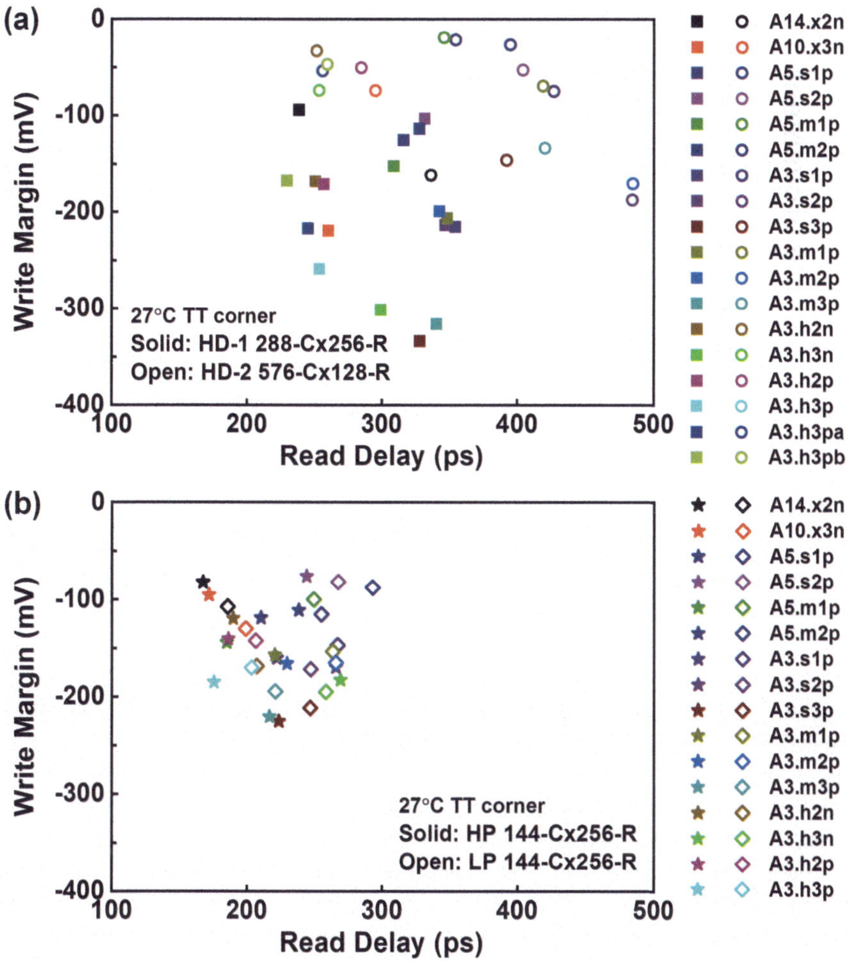

Fig. 6.26 Pareto distribution of WM and RD with (**a**) HD-1 (288-C×256-R; 1024×72M4) and HD-2 (576-C×128-R; 1024×72M8), (**b**) HP (144-C×256-R; 1024×36M4) and LP (144-C×256-R; 1024×36M4) designs

seq and mono CFET SRAMs, the RD and WD do not improve due to larger R_{BL} and C_{BL} caused by the narrower BL and longer via connected from BL to bottom PG devices, respectively. The energy is mostly dominated by the C_{BL} and the operating time. A10.x3n has less RE and WE compared to A14.x2n due to similar read time but less FEOL C_{BL}. For A5 and A3 seq and mono CFET SRAM subarrays, the energies may decrease due to the scaled logic circuits but also increase by large C_{BL}.

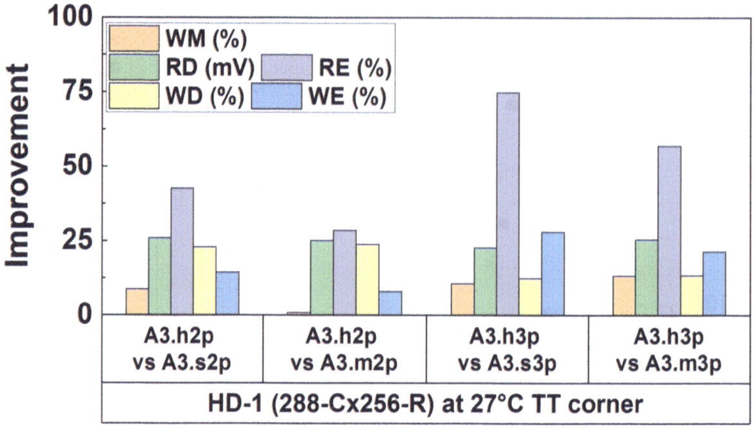

Fig. 6.27 Subarray-level improvements of margin, delay, and energy for read and write operations by HD-1 A3 hybrid CFET SRAM subarray designs

As discussed in Chap. 5, the HD A3 hybrid NS-on-NS CFET SRAM bitcell exhibits significant reductions of 25% in C_{BL} and 35% in C_{WL} primarily by shortening the via height as compared to A3 seq and mono counterparts. Therefore, HD-1 A3.h2p/A3.h3p subarray demonstrates significant PP improvements of up to 43/75 mV, 26/23%, 9/11%, 23/13%, and 15/28% in WM, RD, RE, WD, and WE, respectively, compared to A3.s2p/A3.s3p (Fig. 6.27). The improvements also occur among other designs, except for showing higher energy than its A3 Fin-on-Fin CFET counterparts owing to larger device capacitance. However, it does not fully outperform its A5 counterparts since the increase in R_{BL} with bitcell area scaling becomes a limiting factor with HD-1 subarray design. In conclusion, the optimized subarray designs for best PP and sufficient margin in A5 are A5.m2p for HD-1, HD-2, HP, and LP options. Similarly, in A3, the preferred designs include HD-1 A3.h3pb, HD-2 A3.h2p, HP A3.h2p, and LP A3.h3p subarrays. Notably, stacked Fin CFET SRAM subarrays are excluded in this scope due to the fragility of read stability.

The SRAM subarray area scaling in ultra-scaled nodes is primarily achieved through cell height scaling, and thereby R_{WL} is typically decreased at the expense of larger R_{BL}. This poses a higher risk of write failure, increased energy consumption, and area overhead of write assist techniques. Therefore, it motivates the development of BEOL optimization [8], alternative routing design [7–9], and subarray size reconfiguration [7] to address these challenges. Notably, the alternative routing designs for SRAM bitcells have been explored in Chaps. 4 and 5.

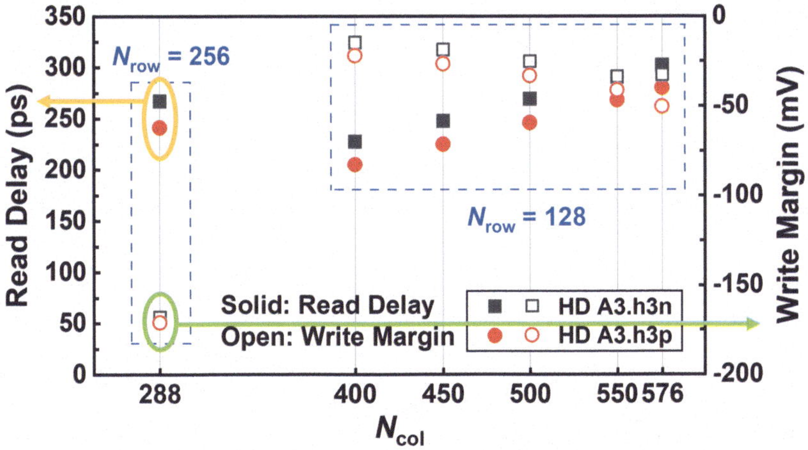

Fig. 6.28 RD and WM with the subarray size of 288-C×256-R and 400-C to 576-C×128-R for HD A3 hybrid CFET SRAM subarray

6.6.1 Subarray Size Optimization for HD Option

By rearranging the subarray size from HD-1 to HD-2 design, while maintaining the same capacity as HD-1 but with longer WL and shorter BL, the WM of HD-2 A3 hybrid CFET SRAM subarray is improved by 138 mV at the cost of 16/29% degradation in RD/WD (Fig. 6.28). Although there is an increase in delay and energy consumption due to the larger total R_{WL} of the longer WL, the potential benefit of the write operation justifies the attention. Note that the realistic write-assist techniques are not considered in this analysis [10]. Figure 6.28 shows that the WM degrades as N_{col} increases under 128-R due to slower WL activation speed compared to BL data transfer speed. To prevent read speed degradation and achieve more WM improvement compared to 576-C×128-R, N_{col} should be at least lower down from 576 to 500 when N_{row} shrinks from 256 to 128. In Chap. 7, we will illustrate that the subarray is not the dominant factor in the overall macro-level results. This implies that purely continuing the traditional bitcell scaling is not sufficient for higher-level PPA enhancement. Consequently, an innovative approach to optimizing subarray-level circuit design for substantial macro-level improvement will be presented in Chaps. 8 and 9.

6.6.2 BEOL Optimization Guideline for HD Option

As discussed in the previous paragraph, node-to-node area scaling may not guarantee PP benefits due to the increased parasitic RC problem. To achieve more PP benefits in the deeply scaled technology, BEOL optimization for lowering R_{BL}

Table 6.3 Test cases of different BEOL R_{BL} and C_{BL}

Cases	R_{BL}	C_{BL}
Baseline	1×	1×
Case 1	0.5×	1×
Case 2	0.25×	1×
Case 3	1×	0.5×
Case 4	1×	0.25×
Case 5	0.5×	0.5×
Case 6	0.25×	0.25×

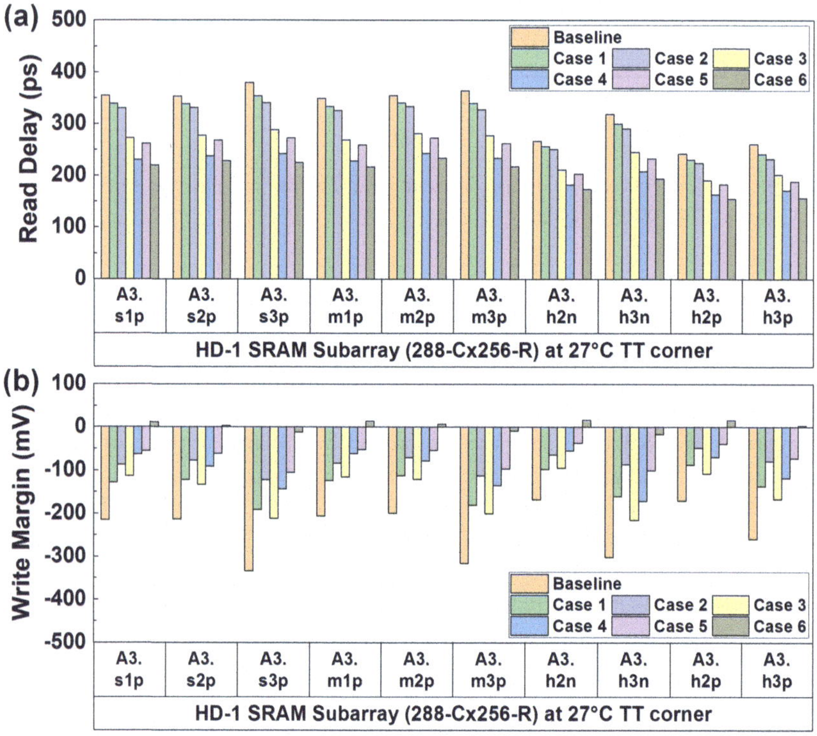

Fig. 6.29 (**a**) RD and (**b**) WM comparison with different BEOL R_{BL} and C_{BL} for HD-1 A3 CFET SRAM subarrays

and C_{BL} would be indispensable. Hence, the corresponding BEOL optimization guidelines for improving RD and WM based on different BEOL R_{BL} and C_{BL} test cases (Table 6.3) is provided based on HD-1 A3 CFET SRAM subarray designs in Fig. 6.29. The read speed improvement is dominated by C_{BL} reduction since case 4 and 6 give the best but similar results. On the other hand, reducing both R_{BL} and C_{BL} is crucial to WM improvement since case 5 and 6 provide the best results. In comparison to HD-2 A3.h3p, HD-1 A3.h3p exhibits superior WM performance under case 6 conditions, characterized by a reduction in both R_{BL} and C_{BL} to less than 0.25×. This proposed approach may serve as a viable alternative to the HD-2 subarray design, contingent upon the development of RC reduction techniques. Several approaches have been reported to reduce RC such as flying BL (which can reduce both R_{BL} and C_{BL}) [1], hybrid-height metal [2], airgap [2], and high aspect ratio [2] techniques. Further developments of BEOL optimization with realistic RC interaction should be tackled in the future by following the RC lowering instructions given in this work.

Takeaways

- **Subarray Size Optimization:**
  - Shortening BL length while enlarging WL length improves WM but increases read delay at subarray level.
  - Exploring a sweet spot between these trade-offs is crucial for future work, requiring more subarray design options.
- **BEOL Optimization Guideline:**
  - Reducing C_{BL} for improving RD.
  - Reducing both R_{BL} and C_{BL} to improve WM.

6.7 Conclusions

This chapter delves into the DTCO PPA analysis of SRAM subarrays across A14 to A3, encompassing stress, geometry, bitcell type, subarray configuration, and RC impacts. In the scaled nodes beyond A14, V_{DD} scaling halts due to limited RSNM, and upscaled N_{row} is constrained by poor WM. Aggressive scaling of bitcell area leads to increased routing complexity, resulting in degraded parasitic RC that hampers node-to-node PP gains. To maximize the full node-to-node PPA benefit for A3 CFET SRAM subarrays, we demonstrate the utilization of an alternative transistor structure (hybrid CFET in Chap. 5), subarray size adjustment, W_{NS} optimization (specifically for HD A3.h3p), and BEOL design guidelines. The optimized design choice varies depending on individual design perspectives and trade-offs, as discussed in the following paragraphs.

The A3 hybrid CFET SRAM bitcell type emerges as the optimized choice for bitcell design across all scenarios, surpassing both seq and mono counterparts. It demonstrates notable reductions of 41%, 25%, and 35% in R_{WL}, C_{BL}, and C_{WL}, respectively, owing to narrower WL and shorter via height. Consequently, the HD A3.h3p (A3 hybrid FS-on-FS CFET SRAM bitcell with P-PG) subarray achieves enhancements of up to 75 mV in WM and 26% in RD without incurring any bitcell area penalty. These findings motivate thorough exploration into the process development of hybrid CFET technology.

In the context of optimizing subarray configuration optimization, HD-2 (576-C×128-R) emerges as the optimized choice when considering the WM perspective, while HP and LP (144-C×256-R) are preferred from the PP perspective. Section 6.6 shows that rearranging the subarray size from HD-1 (288-C×256-R) to HD-2 could potentially lead to speed and energy degradation. For instance, the HD-2 A3.h3p subarray enhances WM/WD by 206 mV/11% but incurs 1%/4.6×/2.1× deterioration in RD/RE/WE compared to its HD-1 counterpart. This trade-off suggests that compromising on delay and energy consumption might be essential to guarantee write functionality. Although HD-1 subarrays generally offer better PP than HD-2 counterparts, the inferior WM raises significant concerns regarding write failure probabilities. Therefore, it is proposed to upscale the bitcell size of A3.h3p to widen the widths of W_{NS}/W_{BL} from 11/12 to20/21 nm, resulting in a new bitcell design denoted as A3.h3pb. The objective is to enhance WM, particularly targeting HD-1 subarrays. It is important to note that the optimized bitcell choice is A3.h2p for HD-2 subarrays to avoid significant energy penalties observed in A3.h3p, despite marginal WM improvements. Comparing the WM optimization between the HD-2 A3.h2p subarray and the HD-1 A3.h3pb subarray, it is evident that the former design outperforms the latter by 117 mV in WM. Thus, resizing the subarray emerges as the most effective approach to enhance WM. In seeking an alternative approach to enhance the WM of the HD-1 subarray to match that of the HD-2 subarray, this study outlines a BEOL optimization guideline in Sect. 6.6.2. It highlights an increase in RD primarily resulting from a decrease in C_{BL}, while an enhanced WM is achieved through reductions in both R_{BL} and C_{BL}. Specifically, to surpass the HD-2 counterparts, R_{BL} and C_{BL} in the HD-1 A3.h3p subarray must be decreased to at least 0.25 times the original RC values. This proposed methodology offers a viable option to attain comparable WM to HD-2 subarray design, subject to the advancement of RC reduction techniques.

Section 6.5.1 highlights that optimized metrics are predominantly observed when H_{fin} and *NNS* are aligned with the logic dimension. However, the optimized *NNS* should be set to 2 rather than 3 to achieve optimized WM in A3.s3p, A3.m3p, A3.h3n, and A3.h2p subarrays. Additionally, optimized RD and RE are attained when *NNS* is set to 2 but not 3 for A3.m2p, A3.m3p, A3.h2p, and A3.h3p subarrays. Therefore, assuming an *NNS* of 3 in these A3 CFET SRAM subarrays to align with the logic dimension may result in potential penalties in WM, RD, and RE. The advantages and disadvantages of choosing the subarray configuration, including HD-1, HD-2, HP, and LP, for each metric are presented in Table 6.2.

In conclusion, the optimized subarray designs to achieve the highest subarray density are A3.s3p, A3.m3p, A3.h3n, and A3.h3p bitcell choice. Moreover, the optimized subarray designs for achieving superior PP and ensuring adequate read and write margins in A3 consist of HD-1 A3.h3pb, HD-2 A3.h2p, HP A3.h2p, and LP A3.h3p subarrays. It is noteworthy that stacked Fin CFET SRAM subarrays are not considered within this framework due to concerns regarding read stability fragility. In the subsequent chapter (Chap. 7), further macro-level PPA analysis will be conducted to offer a clearer overview of optimized bitcell and subarray design options.

References

1. Chang, J., et al. (2017). 12.1 A 7 nm 256 Mb SRAM in high-k metal-gate FinFET technology with write-assist circuitry for low-VMIN applications. In *2017 IEEE International Solid-State Circuits Conference (ISSCC)* (pp. 206–207). https://doi.org/10.1109/ISSCC.2017.7870333
2. Farokhnejad, A., et al. (2022). Evaluation of BEOL scaling boosters for sub-2 nm using enhanced-RO analysis. In *2022 IEEE International Interconnect Technology Conference (IITC)* (pp. 136–138). https://doi.org/10.1109/IITC52079.2022.9881286
3. Ha, D., et al. (2017). Highly manufacturable 7 nm FinFET technology featuring EUV lithography for low power and high performance applications. In *2017 Symposium on VLSI Technology* (pp. T68–T69). https://doi.org/10.23919/VLSIT.2017.7998202
4. Jang, D., et al. (2017). Device exploration of NanoSheet transistors for sub-7-nm technology node. *IEEE Transactions on Electron Devices, 64*(6), 2707–2713. https://doi.org/10.1109/TED.2017.2695455
5. Kim, T. H., et al. (2020). An embedded level-shifting dual-rail SRAM for high-speed and low-power cache. *IEEE Access, 8*, 187126–187139. https://doi.org/10.1109/ACCESS.2020.3030099
6. Lee, S., et al. (2022). Novel modeling approach to analyze threshold voltage variability in short gate-length (15–22 nm) nanowire FETs with various channel diameters. *Nanomaterials, 12*(10). ISSN: 2079-4991. https://doi.org/10.3390/nano12101721
7. Liu, H.-H., et al. (2023). CFET SRAM DTCO, interconnect guideline, and benchmark for CMOS scaling. *IEEE Transactions on Electron Devices, 70*(3), 883–890. https://doi.org/10.1109/TED.2023.3235701
8. Liu, H.-H., et al. (2023). CFET SRAM with double-sided interconnect design and DTCO benchmark. *IEEE Transactions on Electron Devices* 1–8. https://doi.org/10.1109/TED.2023.3305322
9. Liu, H.-H., et al. (2023). DTCO of sequential and monolithic CFET SRAM. In R.-H. Kim & N. V. Lafferty (Eds.), *DTCO and computational patterning II* (Vol. 12495, p. 124950Z). International Society for Optics and Photonics. SPIE. https://doi.org/10.1117/12.2657524
10. Liu, H.-H., et al. (2022). Extended methodology to determine SRAM write margin in resistance-dominated technology node. *IEEE Transactions on Electron Devices, 69*(6), 3113–3117. https://doi.org/10.1109/TED.2022.3165738
11. Matsukawa, T., et al. (2012). Suppressing Vt and Gm variability of FinFETs using amorphous metal gates for 14 nm and beyond. In *2012 International Electron Devices Meeting* (pp. 8.2.1–8.2.4). https://doi.org/10.1109/IEDM.2012.6479002
12. Schuddinck, P., et al. (2019). Device-, Circuit-Block-level evaluation of CFET in a 4 track library. In *2019 Symposium on VLSI Technology* (pp. T204–T205). https://doi.org/10.23919/VLSIT.2019.8776513
13. Schuddinck, P., et al. (2022). PPAC of sheet-based CFET configurations for 4 track design with 16 nm metal pitch. In *2022 IEEE Symposium on VLSI Technology and Circuits (VLSI Technology and Circuits)* (pp. 365–366). https://doi.org/10.1109/VLSITechnologyandCir46769.2022.9830492
14. Seevinck, E., List, F. J., & Lohstroh, J. (1987). Static-noise margin analysis of MOS SRAM cells. *IEEE Journal of Solid-State Circuits, 22*(5), 748–754. https://doi.org/10.1109/JSSC.1987.1052809
15. Weckx, P., et al. (2019). Novel forksheet device architecture as ultimate logic scaling device towards 2 nm. In *2019 IEEE International Electron Devices Meeting (IEDM)* (pp. 36.5.1–36.5.4). https://doi.org/10.1109/IEDM19573.2019.8993635

Chapter 7
Macro-level Performance-Power-Area Benchmark of CFET SRAM Macro

Chapter 6 presents a comprehensive subarray-level power, performance, and area (PPA) analysis for various static random-access memory (SRAM) bitcell and subarray design options. The evaluation spans from 14-Å-compatible technology (A14), 10-Å-compatible technology (A10), 5-Å-compatible technology (A5), to 3-Å-compatible technology (A3), including nanosheet (NS) [4], forksheet (FS) [8], sequential (seq) complementary field-effect transistors (CFET) [1, 3], mono CFET [1, 3], and hybrid CFET [2] SRAM bitcell designs. The necessity of exploring macro-level evaluation arises from the significant influence of inter-subarray interconnect and logic design on the macro-level PPA results of SRAM macros [5, 6]. Thus, this chapter provides design-technology co-optimization (DTCO) PPA analysis at the macro level to identify the most promising SRAM bitcell design in deeply scaled nodes. This analysis is performed utilizing the subarray-level PPA data provided in Chap. 6, along with the PPA simulation framework outlined in Chap. 2.

The PPA analysis of the SRAM macro incorporates contributions from both subarray and inter-subarray components. Local interconnects, namely bitline (BL) and wordline (WL), play a crucial role at the subarray level. The node-to-node gain in power and performance (PP) is hindered by undesirable increases in resistance and capacitance (RC) for both BL and WL. This is attributed to the scaling of SRAM bitcell area and the utilization of stacked architecture, as described in Chap. 6 [1–3]. As cell height scales with technology, the degradation of write margin (WM) occurs due to increased BL resistance (R_{BL}) [4]. Addressing this issue necessitates the implementation of an additional design of subarray size or RC reduction technique to guarantee a successful write operation (Chap. 6). Conversely, macro-level PP heavily relies on the overhead of inter-subarray interconnects, benefiting significantly from subarray area scaling. Hence, high-density (HD) A5 CFET SRAM macro demonstrates a 53%–76% improvement in macro-level energy-delay-area product (EDAP) compared to A14 NSFET counterpart. This improvement stems from reducing inter-subarray interconnect overhead through node-to-node

H.-H. Liu, F. Catthoor, *Circuit-Technology Co-Optimization of SRAM Design in Advanced CMOS Nodes*, https://doi.org/10.1007/978-3-031-76109-6_7

subarray area scaling, as delineated in this chapter. It is important to emphasize that the comprehensive evaluation of the impact of bitcell and subarray area scaling is essential and can be thoroughly addressed solely at the macro level. Although A3 hybrid FS-on-FS CFET SRAM macro continues to offer a macro-level EDAP improvement by 87% over A14, the node-to-node PPA gain seems to reach a plateau beyond A5. Given the added process complexities and expenses involved in the hybrid CFET structure—such as wafer flipping, layer transfer bonding, substrate removal, and double MINT layer fabrications—the incremental benefit when transitioning from A5 to A3 is minimal and not particularly worthwhile. This observation prompts further exploration of alternative PPA enhancement techniques, such as alternative SRAM subarray design, as introduced in Chaps. 8 and 9.

This chapter focuses on presenting the optimized SRAM bitcell, subarray, and macro designs derived from macro-level PPA, incorporating data input from the subarray level obtained in Chap. 6. The analysis comprehensively examines the impacts of inter-subarray interconnect, pin position, subarray size, bitcell configuration, data capacity, and bitcell scaling. It is important to highlight that the macro-level results were achieved with the support of Zhenlin Pei and Chenyun Pan from The University of Texas at Arlington. Notably, the collaborative work referenced in [5–7][1] extensively addresses inter-subarray interconnect optimization, covering various macro aspects. However, this doctoral research diverges from those topics. Instead, this chapter focuses solely on clarifying the relationship between macro-level PPA outcomes and the available subarray design alternatives. Section 7.1 briefly outlines the design options for the macro and back-end-of-line (BEOL) assumptions. The study employs a macro architecture comprising surrounding cores and macros [5–7]. The macro follows a hierarchical design, including bank, mat, subarray, and bitcell arrays, but excludes the tag array. Section 7.2 breaks down the macro-level PPA into four contributors: subarray, tree, core-to-data, and control/driver. The significance of each component at macro level is thoroughly examined. Section 7.3 discusses the optimized inter-subarray interconnect routing and pin position design. Section 7.4 delves into the impacts of subarray size and bitcell configuration designs at macro level, building upon the discussion provided in Chap. 6. The examination of macro capacities is presented in Sect. 7.5. Section 7.6 illuminates the influence of bitcell/subarray scaling on macro-level PPA results. Finally, Sect. 7.7 concludes this chapter. The HD-2 subarray, A3.h3pb bitcell type, centre pin access, and E-tree interconnect stand out as the best design options, offering minimal write failure risk and favourable macro-level PPA results.

7.1 Experimental Setup of SRAM Macro Design

Figure 7.1 depicts four unique SRAM macro and logic core designs to be explored in this chapter, with corresponding BEOL assumptions listed in Table 7.1. Assuming

[1] Acknowledgment is given to Zhenlin Pei and Chenyun Pan for the Macro-level PPA analysis framework and inter-subarray interconnect advancements.

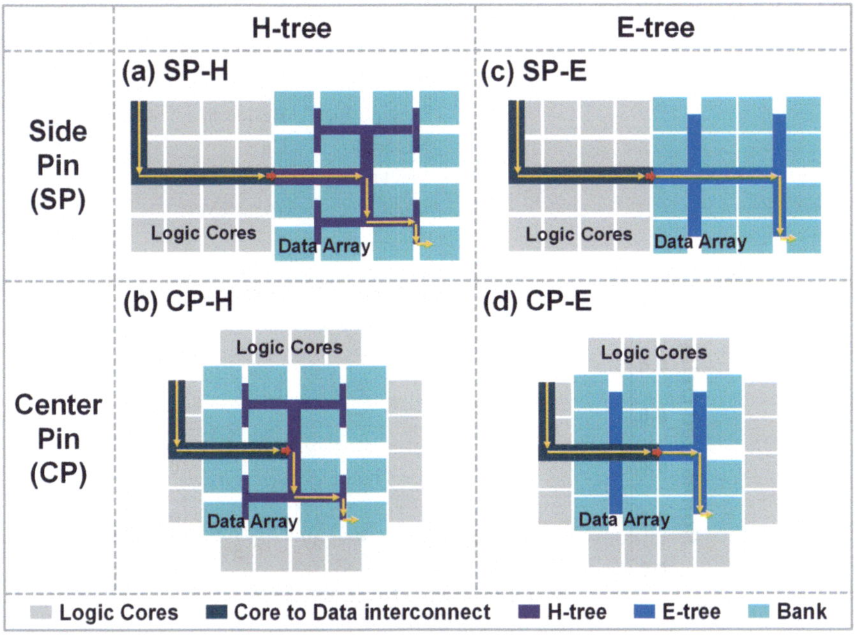

Fig. 7.1 SRAM macro design options with distinct inter-subarray interconnect assumptions in this work. (**a**) side pin with H-tree (SP-H), (**b**) centre pin with H-tree (CP-H), (**c**) side pin with E-tree (SP-E), and (**d**) centre pin with E-tree (CP-E) technologies

identical A14 design rules for BEOL is undertaken for simplicity, given the similar leakage and performance for each simulation model [1–4, 8] and unchanged geometry of M5 metal layer from A14 to A3 (Table 5.4 in Chap. 5). These designs are categorized according to the configurations of inter-subarray types, specifically H-tree or E-tree, and pin positions, denoted as side pin (SP) or centre pin (CP). The identified technologies include side pin with H-tree (SP-H), centre pin with H-tree (CP-H), side pin with E-tree (SP-E), and centre pin with E-tree (CP-E) technologies. These designs serve to evaluate the impacts of distinct inter-subarray interconnect overhead on macro-level PPA. Furthermore, the SRAM macro investigates data capacities of 4, 16, and 128 MB, along with four distinct standard subarray designs: HD-1, HD-2, HP, and LP, as detailed in Chap. 6. Generally, 16MB, HD-1 (288 columns and 256 rows, denoted as 288-C$\times$256-R), and A14 NSFET SRAM (A14.x2n) technologies are employed as constant parameters for simplicity when testing various variables. A comparable trend is anticipated across various bitcell, subarray, and data capacity designs. In this chapter, the CACTI simulator conducts macro-level simulations at the nominal temperature (27 °C) and typical–typical (TT) corner, as outlined in Sect. 2.1.4.2. Additional insights into the simulation methodology, spanning from bitcell to macro level, can be found in Chap. 2.

Table 7.1 BEOL assumptions for inter-subarray interconnect

Technology	A14 NSFET
Routing style	H-tree; E-tree
Pin position	Side Pin (SP); Centre Pin (CP)
Core position	Beside array; Surround array
Material	Cu
Thickness	28 nm (M5)
Width	16 nm (M5)

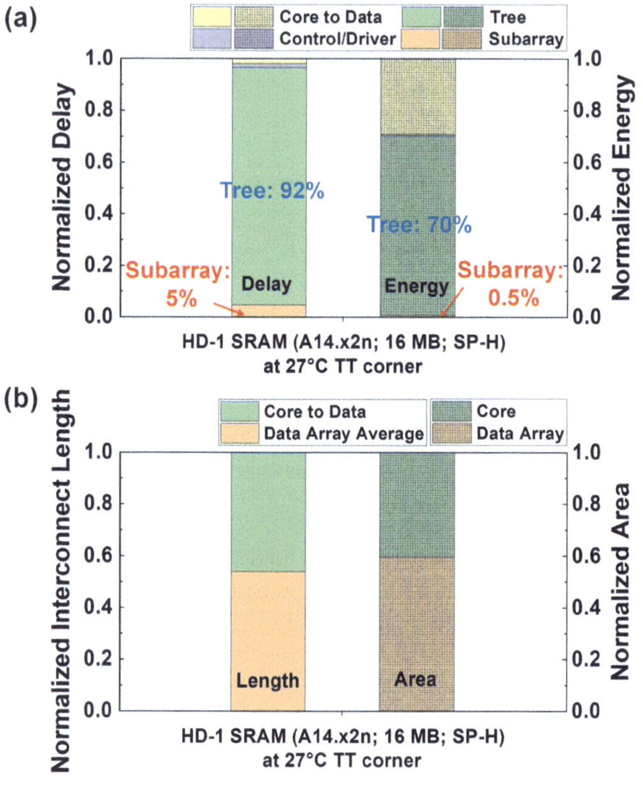

Fig. 7.2 Breakdown of normalized (**a**) delay and energy; (**b**) interconnect length and area for 16 MB HD-1 A14.x2n SRAM macro based on SP-H technology

7.2 Breakdown of Macro-level PPA Contribution in SRAM Macro

In the simulation framework, the macro-level PPA of the SRAM macro can be dissected into four distinct components: core to data, tree, control/driver, and subarray. Figure 7.2 illustrates the breakdown of normalized delay, energy, interconnect length, and area for the 16MB HD-1 SRAM macro utilizing A14.x2n bitcell design and SP-H technology. This analysis aims to determine the components with the most substantial influence on the macro-level PPA. It is important to note that in academic work, SP-H is commonly assumed for inter-subarray interconnect and pin position [5, 6].

In Fig. 7.2a, the contribution of the subarray and tree occupies 5%/0.5% and 92%/72% of the total delay and energy, respectively. The impact of the tree overhead surpasses that of the subarray, attributed to increased interconnect resistance in advanced technology nodes, as corroborated by previous studies [5, 6]. Thus, it is evident that bitcell/subarray scaling lacks efficacy as a macro-level PPA booster beyond A14, given the current SP-H assumption. The suggested approach for bolstering PPA lies in advancing techniques that mitigate inter-subarray interconnect overhead, exemplified by SP-E, CP-H, and CP-E [5, 6]. It is noteworthy that bitcell/subarray scaling remains crucial for curtailing inter-subarray interconnect overhead, representing an indispensable avenue for ongoing development [2].

Takeaways

- **Challenges in the Interconnect-Centric Technology Nodes:** Inter-subarray interconnect has a greater impact on macro-level factors compared to subarray contributions. Consequently, traditional bitcell scaling is less effective in boosting the PPA beyond A14.
- **Motivated Design Strategy:** Find an alternative array-level design to reduce the inter-subarray interconnect overheads.

7.3 Impacts of Inter-subarray Routing Design and Pin Position

Figure 7.3 demonstrates the effects of utilizing four configurations for inter-subarray interconnect routing and pin position designs, namely SP-H, SP-E, CP-H, and CP-E. It provides a detailed analysis of each component's contribution to macro-level delay, energy consumption, interconnect length, and area. This investigation focuses on HD-1 subarray design, A14.x2n bitcell design, and 16MB macro capacity for simplicity. In Fig. 7.3a, CP-E technology emerges as the optimized design option, reducing tree contribution to both delay and energy by 57% compared to SP-H counterpart. Figure 7.3b shows that CP-E technology achieves a 55% reduction in interconnect length and an 83% decrease in area for the macro. Consequently, macro-level delay, energy, interconnect length, and area can be enhanced by 53%, 45%, 37%, and 41%, respectively. The E-tree technology has been observed to operate faster than its H-tree counterpart. This improvement can be attributed to two main factors:

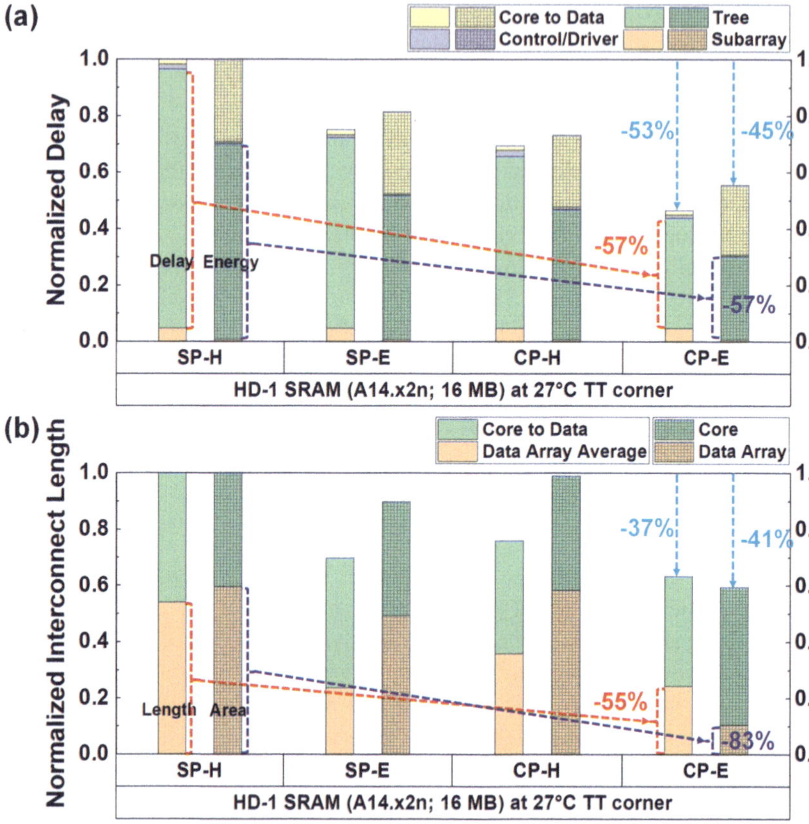

Fig. 7.3 Breakdown of normalized (**a**) delay and energy; (**b**) interconnect length and area for 16 MB HD-1 A14.x2n SRAM macros based on SP-H, SP-E, CP-H, and CP-E technologies

- **Reduced Area Overhead**: For a given cache size, technology, and workload, the array utilizing the H-tree has a larger area overhead compared to the E-tree. As a result, the E-Tree occupies less space.
- **Shorter Average Wire Length**: The average wire length of the E-Tree depends on the probability of access to each subarray and the distance from the root pin to the subarray. In contrast, the H-tree's wire length to all subarrays is equal to the distance to the furthest subarray. Therefore, the E-tree has a shorter average wire length, improving both space and timing.

Figure 7.4a illustrates that the enhancements in delay and energy exhibit a comparable trend. The subsequent discussion will specifically focus on the percentage of delay improvement. By utilizing CP-E technology, EDAP improvement can reach up to 85% (factor of 6.7×) compared to the conventional SP-H counterpart (Fig. 7.4b). Notably, further advancements in inter-subarray interconnect delay and

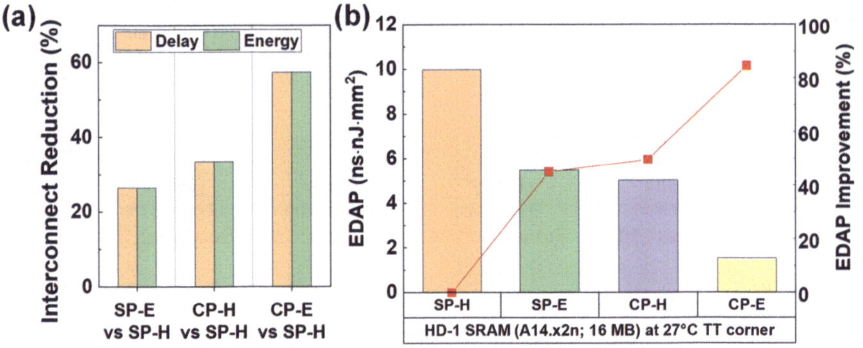

Fig. 7.4 (**a**) Relative inter-subarray interconnect reduction by compared to SP-H technology according to Fig. 7.3. (**b**) EDAP of different inter-subarray interconnect designs and their improvement compared to SP-H technology

energy overhead are possible by using alternative interconnect materials, such as graphene, and adjusting the aspect ratio and width-scaling factor [6].

Takeaways

- **Proposed CP-E Technologies:** By relocating the access pin to the array centre and employing E-tree routing design, significant enhancements in macro-level PPA can be achieved due to reduced interconnect overheads.

7.4 Impacts of SRAM Subarray Size and Bitcell Configuration

Figure 7.5 illustrates the impacts of selecting different subarray designs (HD-1, HD-2, HP, and LP) based on A14.x2n bitcell design, 16 MB macro capacity, and the optimized CP-E technology. The significant enhancement of WM through HD-2 subarray design presents a viable solution for mitigating write failures when compared to HD-1, HP, and LP counterparts, as elaborated in Sect. 6.5.1. However, the adoption of HD-2 subarray design introduces increased inter-subarray interconnect and area overhead compared to the equivalent subarray capacity of HD-1 counterpart, owing to its longer WL design (Fig. 7.5b). Hence, the overall macro-level delay and energy consumption of HD-2 subarray surpass those of HD-1 counterpart (Fig. 7.5a). Among the four subarray designs, HD-1 option stands out due to its compact bitcell design, offering minimal interconnect length and area overhead.

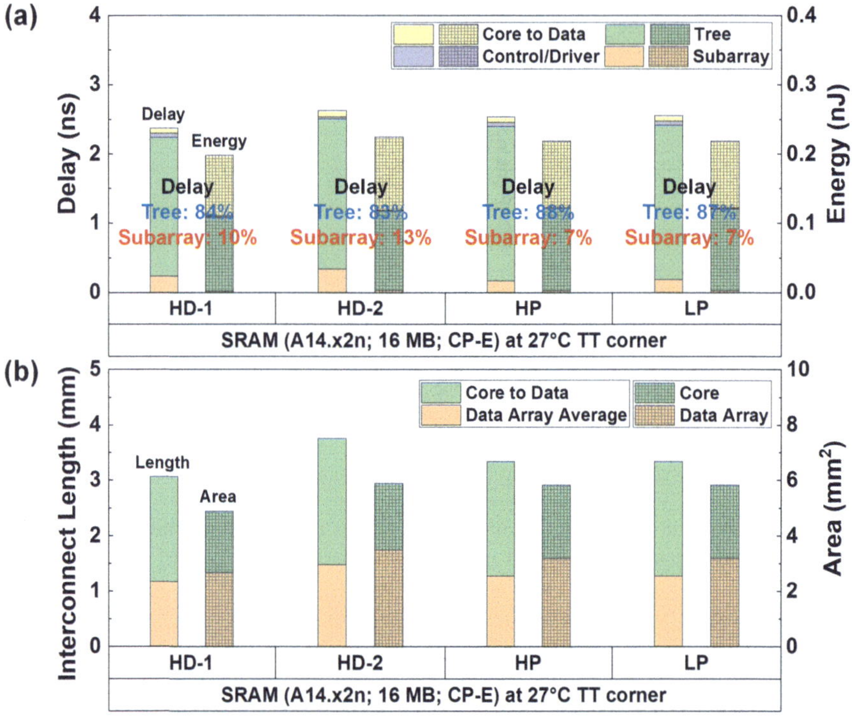

Fig. 7.5 Breakdown of normalized (**a**) delay and energy; (**b**) interconnect length and area for 16 MB HD-1, HD-2, HP, and LP A14.x2n SRAM macros based on CP-E technology

It is imperative to emphasize that the design should exhibit consideration for subarray-level WM to ensure a successful write operation while maintaining good macro-level PPA. If the HD-1 subarray design is chosen, the challenge of poor WM persists and necessitates addressing through additional RC reduction techniques, as detailed in Sect. 6.6. Otherwise, the stronger write-assist technique is required to prevent the write failure at the cost of higher energy and area. This write-assist technique, which has not yet been implemented in the current subarray designs, may incur energy and area overhead in HD-1 subarray more than its HD-2 counterpart. Alternatively, the recommended approach is to opt for the compromised option, namely HD-2 subarray design, to achieve an improved WM condition while minimizing potential energy and area overhead through a practical subarray design with write-assist technique.

The differences between the subarray designs (HD, HP, and LP) appear insignificant, primarily because the peripheral designs are primarily tailored for HD applications in this work. In contrast, for HP and LP applications, optimization mainly focuses on transistor sizing and subarray size reorganization. This presents a challenge in determining how to implement these subarrays for various cache levels. Thus, it is advisable to undertake specialized circuit-level designs for the peripheries in HP and LP subarrays to achieve noticeable distinctions at the macro level in future endeavours. However, the results presented in this chapter can still serve as a general trend for optimizing SRAM macro design in advanced technology.

Takeaways

- **Trade-offs Between WM and Macro-Level EDAP:**
  - **HD-1 (288-C×256-R) Subarray:** Exhibits superior EDAP but inferior WM. Requires stronger write-assist techniques, potentially leading to additional area and energy overhead.
  - **HD-2 (576-C×256-R) Subarray:** Demonstrates superior WM but inferior EDAP. Opting for this compromise option minimizes the demand for write-assist circuits.
- **Motivated Design Strategy:** Given that the differences in HD, HP, and LP designs are not distinguishable solely by adjusting bitcell ratios and subarray sizes, future work necessitates further circuit-level techniques aimed at HP and LP applications.

7.5 Impact of SRAM Data Capacity

Figure 7.6 depicts the effects of varying macro capacities (4, 16, and 128 MB) on delay, energy, interconnect length, and area utilizing HD-1 subarray design, A14.x2n bitcell design, and CP-E technology. As the data capacity increases, the impact of tree interconnect on macro-level PPA becomes more prominent. Moreover, the subarray's contribution remains significantly lower than that of the tree across varying data capacities. This underscores the necessity to prioritize techniques aimed at reducing inter-subarray interconnects for macro-level PPA enhancement.

Takeaways

- **Impact of Increased Data Capacity in Modern SoC Design:** The interconnect significantly influences macro-level PPA outcomes, particularly as data array capacity increases. This underscores the importance of investigating methods to minimize inter-subarray interconnect overhead.

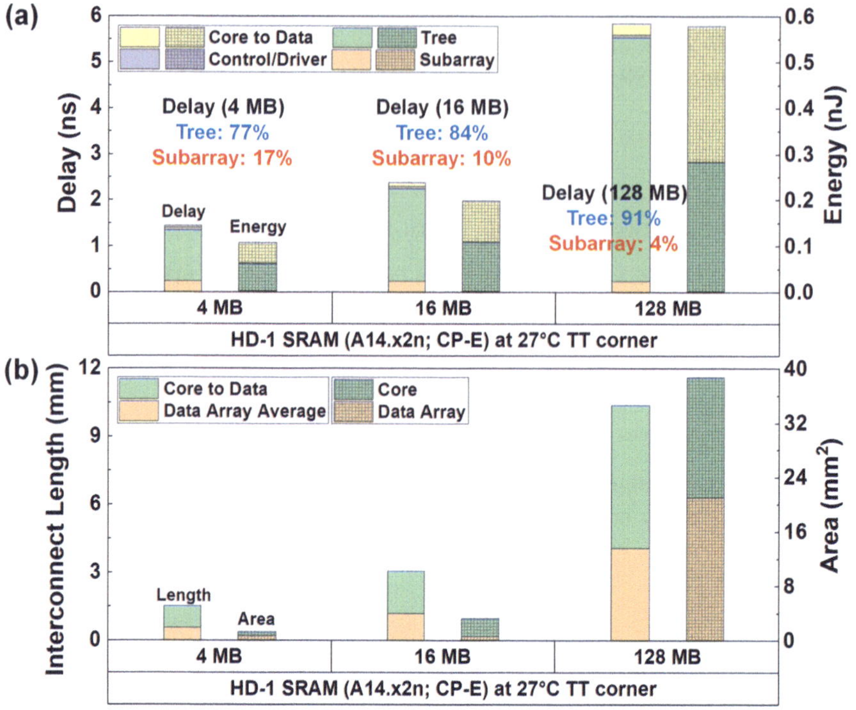

Fig. 7.6 Breakdown of normalized (**a**) delay and energy; (**b**) interconnect length and area for 4, 16, and 128 MB HD-1 A14.x2n SRAM macros based on CP-E technology

7.6 Impact of SRAM Bitcell/Subarray Scaling

Figure 7.7 illustrates the impacts of subarray scaling employing HD-1, 16 MB, and CP-E technology on A14 to A3 SRAM bitcell design options, as outlined in Chap. 6. A5.m2p (A5 mono NS-on-NS CFET SRAM bitcell) and A3.h3p (A3 hybrid FS-on-FS CFET SRAM bitcell) device options are chosen as representatives for A5 and A3, respectively. This selection is based on their superior PPA results at the identical technology node. Notably, stacked Fin CFET SRAM bitcells are not considered due to the poor read stability, as reported in Sect. 6.3. A systematic comparison is conducted between these configurations and A14.x2n option across multiple metrics. A5.m2p bitcell design exhibits a notable enhancement in macro-level delay (18%), energy (22%), interconnect length (19%), and area (39%) compared to A14.x2n counterpart. This observation emphasizes the significant impact of bitcell/subarray scaling on inter-subarray interconnect and macro area. Thus, bitcell/subarray scaling offers supplementary support, with the reduction of inter-subarray interconnect as the most effective strategy for PPA improvement in resistance-dominated nodes. On the other hand, the transition from A5.m2p to A3.h3p bitcell design yields noteworthy improvements in macro-level delay (31%),

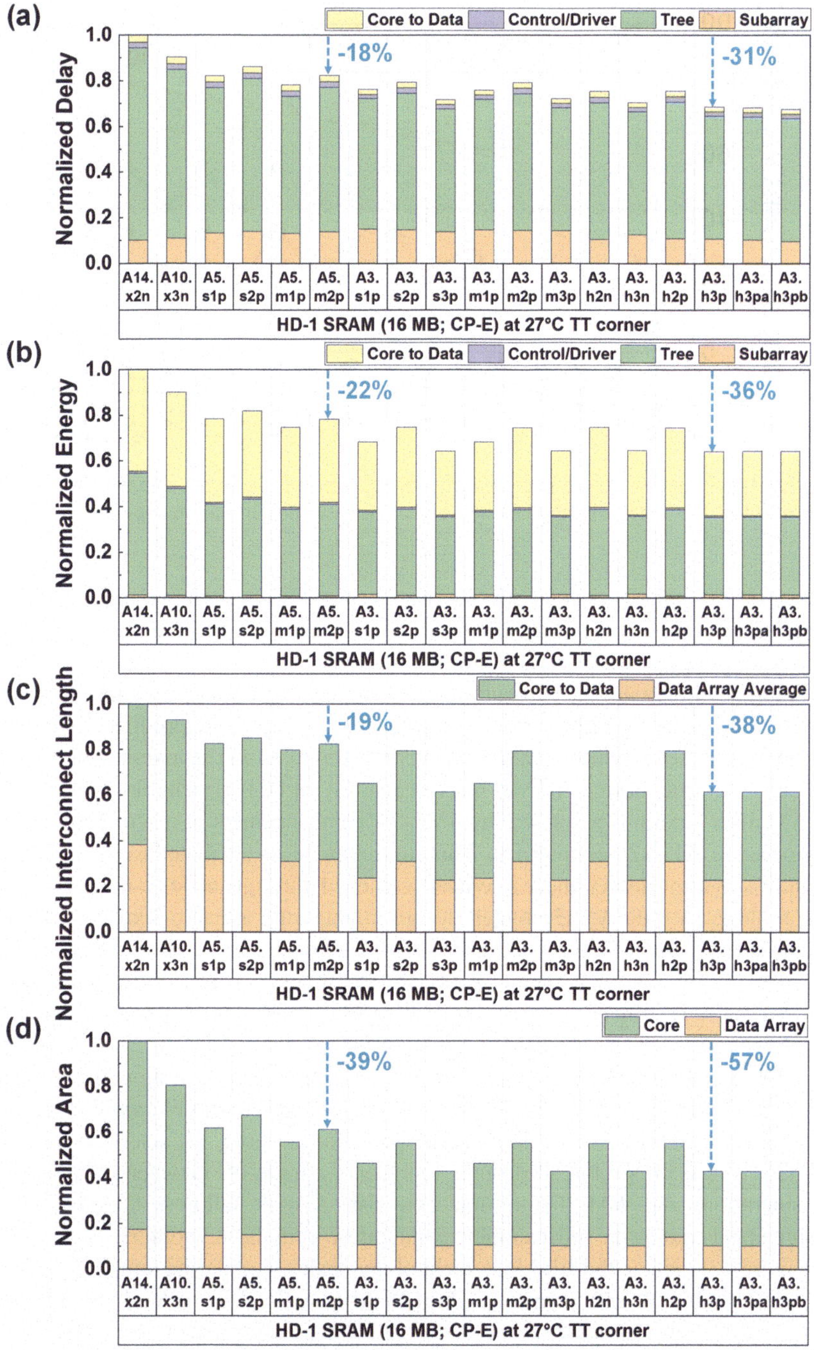

Fig. 7.7 Breakdown of normalized (**a**) delay, (**b**) energy, (**c**) interconnect length, and (**d**) area for 16 MB HD-1 SRAM macro based on CP-E technology

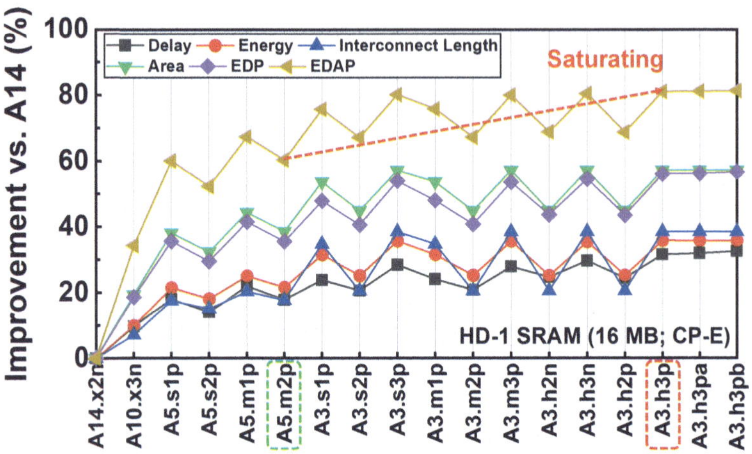

Fig. 7.8 Improvements of each node in delay, energy, interconnect length, area, EDP, and EDAP compared to A14.x2n for 16 MB HD-1 SRAM macros based on CP-E technology

energy (36%), interconnect length (38%), and area (57%) in comparison to A14.x2n counterpart. Although A3 continues to advance node-to-node PPA enhancement, the scaling advantages seem to reach saturation after A5.

Figure 7.8 illustrates the percentage improvement achieved by each technology in terms of delay, energy, interconnect length, area, energy-delay product (EDP), and energy-delay-area product (EDAP) when compared to A14. The assessment is based on HD-1 subarray design, 16 MB macro capacity, and CP-E technology. A5 CFET SRAM macro exhibits a noteworthy 53%–76% enhancement in macro-level EDAP compared to A14.x2n counterpart. This improvement is attributed to the reduction of inter-subarray interconnect overhead achieved through node-to-node subarray area scaling. Despite A3.h3p bitcell design maintaining a macro-level 87% EDAP improvement over A14, the node-to-node PPA gain appears to plateau beyond A5. It is noteworthy that A3.h3pa and A3.h3pb bitcell designs exhibit identical macro-level EDAP to A3.h3p counterpart yet offer superior WM at the expense of increased bitcell/subarray area, as detailed in Chap. 6, Sect. 6.5.1.3. Therefore, A3.h3pa and A3.h3pb bitcell designs emerge as the most favourable bitcell designs, showcasing optimized macro-level EDAP enhancement in comparison to A14, coupled with an improved WM relative to A3.h3p counterpart. Furthermore, the selection between A3.h3pa and A3.h3pb bitcell designs represents a trade-off between delay/energy considerations and WM. To minimize the risk of write failures, A3.h3pb bitcell design stands out as the most suitable candidate within the A3 context.

Figure 7.9 depicts the EDAP as technology scales, comparing subarray designs (HD-1, HD-2, HP, LP) across 4, 16, 128 MB, and CP-E technology. In most cases, the HP and LP subarray designs exhibit a significantly greater macro-level EDAP than HD-1 and HD-2 counterparts. This can be attributed to the increased interconnect overhead between subarrays, resulting from the reduced capacity of

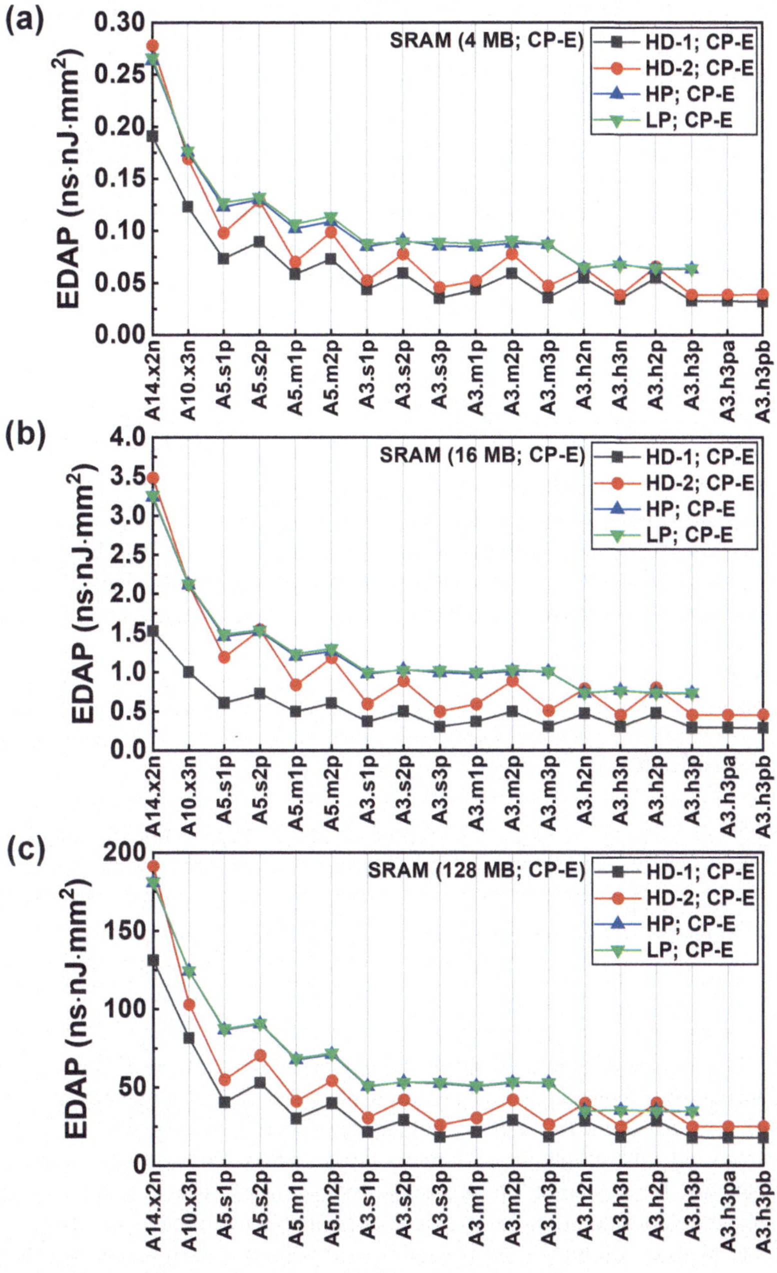

Fig. 7.9 EDAP benchmarks for 4, 16, and 128 MB HD-1, HD-2, HP, and LP SRAM macros based on CP-E technology from A14 to A3

subarray (144-C×256-R). Among all the designs, HD-1 subarray design provides the most favourable macro-level EDAP when compared to its counterparts due to the square-like subarray structure. Although HD-1 option seems promising, its poor WM necessitates greater power and area for a write-assist technique not addressed in this study. To strike a balance between PPA and write failure risk, HD-2 subarray design emerges as the optimized candidate with the least write-assist requirement compared to HD-1, HP, and LP counterparts. As aligned with the previous figures, the macro-level EDAP results demonstrate that technology scaling significantly benefits from A14 to A5 but starts saturating beyond A5.

Takeaways

- **Challenges in Traditional Bitcell Area Scaling:** Introducing the CFET architecture in A5 bitcell designs initially results in significant overall macro-level gains. However, beyond A5, these gains become saturated due to higher process complexity and increased cost concerns.
- **Motivated Design Strategy:** Shift towards higher-level design to achieve significant macro-level PPA improvements with a focus on reducing inter-subarray interconnect overheads.

7.7 Conclusions

This chapter focuses on macro-level PPA analysis, considering the impacts of inter-subarray interconnect, pin position, subarray size, bitcell configuration, data capacity, and bitcell/subarray scaling. In multiple PPA breakdown plots, it is evident that the impact of inter-subarray interconnect significantly outweighs macro-level factors than that of subarray contributions. Moreover, this trend intensifies with increased macro capacity, emphasizing the need to explore techniques for reducing inter-subarray interconnect overhead. Among the options of side pin (SP), centre pin (CP), H-tree, and E-tree, the optimized tree and pin design is CP-E technology. CP-E technology demonstrates a 57% improvement in delay and energy compared to SP-H counterpart due to lower inter-subarray overhead. Adhering to conventional approaches, bitcell configuration, subarray design, and bitcell/subarray scaling can contribute to the reduction of inter-subarray overhead. These factors stand as supplementary considerations alongside the optimization of the macro design.

In the analysis of bitcell configuration and subarray design, HD-1 subarray design exhibits superior macro-level EDAP compared to HD-2, HP, and LP counterparts. However, this analysis overlooks the potential drawback of inadequate WM, resulting in increased energy and area demands due to the utilization of write-assist techniques. As a result, HD-2 subarray design emerges as a promising alternative candidate owing to its minimal reliance on write-assist circuits, attributed to its superior WM characteristics. The overall gains of macro-level EDAP increase by up to 76% as the bitcell design progresses from A14 to A5. This underscores the importance of scaling at the bitcell level in minimizing inter-subarray interconnect overhead. However, beyond A5, this macro-level EDAP improvement saturates, reaching a maximum of 87% improvement compared to A14. The A3.h3pb bitcell emerges as the top choice among A3 variants, featuring a hybrid FS-on-FS CFET SRAM design with a 21 nm BL width. Its standout feature lies in its excellent WM and superior macro-level EDAP outcomes. However, the additional process steps involved in hybrid CFET, including wafer flipping, bonding, substrate removal, and double MINT layer fabrication, significantly increase process complexities and costs compared to their seq and mono counterparts. Moreover, the benefit of macro-level EDAP improvement becomes insignificant when transitioning from A5 to A3 bitcell designs. In light of these challenges and the marginal node-to-node gain, it is warranted to carefully evaluate the economic viability of applying the hybrid CFET technique and continuing scaling beyond A5. This observation prompts a deeper investigation into alternative PPA enhancement techniques, such as the alternative SRAM subarray designs introduced in Chaps. 8 and 9.

References

1. Liu, H.-H., et al. (2023). CFET SRAM DTCO, interconnect guideline, and benchmark for CMOS scaling. *IEEE Transactions on Electron Devices, 70*(3), 883–890. https://doi.org/10.1109/TED.2023.3235701
2. Liu, H.-H., et al. (2023). CFET SRAM with double-sided interconnect design and DTCO benchmark. *IEEE Transactions on Electron Devices* 1–8. https://doi.org/10.1109/TED.2023.3305322
3. Liu, H.-H., et al. (2023). DTCO of sequential and monolithic CFET SRAM. In R.-H. Kim & N. V. Lafferty (Eds.), *DTCO and computational patterning II* (Vol. 12495). International society for optics and photonics. SPIE. https://doi.org/10.1117/12.2657524
4. Liu, H.-H., et al. (2023). Extended methodology to determine SRAM write margin in resistance-dominated technology node. *IEEE Transactions on Electron Devices, 69*(6), 3113–3117. https://doi.org/10.1109/TED.2022.3165738
5. Pei, Z., et al. (2023). Emerging interconnect exploration for SRAM application using non-conventional H-tree and center-pin access. In *2023 24th International Symposium on Quality Electronic Design (ISQED)* (p. 1). https://doi.org/10.1109/ISQED57927.2023.10129316
6. Pei, Z., et al. (2023). Graphene-based interconnect exploration for large SRAM caches for ultrascaled technology nodes. *IEEE Transactions on Electron Devices. 70*(1), 230–238. https://doi.org/10.1109/TED.2022.3225512
7. Pei, Z., et al. (2023). Technology/memory co-design and co-optimization using E-tree interconnect. In *Proceedings of the Great Lakes Symposium on VLSI 2023*. GLSVLSI '23 (pp. 159–162). Association for Computing Machinery. ISBN: 9798400701252. https://doi.org/10.1145/3583781.3590311
8. Weckx, P., et al. (2019). Novel forksheet device architecture as ultimate logic scaling device towards 2 nm. In *2019 IEEE International Electron Devices Meeting (IEDM)* (pp. 36.5.1–36.5.4). https://doi.org/10.1109/IEDM19573.2019.8993635

Part IV
Beyond Traditional SRAM Bitcell-Level Scaling – Future Design Direction for SRAM Data Array Towards A14

Chapter 8
Write-Ability-Aware SRAM Hierarchical Subarray Design Optimization

In Chap. 7, the saturation of macro-level energy-delay-area product (EDAP) enhancement becomes evident with static random-access memory (SRAM) technology scaling. It is imperative to identify an alternative node-to-node performance, power, and area (PPA) booster, moving beyond the conventional approach of SRAM bitcell scaling. This chapter and the upcoming Chap. 9 redirect attention towards the design of SRAM subarrays to achieve macro-level PPA improvement. In sub-10 nm nodes, the SRAM macro experiences a shift towards dominance by inter-subarray interconnects and logic elements (Chap. 7) [7, 16, 17]. Increasing the SRAM subarray size yields notable improvements (56%) in macro-level EDAP compared to smaller subarrays, as depicted in Fig. 8.1. However, the design of a larger subarray introduces the write failure risk (WFR) due to increased resistance (R) and capacitance (C) in both bitlines (BL) and wordlines (WL) [6, 8, 9]. To address the WFR, this chapter investigates write-ability-aware SRAM subarray optimization across a wide range of subarray sizes, particularly in 14-Å-compatible nanosheet (NS) technology (A14) [4]. This A14 technology serves as a representative node in ultra-scaled node for the sake of simplicity. Further research is necessary to substantiate these findings for the nodes beyond A14. It is essential to emphasize that the dedicated SRAM subarray designs, periphery designs, obtained outcomes, and design framework have potential applicability to nodes beyond A14. Furthermore, this simulation-based exploration is crucial for ensuring successful write operations of SRAM subarray in deeply scaled technologies, given the current absence of silicon data for validation.

To design a larger subarray size, the well-known divided wordline (DWL) [21] and divided bitline (DBL) [5] techniques are utilized at the cost of subarray and macro area penalty. These techniques offer reduced effective resistance and capacitance (RC) for BL and WL. By applying the simulation circuits and methodologies outlined in Chap. 2, this chapter provides detailed discussions on write failure criteria. The limitations of a successful write operation imposed by leakage through pass-gate (PG) device (PG leakage) and boost capacitors (C_{boost}) in both negative

H.-H. Liu, F. Catthoor, *Circuit-Technology Co-Optimization of SRAM Design in Advanced CMOS Nodes*, https://doi.org/10.1007/978-3-031-76109-6_8

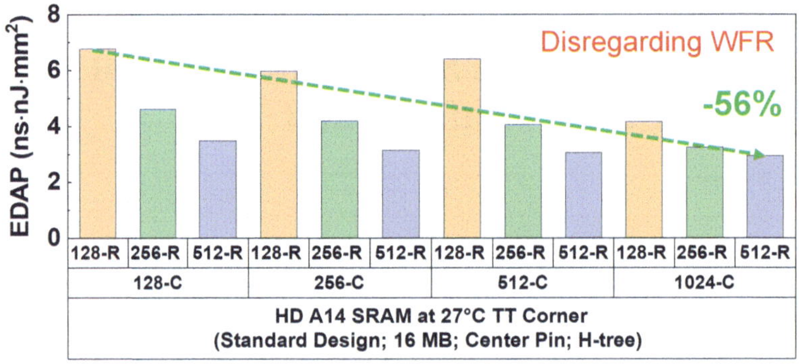

Fig. 8.1 Macro-level EDAP benchmark for different subarray sizes of HD A14 NSFET SRAM data array with Std. subarray design (disregarding WFR), data capacity of 16 MB, centre pin access, and H-tree technology based on typical-typical (TT) corner and nominal temperature (27 °C). C: column; R: row

bitline (NBL) [13, 14] and boosted wordline (BWL) [20] write-assist techniques are analysed. For simplicity, the SRAM subarrays with and without DWL and DBL techniques are called "standard (Std.)" and "divided" designs, respectively, in the following discussions. The investigation into optimized choices for local block (LB) size and parasitic RC for WL and BL is conducted for each subarray size, considering both Std. and conventional (Conv.) divided subarray designs. Although the Conv. divided subarray design significantly enhances the write margin (WM) and operating write speed, it incurs a noticeable subarray area penalty compared to the Std. counterpart. This subarray area penalty will be effectively addressed with a proposed divided subarray design in the upcoming Chap. 9.

This chapter focuses on the electrical-level outcomes at the subarray level, presenting power and performance (PP) enhancements relevant to both the Conv. (Chap. 8) and proposed (Chap. 9) divided subarray designs, in comparison to the Std. counterpart. Section 8.1 outlines the potential benefits of designing a large-size subarray to achieve 56% macro-level EDAP enhancement. However, these approaches in Std. subarray design is constrained by the poor WM value. Section 8.2 details the SRAM subarray circuit and bitcell design of divided techniques to enable large-size subarray by reducing the effective RC of BL and WL. Section 8.3 analyses the criteria for a successful write operation based on various sources of write failure. It is crucial to emphasize that the criteria governing the write failure boundary are represented by specific WM values for a precise and comprehensive overview. Section 8.4 investigates the impact of RC while optimizing the LB size within the SRAM subarray. Section 8.5 explores potential design directions and challenges for enabling larger-sized designs without incurring significant area penalties. Finally, Sect. 8.6 concludes this chapter by presenting the framework for write-ability-aware SRAM subarray optimization. This framework stands as a crucial reference for the proposed divided subarray design in Chap. 9.

8.1 Poor Write Margin Limits Large-size SRAM Subarray Design

In interconnect-dominant nodes, the SRAM subarray area becomes a critical concern due to its impact on inter-subarray interconnect overhead. This inter-subarray interconnect, in turn, exerts significant influence over the overall macro-level PPA [7, 16, 17]. Figure 8.1 presents an insightful graph of the macro-level EDAP for a 16 MB data array consisting of subarrays with varying number of row (N_{row} or -R) and column (N_{col} or -C) based on high-density (HD) 6-transistors (6T) SRAM bitcell. As depicted in Fig. 8.1, a significant 56% reduction in macro-level EDAP is achieved when transitioning from a small- to a large-size Std. subarray design. Notably, these macro-level results have not yet considered the potential WFR [9] within large Std. subarray design. This EDAP reduction is primarily attributed to the decreased inter-subarray interconnect congestion [16]. Remarkably, given the dominance of inter-subarray interconnects, the overall macro-level EDAP continues to decrease even as subarray-level PP increases with bigger subarray size [16]. This observation propels cache design towards larger SRAM subarrays to attain further macro-level PPA improvements.

However, enabling a large-size SRAM subarray poses challenges due to the significant parasitic RC overhead on both BL and WL [6, 8, 9]. This increased RC may cause write failure and longer write time as the subarray's N_{row} and N_{col} increase in the Std. subarray design (Fig. 8.2) [9]. The WFR challenges can be mitigated by introducing the DWL [21] and DBL [5] techniques due to the effective RC reduction at the cost of subarray area overhead. However, both the Std. and Conv. divided techniques remain constrained by various sources of write failures, including PG leakage [9], maximum C_{boost}-generated voltage by NBL and BWL write-assist techniques, reliability, and substrate current issue explained in Sect. 2.5.

In Fig. 8.2, the 6-sigma (6σ) Monte Carlo (MC) simulation is performed with random threshold voltage (V_T) distribution with V_T mismatch index (A_{VT}) of 1 mV·μm at the worst-case temperature (80 °C) and nominal supply voltage (V_{DD}) (0.7 V) to avoid underestimating the WFR. The A_{VT} of 1 mV·μm is artificially assumed since it should be smaller than the reported values [3, 12] in more advanced technology. Note that the occurrence of WFR is assumed when the WM falls below V_T of the A14 NS device (~300 mV) [9]. This condition is usually considered as an initial reference point before conducting a detailed examination based on simulation or silicon data. However, the determination of V_T value depends on the chosen methodology [15], introducing a lack of consistency in the approach with distinct methodologies. Hence, a thorough examination of the write failure boundary is imperative, ideally presented as a certain value associated with WM simulation.

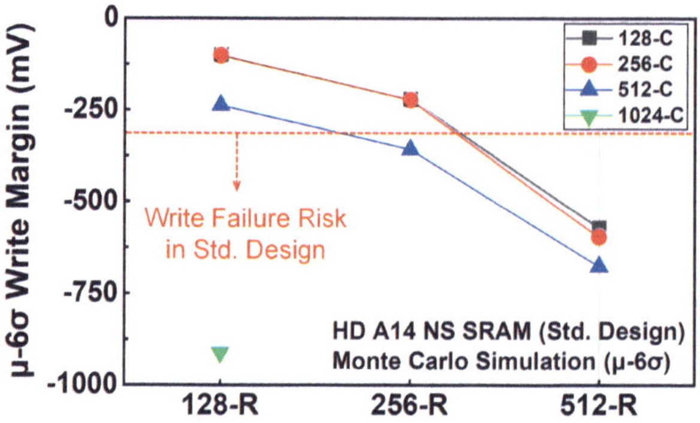

Fig. 8.2 Worst-case WM of HD A14 NSFET SRAM subarray with Std. design based on 6σ MC simulation at the worst-case temperature (80 °C)

Takeaways

- **Target Research Goal:** Enlarge the subarray size to minimize interconnect overheads, enhancing Marco-level PPA results.
- **Challenges in Standard Subarray Design:** The RC of BL and WL rises with subarray size enlargement, leading to degraded WM and increased WFR.
- **Motivated Design Strategy:** Develop an alternative subarray design to enhance WM while preserving potential PPA improvements at the macro level.

8.2 SRAM Bitcell and Subarray Circuit Designs with DWL and DBL Techniques

8.2.1 Example of SRAM Subarray Design

Figure 8.3 illustrates a circuit diagram example for SRAM subarray utilizing DWL and DBL techniques [5, 21] under the size of 128 columns (-C) and 256 rows (-R) (128-C×256-R). The SRAM subarray with Conv. divided design incorporates identical circuit components as Std. design [6–9], including SRAM bitcell array, WL decoder, precharge, sense amplifier, write driver, and column multiplexer (MUX) (4–1; M4). Additionally, this Conv. divided subarray design integrates extra AND logic gate and transmission gate (TG) to facilitate the selection of signals within

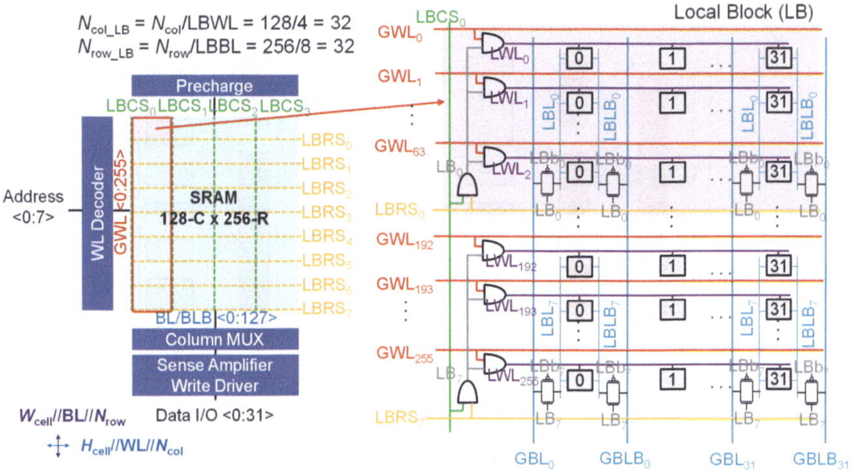

Fig. 8.3 Schematic view of SRAM subarray circuit with Conv. divided technique (subarray size = 128-C×256-R; LBWL = 4; LBBL = 8)

each LB. In Chap. 9, a clear layout example and more detailed comparison between different subarray designs will be presented for considering the area penalty.

Connecting the AND and TG to the global WL (GWL) and global BL (GBL), respectively, divides the SRAM bitcell array into discrete local blocks (LBs), each with its own local WL (LWL) and local BL (LBL). In this illustration, the number of LBs in the WL (LBWL) and BL (LBBL) directions is assumed to be 4 and 8, respectively. Activating each LB is achieved by selecting the corresponding LB column select (LBCS) and row select (LBRS) signals. The number of columns ($N_{\mathrm{col-LB}}$) and rows ($N_{\mathrm{row-LB}}$) in each LB is calculated by dividing N_{col} and N_{row} by LBWL and LBBL, respectively. Full swing on GBL and LBL is assumed in this initial work, which can potentially be extended to full-swing GBL and low-swing LBL for lower energy design in the future.

8.2.2 SRAM Bitcell Design

Figure 8.4 illustrates the HD A14 NSFET SRAM bitcell layout design for both Std. and Conv. divided subarray designs. Moreover, the routing design has been explored specifically for the Conv. divided technique, where LBL, LWL, GBL, and GWL are located at MINT, M1, M2, and M3 layers, respectively. There is the freedom to design GBL in different widths to find the best WM since it is the only rail in M2 layer. Figure 8.4 shows the impact of GBL width on the WM, presented at the nominal temperature of 27 °C and the typical-typical (TT) corner. The WM improves as the GBL width increases due to the lower GBL resistance (R_{GBL}),

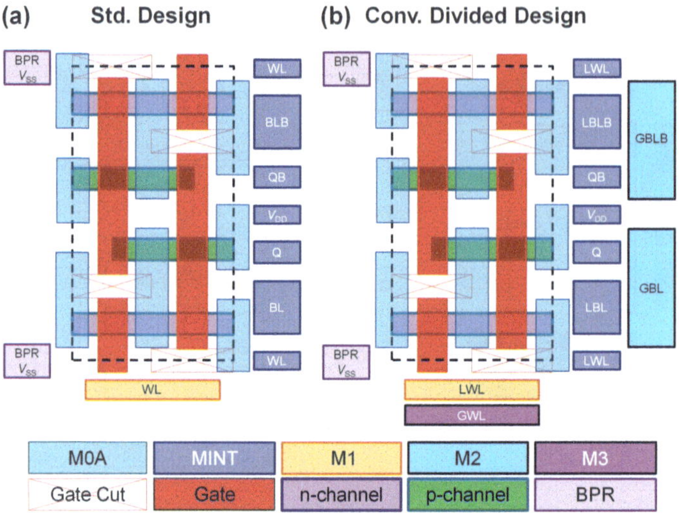

Fig. 8.4 Schematic views of HD A14 6T NSFET SRAM bitcell layout design for (**a**) Std., and (**b**) Conv. divided subarray designs

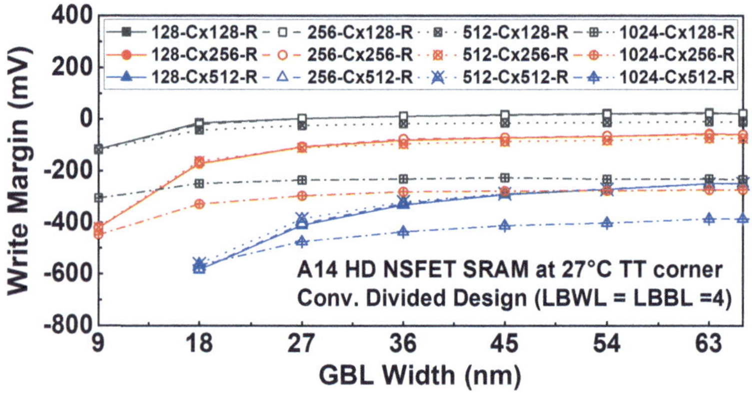

Fig. 8.5 WM with varying GBL width and subarray sizes of Conv. divided subarray designs (LBWL = LBBL = 4)

and the improvements gradually saturate caused by the increased GBL capacitance (C_{GBL}). To be consistent and leave out the impact of GBL width, the same GBL width of 63 nm is selected for each subarray size to achieve the best WM for Conv. divided design (Fig. 8.5).

8.3 Write Failure Examination

In this section, an examination of write failures is analysed to determine the upper limits of the SRAM subarray size. This analysis is conducted through the simulation circuit and methodology outlined in Sect. 2.5 of Chap. 2. Moreover, the static noise margins for hold and read have been discussed in Chap. 6. Regarding characteristics in divided subarray design, NBL and BWL write-assist techniques are chosen to avoid stability degradation of half-selected bitcell (Sect. 2.4 of Chap. 2). The assessment of write failure criteria, involving varied N_{col} and N_{row}, is cross-validated through the WM (80 °C and 0.7 V) and the two proposed write delay (WD) (−40 °C and 0.7 V) 6σ MC simulations. These WD simulations are based on both conceptual (Figs. 8.6 and 8.7 for testing PG leakage) and realistic (Figs. 8.8, 8.9, and 8.10 for testing limited C_{boost}-generated voltages) write-assist circuits, as discussed in Sect. 2.5 of Chap. 2.

8.3.1 PG-leakage-caused Write Failure Test

In Figs. 8.6 and 8.7, the worst-case ($\mu - 6\sigma$) WM is initially established as the conceptual write-assist voltage source for the worst-case WD simulation in both Std. and Conv. divided subarray designs. To evaluate the impact of varying N_{row} and N_{col} on their corresponding changes in write-assist voltage, the circuits are designed to conceptually mimic the NBL and BWL techniques (refer to Chap. 2). As the N_{col}/N_{row} increases while maintaining a constant N_{row}/N_{col}, the WM degrades implying the requirement for stronger write-assist voltage. The PG leakage occurs when over-boosting voltage level of NBL (V_{NBL}) or BWL (V_{BWL}) technique, which causes larger WD and eventually write failure [9]. Thus, the upper limit of N_{col} shifts negatively towards a larger N_{row}, and it also becomes higher in Conv. divided subarray design due to less parasitic RC of BL and WL in an activated LB. This test shows that even if the WM can be successfully obtained by the simulation, the WD operation may not work properly due to the PG leakage under the worst-case write-assist voltage setting [9]. It also demonstrates that the Conv. divided subarray design is beneficial and essential to push the upper limit of the subarray size compared to the Std. counterpart. The results of this section are summarized in Table 8.1. Note that they are not yet the final conclusion of the subarray size limitation, other WFRs are discussed in the following sections.

8.3.2 Limited C_{boost}-generated V_{NBL} and V_{BWL} Write Failure Test

In this section, the configurations of 128-C, 256-C, 512-C, and 1024-C are selected for detailed examination in conjunction with the scenarios of 128-R, 256-R, and

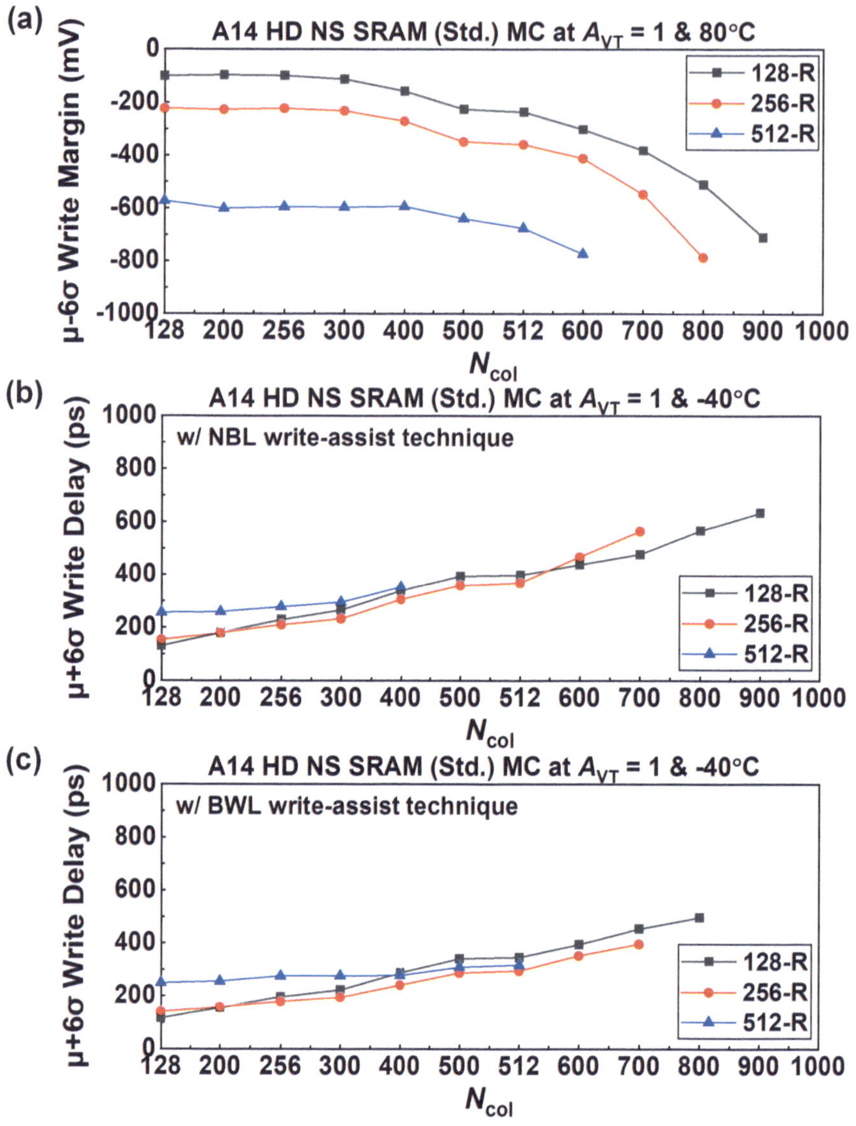

Fig. 8.6 PG-leakage-caused write failure test by varying Nrow and Ncol of Std. subarray design. (**a**) $\mu - 6\sigma$ WM at 80 °C; $\mu + 6\sigma$ WD at −40 °C with conceptual (**b**) NBL and (**c**) BWL write-assist circuits

512-R according to Table 8.1. The limitation of the C_{boost}-generated write-assist voltage ($V_{\mathrm{NBL-max}}$ and $V_{\mathrm{BWL-max}}$) based on varying C_{boost}, N_{row}, and N_{col} of Std. subarray design with realistic NBL and BWL circuits are given in Fig. 8.8 (128-R), Fig. 8.9 (256-R), and Fig. 8.10 (512-R). As C_{boost} (ideal capacitor is assumed) increases, the $|V_{\mathrm{NBL}}|$ and $|V_{\mathrm{BWL}}|$ become larger resulting in faster write speed.

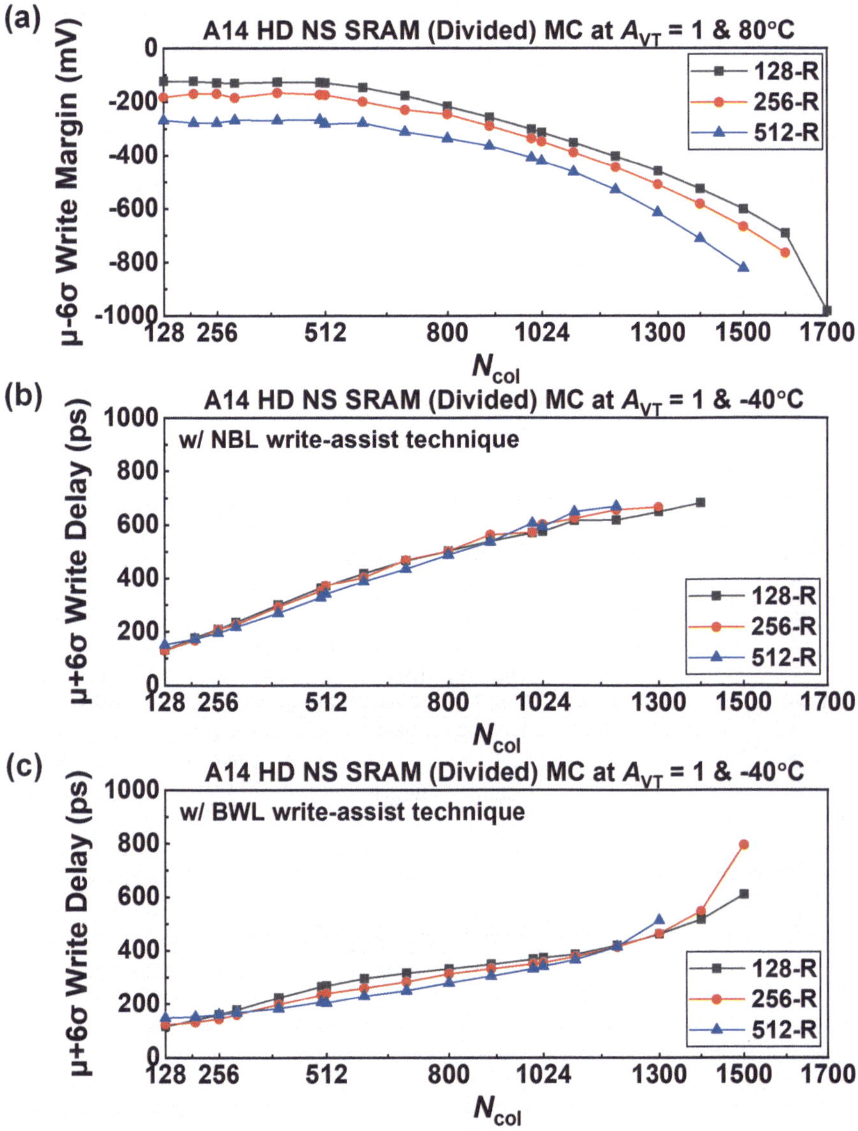

Fig. 8.7 PG-leakage-caused write failure test by varying Nrow and Ncol of Conv. divided subarray design (LBWL = 4; LBBL = 4/8/16 for 128-R/256-R/512-R). (**a**) $\mu - 6\sigma$ WM at 80 °C; $\mu + 6\sigma$ WD at −40 °C with conceptual (**b**) NBL and (**c**) BWL write-assist circuits

However, the enhancement of write-assist voltage and write speed starts saturating when C_{boost} is sufficiently large (reaches the threshold C_{boost}, $C_{\text{boost-th}}$). The power and area overhead become greater than the gain in write speed beyond $C_{\text{boost-th}}$. To ensure a successful write operation, the C_{boost}-generated "V_{NBL}" or "V_{DD}-V_{BWL}"

Table 8.1 The ideal upper limits of N_{col} at different N_{row} for Std. and Conv. divided subarray designs based on PG-leakage-caused write failure test. It would change after considering other write failure constraints

Subarray design	N_{row}	Ideal upper limit of N_{col}		
		WM	WD (NBL)	WD (BWL)
Std.	128-R	900-C	900-C	800-C
	256-R	800-C	700-C	700-C
	512-R	600-C	400-C	512-C
Conv. divided	128-R	1700-C	1400-C	1500-C
	256-R	1600-C	1300-C	1500-C
	512-R	1500-C	1200-C	1300-C

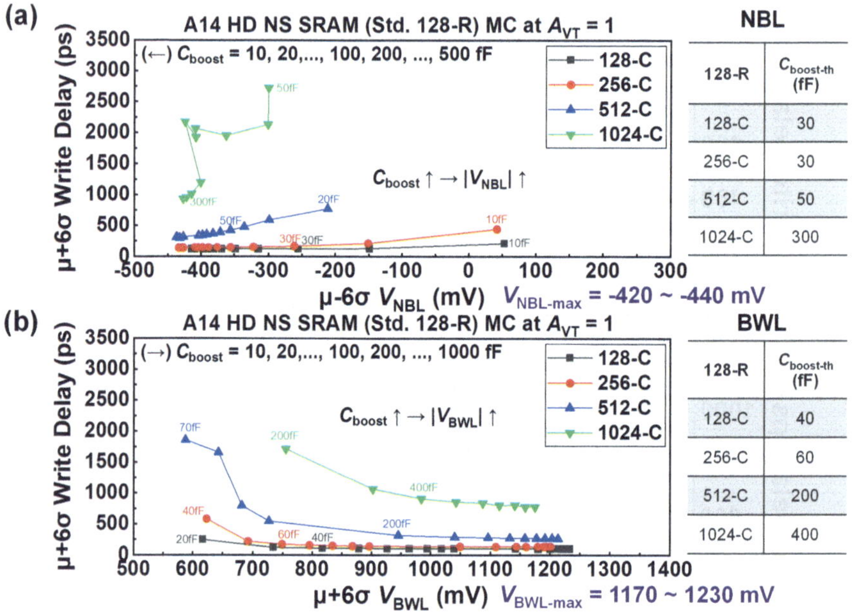

Fig. 8.8 C_{boost}-generated write-assist voltage (V_{NBL} and V_{BWL}) limitation test by varying C_{boost} and N_{col} at fixed N_{row} = 128 of Std. subarray design with realistic (**a**) NBL and (**b**) BWL write-assist circuits

should at least match or exceed the WM value (reported in the next section). As long as a functional write is guaranteed, the practical C_{boost} design should not go beyond $C_{boost-th}$. This test explores the $C_{boost-th}$, $V_{NBL-max}$, and $V_{BWL-max}$ under different subarray sizes. Note that the V_{NBL} and V_{BWL} are extracted from the write driver and WL voltage, respectively, when data flips. According to the summary of 512-C in each table, the area and energy overhead from C_{boost} for BWL may be higher than NBL due to the higher C_{boost} requirement. The reason might be that BWL is more sensitive to changes in N_{col}, which represents WL lengths, than NBL.

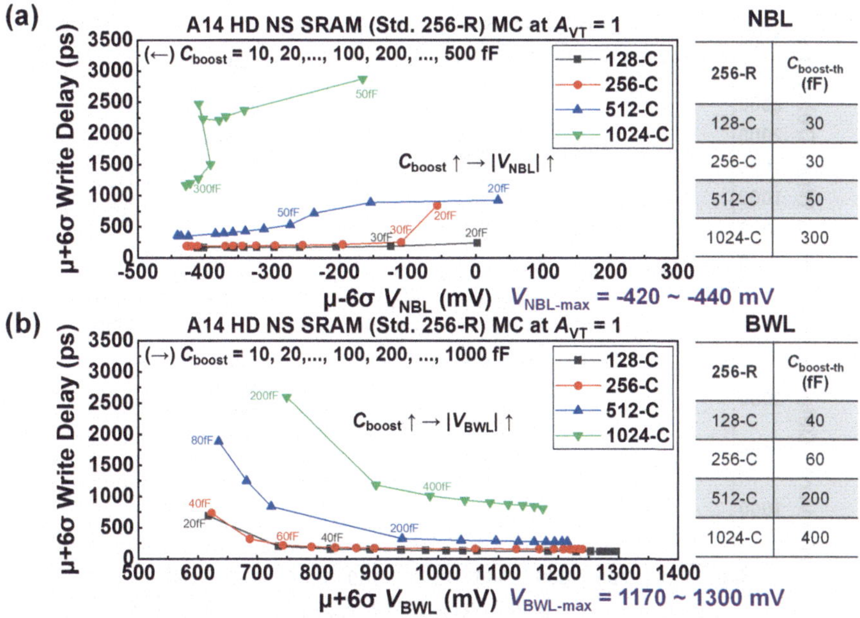

Fig. 8.9 C_{boost}-generated write-assist voltage (V_{NBL} and V_{BWL}) limitation test by varying C_{boost} and N_{col} at fixed N_{row} = 256 of Std. subarray design with realistic (**a**) NBL and (**b**) BWL write-assist circuits

In Figs. 8.8a, 8.9a, and 8.10a, 128-R, 256-R, and 512-R subarrays with NBL write-assist circuits exhibit varying minimum required C_{boost} values for successful write operations: 10, 20, and 30 fF for 128-C; 10, 20, and 30 fF for 256-C; 20, 20, and 30 fF for 512-C; and 50, 50, and 70 fF for 1024-C, respectively. This observation underscores that longer WL or BL length in the subarray leads to prolonged write times, necessitating a stronger NBL write-assist technique to ensure write ability. Furthermore, changes in V_{NBL} plateau at approximately −420 mV across all cases, indicating that increasing C_{boost} beyond the threshold value that achieves $V_{NBL-max}$ would be futile and ineffectual for further improving write speed. Conversely, enhancements in write delay within 128-R, 256-R, and 512-R subarrays decelerate as C_{boost} approaches 30, 30, and 50 fF for 128-C; 30, 30, and 50 fF for 256-C; 50, 50, and 70 fF for 512-C; and 300 fF across all cases for 1024-C. These capacitance values are proposed as the $C_{boost-th}$ to prevent unnecessary additional area and energy overheads stemming from larger C_{boost}. For 128-R, 256-R, and 512-R subarrays, increasing C_{boost} from the minimum required value to the $C_{boost-th}$ yields reductions in write delay by 42%, 23%, and 32% for 128-C; 62%, 70%, and 64% for 256-C; 45%, 43%, and 49% for 512-C; and 63%, 56%, and 46% for 1024-C. It is evident that achieving better performance, particularly for the 1024-C scenario, necessitates a significantly larger C_{boost}-generated $|V_{NBL}|$ at the expense of increased area and energy penalties. This trade-off between PPA warrants

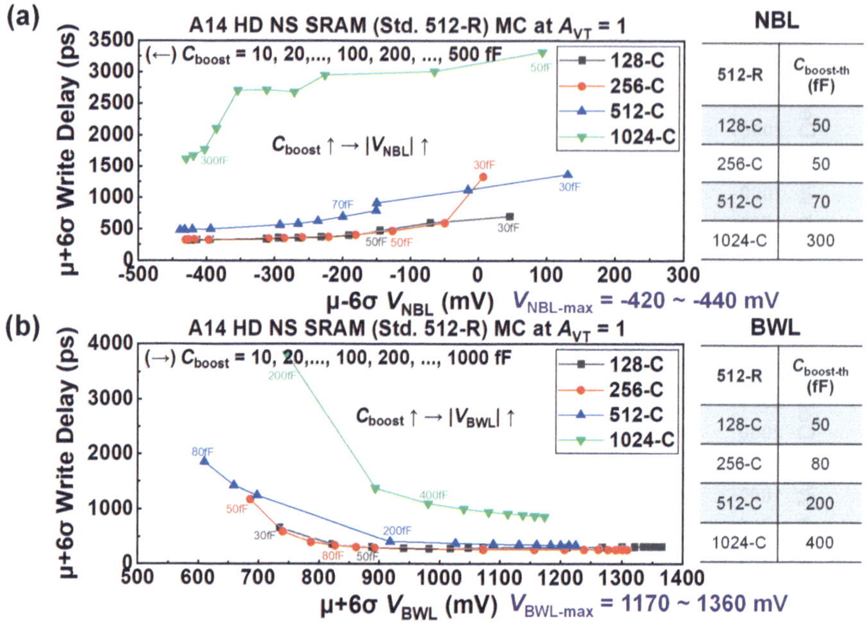

Fig. 8.10 C_{boost}-generated write-assist voltage (V_{NBL} and V_{BWL}) limitation test by varying C_{boost} and N_{col} at fixed N_{row} = 512 of Std. subarray design with realistic (**a**) NBL and (**b**) BWL write-assist circuits

comprehensive exploration in future endeavours, particularly through tailored full subarray layout designs incorporating realistic NBL write-assist circuitry.

In Figs. 8.8b, 8.9b, and 8.10b, 128-R, 256-R, and 512-R subarrays with BWL write-assist circuits illustrate varying minimum required C_{boost} values for successful write operations: 20, 20, and 30 fF for 128-C; 40, 40, and 50 fF for 256-C; 70, 80, and 80 fF for 512-C; and 200, 200, and 200 fF for 1024-C, respectively. These findings demonstrate that the minimum required C_{boost} for BWL scenario is higher than that of NBL counterpart, particularly evident in the cases of 512-C and 1024-C subarrays. Furthermore, the suggested $C_{boost-th}$ is 40, 40, and 50 fF for 128-C; 60, 60, and 80 fF for 256-C; 200, 200, and 200 fF for 512-C; and 400, 400, and 400 fF for 1024-C, respectively, within 128-R, 256-R, and 512-R subarrays. This indicates that employing NBL write-assist technique proves more efficient in ensuring write operations with less area and energy overheads induced by C_{boost} compared to BWL counterpart, particularly for longer-WL subarrays. In 128-R, 256-R, and 512-R subarrays, augmenting C_{boost} from the minimum required value to the $C_{boost-th}$ improves the write speed by 56%, 77%, and 54% for 128-C; 69%, 70%, and 71% for 256-C; 83%, 83%, and 78% for 512-C; and 47%, 61%, and 71% for 1024-C. Notably, the saturation of changes in $V_{BWL-max}$ occurs near 1170 mV, equivalent to −470 mV of WM value, and closely aligns with $V_{NBL-max}$ (−420 mV) in NBL cases. However, the requisite C_{boost} value for BWL case should exceed 500 fF to reach the saturation of $V_{BWL-max}$, particularly for 512-C and 1024-C, surpassing that of NBL counterpart. These observations could potentially result in inferior PPA benchmarks for BWL-triggered write operation in comparison to NBL case, considering the overall impacts caused by additional write-assist circuitry.

8.3.3 Reliability and Substrate Current Discussion

As discussed in the previous subsections, the $V_{NBL-max}$ and $V_{BWL-max}$ exhibit equivalence to the WM of approximately −420 to −440 mV and −470 to −660 mV, respectively, depending on the N_{row} and N_{col}. However, considering the insights from Sect. 2.5 of Chap. 2, the WFR must also account for the restricted maximum allowed operating gate voltage (V_{GS-max}) and the presence of unwanted substrate current. In the context of bias temperature instability (BTI) testing conducted on A14 NS devices [2, 18], the projected V_{GS-max} for the target device is 1.2 V. This projection takes into consideration the active cells to the subarray size, set at 32 over 128 × 128 (representing the worst-case scenario in this chapter). The provided margin of 500 mV above the nominal V_{DD} (0.7 V) of A14 technology is deemed adequate for the implementation of write-assist techniques. Conversely, substrate current is observed only when the negative voltage hovers around 800 mV, as simulated in Sect. 2.5 of Chap. 2. This condition corresponds to a relatively lenient WM constraint of −800 mV. Considering the constraints imposed by V_{GS-max} (for both NBL and BWL) and substrate current (exclusive to NBL), a conservative recommendation is made: the WM should not exceed −420 mV and −470 mV when employing NBL and BWL write-assist techniques, respectively, as summarized in Table 8.2. Hence, the maximum N_{col} considered for testing is 1024-C in A14 divided subarray, determined from the upper limits observed across 128-R, 256-R, and 512-R shown in Table 8.2. Considering incomplete process variation information in A14 and the nodes below, a conservative WFR boundary of −300 mV is adopted to enhance design robustness in the following sections.

Table 8.2 The realistic upper limits of N_{col} at different N_{row} for Std. and Conv. divided subarray designs based on all the write failure constraints. (WF: Write Fail)

		Realistic upper limit of N_{col}	
Subarray design	N_{row}	WM (NBL) (WM >−420 mV)	WD (BWL) (WM >−470 mV)
Std.	128-R	700-C	700-C
	256-R	600-C	600-C
	512-R	WF	WF
Conv. divided	128-R	1200-C	1200-C
	256-R	1100-C	1100-C
	512-R	1024-C	1024-C

Takeaways

- **Write Failure Sources:**
  - **PG Leakage:** Over-boosting the write-assist voltage level in NBL or BWL techniques causes larger write time and eventually write failure.
  - **Limited Write-Assist Voltage:** Increasing C_{boost} raises write-assist voltages and speeds, but this saturates at high C_{boost}, causing unnecessary power and area overhead. Optimal C_{boost} should balance improvement in WM with minimal power and area penalties.
  - **Maximum Operating Gate Voltage:** BTI tests show V_{GS-max} is around 1.2 V, allowing for a 0.5 V write-assist voltage at V_{DD} = 0.7 V for NBL and BWL techniques.
  - **Substrate Current:** In the NBL technique, negative bias on the bitcell's source node may cause substrate current when the bias exceeds -0.8 V.
- **Successful Write Operation Criteria for A14 NS SRAM Subarray Design:**
  - The WM should be more positive than −0.42 V for NBL and −0.47 V for BWL write-assist techniques.
  - This limit also defines the maximum subarray size as shown in Table 8.2.

8.4 SRAM Subarray Optimization with LB Size, RC, and Area Impacts for HD Option

In this section, the subarray is optimized based on varying N_{col} (128, 256, 512, 1024), N_{row} (128, 256, 512), LBWL (4, 8, 16), and LBBL (4, 8, 16) including actual RC and other 12 artificial RC test cases (Table 8.3) by 6σ MC simulation at the worst-case temperature (80 °C) and nominal supply voltage (0.7 V) (Figs. 8.11, 8.12, and 8.13). Figure 8.14 summarizes LB size optimization across all LBWL and LBBL configurations, based on write-ability studies. The comparison is made between the Std. and Conv. divided subarray designs with the baseline RC (#0). Table 8.4 and Fig. 8.15 provide an overview of the optimized results for each subarray size while also considering the write failure limitation. Except for the row decoder and column multiplexer, the peripheral design for each subarray is assumed to be identical for simplicity and efficiency, where there is room for further optimization. Although implementing the divided technique into SRAM subarray can significantly enhance the WM and enlarge the subarray size, the additional logic gates also cost unignorable subarray area overhead (Fig. 8.16). This vital issue will be addressed in the next chapter (Chap. 9). In this section, the emphasis lies on the

Table 8.3 Test cases of different R_{LBL}, C_{LBL}, R_{LWL}, and C_{LWL}. (R_{LBL}: LBL resistance; C_{LBL}: LBL capacitance; R_{LWL}: LWL resistance; C_{LWL}: LWL capacitance)

RC Test Case #	R_{LBL}	C_{LBL}	R_{LWL}	C_{LWL}
0-Baseline	1×	1×	1×	1×
1	0.5×	1×	1×	1×
2	0.25×	1×	1×	1×
3	1×	0.5×	1×	1×
4	1×	0.25×	1×	1×
5	1×	1×	0.5×	1×
6	1×	1×	0.25×	1×
7	1×	1×	1×	0.5×
8	1×	1×	1×	0.25×
9	0.5×	0.5×	1×	1×
10	0.25×	0.25×	1×	1×
11	1×	1×	0.5×	0.5×
12	1×	1×	0.25×	0.25×

electrical-level improvements arising from the divided subarray design, whether it be the Conv. or any other proposed alternative, in contrast to the Std. counterpart. In the preliminary stage, the Conv. divided subarray design [5, 21] will be implemented as a representative illustration of diversified design approaches. As we disregard the area impact at the subarray level within this section, the same outcomes are expected to be applicable to the proposed divided subarray design discussed in Chap. 9.

8.4.1 Subarray and LB Size Optimization with RC Impacts

The RC impacts in PP analysis for both Std. and Conv. divided subarray designs are included since the subarray-level PP is strongly dependent on the RC of local interconnects (BL and WL) [6–9]. If the RC for LBL and LWL (R_{LBL}, C_{LBL}, R_{LWL}, and C_{LWL}) can be improved to a certain level that causes the PP of the Std. subarray design to outperform the divided counterpart in different LB sizes, the divided technique may not require. On the contrary, the divided technique should be involved in a larger subarray size design if it is more effective than pure local interconnect developments. Note that the LB size should also be kept as small as possible with sufficient PP enhancements since the subarray area penalty becomes higher with increased LB size (Fig. 8.16). The optimized (opt.) results for design type, LB size, and RC test case are marked in the red (Std.) and purple (Conv. divided) frames in each figure and listed in Table 8.4.

For both 128-C×128-R and 256-C×128-R subarrays (Fig. 8.11a and b), the WMs of Conv. divided subarray design are always inferior to Std. counterpart due to the parasitic RC introduced by additional logic gates. Moreover, both Std. and Conv. divided subarrays exceed the WFR boundary without necessitating the adoption of divided techniques. Therefore, only the Std. subarray design and potential RC optimization (best case = #2) are required for these subarray sizes. According to

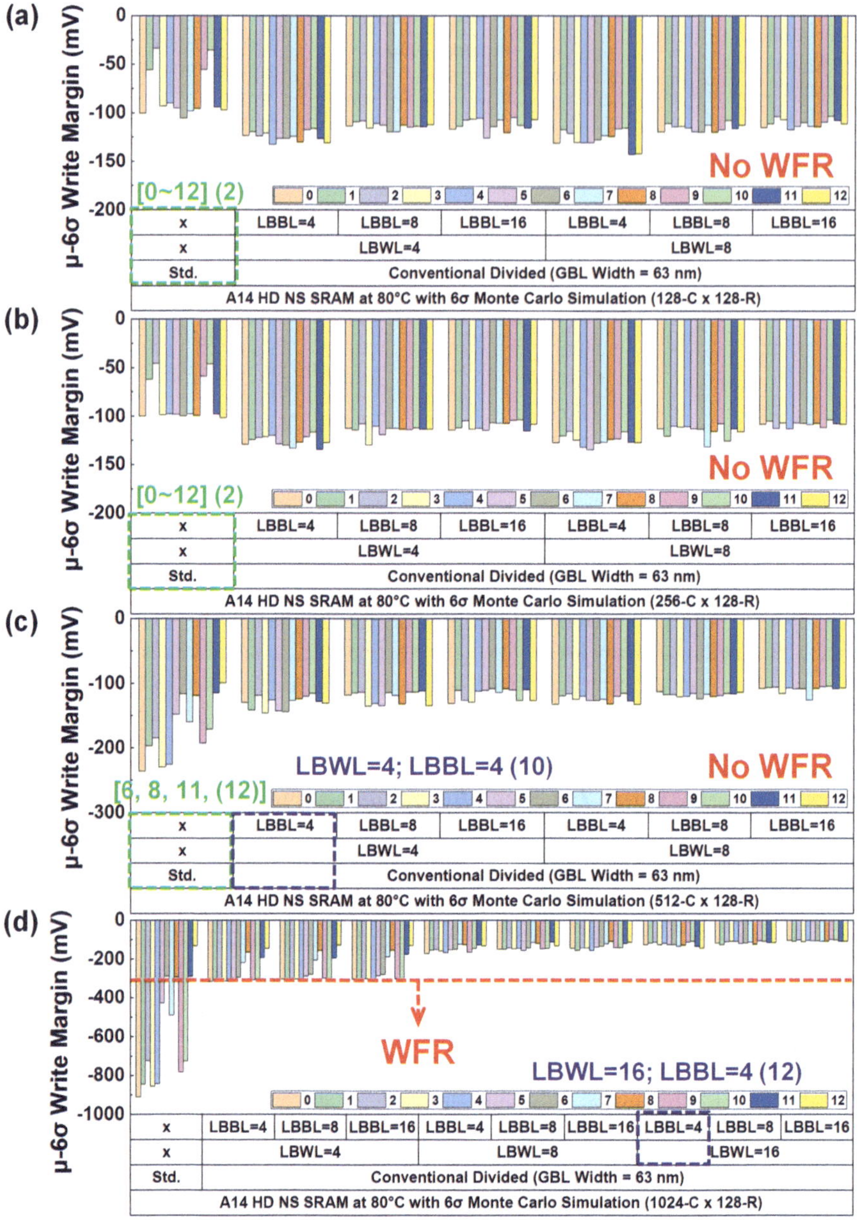

Fig. 8.11 WM with varying LBWL, LBBL, and RC at fixed $N_{\mathrm{col}} = 128$, 256, 512, and 1024, and $N_{\mathrm{row}} = 128$ of Std. and Conv. divided subarray designs as given by Table 8.3. Std. subarray designs are suitable for 128-C×128-R and 256-C×128-R in all RC scenarios. When N_{row} is either 128 or 256, WM is particularly affected by R_{BL}. As N_{col} increases, reducing both R_{WL} and C_{WL} proves more effective in enhancing WM

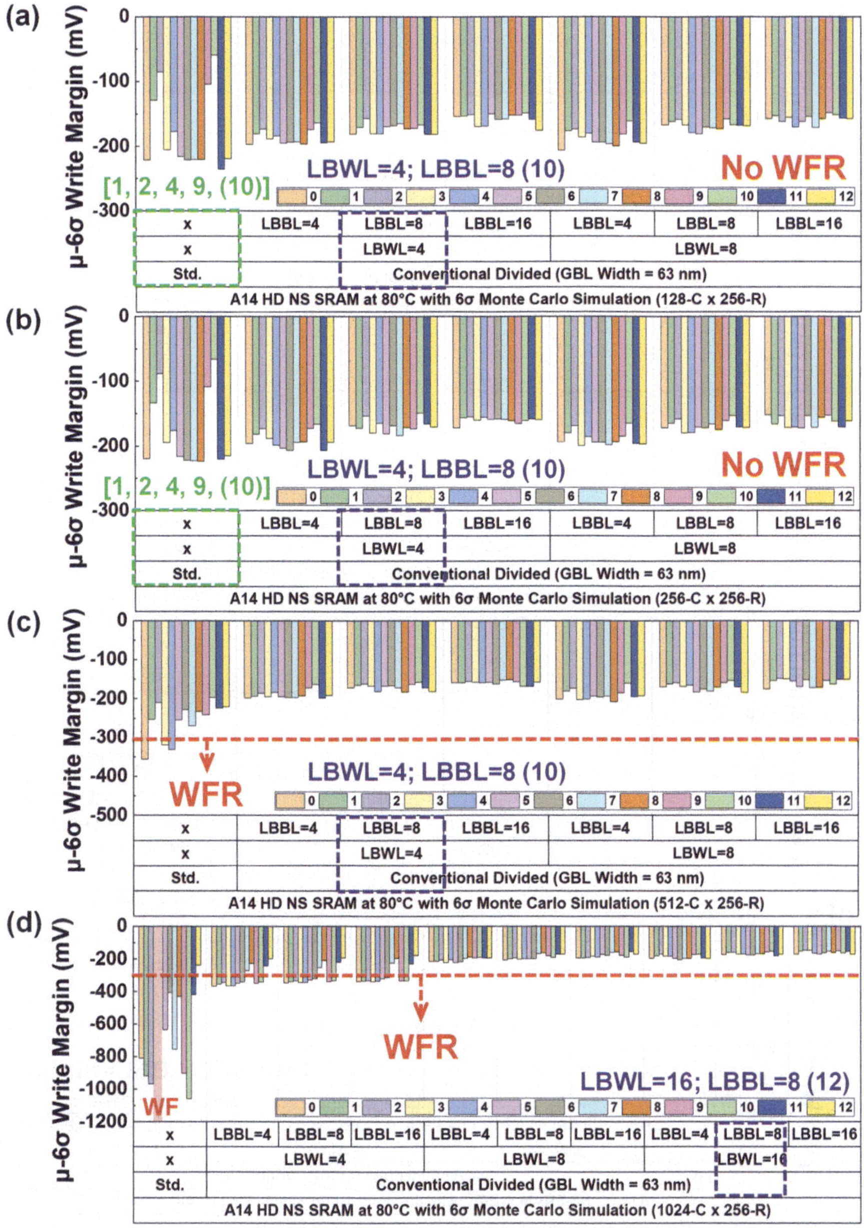

Fig. 8.12 WM with varying LBWL, LBBL, and RC at fixed N_{col} = 128, 256, 512, and 1024, and N_{row} = 256 of Std. and Conv. divided subarray designs as given by Table 8.3. The adoption of the divided technique is crucial when N_{row} reaches 256 and RC optimization is unattainable for Std. subarray design. When N_{col} is set to 128, 256, or 512, the optimized approach for enhancing WM involves decreasing both R_{BL} and C_{BL}. However, when N_{col} increases to 1024, it becomes more effective to decrease both R_{WL} and C_{WL} to improve WM

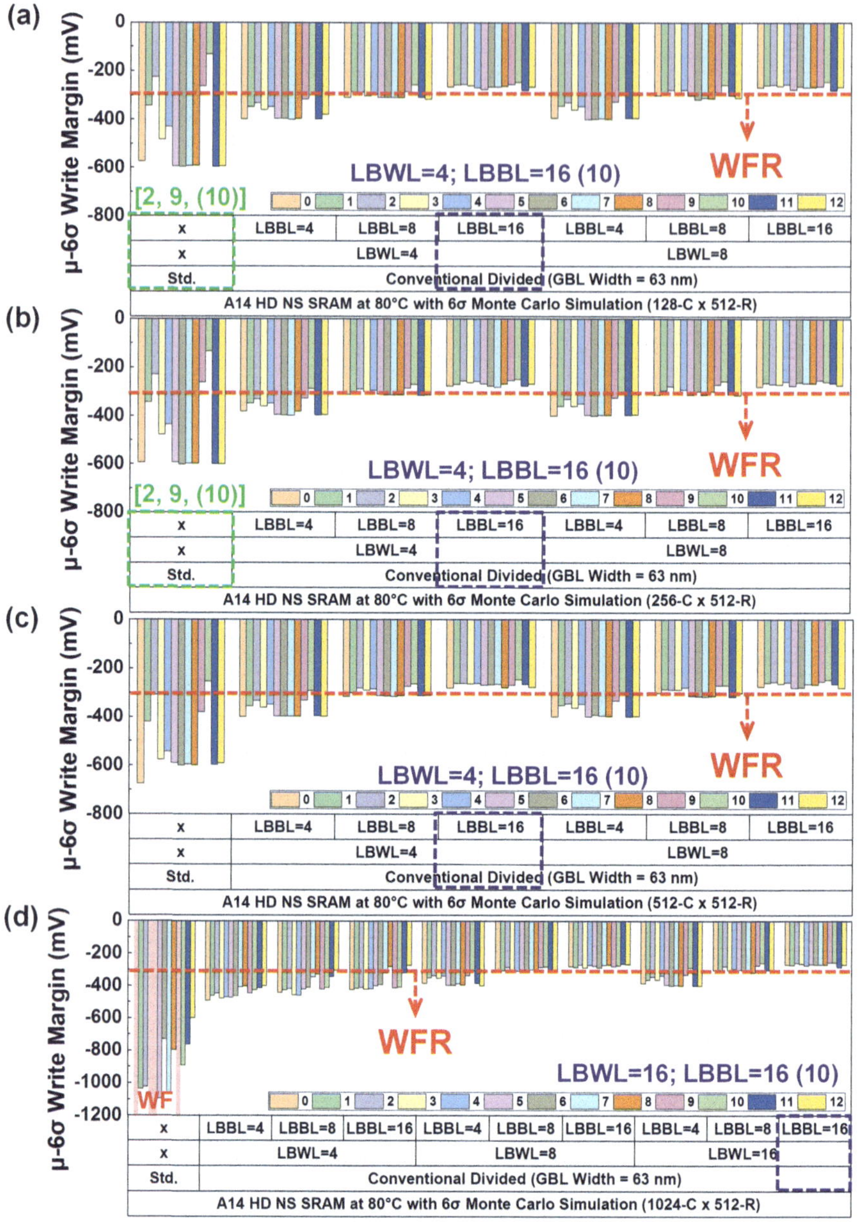

Fig. 8.13 WM with varying LBWL, LBBL, and RC at fixed $N_{\mathrm{col}} = 128$, 256, 512, and 1024, and $N_{\mathrm{row}} = 512$ of Std. and Conv. divided subarray designs as given by Table 8.3. The implementation of divided technique becomes essential due to the increased WFR in Std. subarray design. As N_{row} increases to 512, reducing both R_{BL} and C_{BL} emerges as the most effective method for enhancing WM

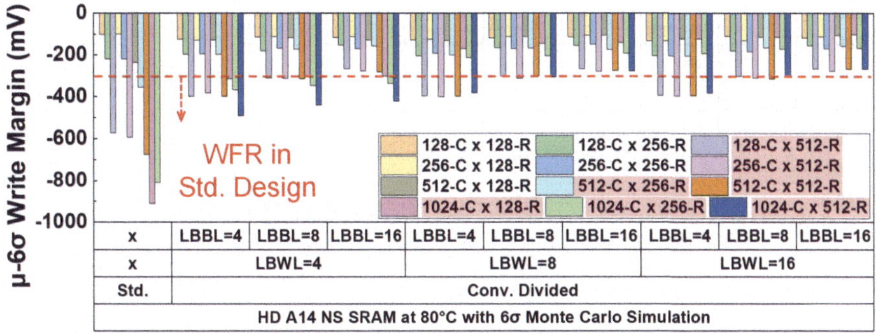

Fig. 8.14 Worst-case WM of HD A14 NSFET SRAM subarray with Conv. divided technique based on 6σ MC simulation at the worst-case temperature (80 °C). Subarrays that experience write operation failures are highlighted in red

Table 8.4 Optimized results of each subarray with the corresponding LB size and RC test case between Std. and divided subarray designs. *The indications favour the Std. design over the Conv. divided design, without encountering WFR issues

N_{row}	N_{col}	Opt. design	LB WL	LB BL	[Opt. case in Std.]* (Best case for both)
128-R	128-C	Std.	×	×	[0∼12] (2)
	256-C	Std.	×	×	[0∼12] (2)
	512-C	Std.	×	×	[6, 8, 11, (12)]
		Divided	4	4	(10)
	1024-C	Divided	16	4	(12)
256-R	128-C	Std.	×	×	[1, 2, 4, 9, (10)]
		Divided	4	8	(10)
	256-C	Std.	×	×	[1, 2, 4, 9, (10)]
		Divided	4	8	(10)
	512-C	Divided	4	8	(10)
	1024-C	Divided	16	8	(12)
512-R	128-C	Std.	×	×	[2, 9, (10)]
		Divided	4	16	(10)
	256-C	Std.	×	×	[2, 9, (10)]
		Divided	4	16	(10)
	512-C	Divided	4	16	(10)
	1024-C	Divided	16	16	(10)

the RC optimization results, minimal changes in WM occur when reducing C_{LBL}, R_{LWL}, and C_{LWL}. However, significant impacts are observed solely with R_{LBL} reduction, resulting in WM improvements of 66 mV and 54 mV for the Std. 128-C×128-R and 256-C×128-R subarrays, respectively, when comparing #2 to #0. Hence, reducing RLBL emerges as the most efficient and sole method to enhance the WM for relatively small Std. subarray sizes. It is worth noting that the alternations in WM for each configuration of LBBL and LBWL are insignificant in Conv. divided

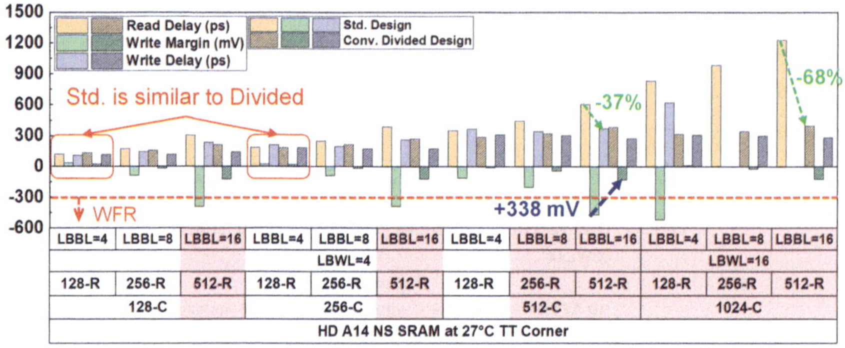

Fig. 8.15 SRAM subarray-level RD, WM, and WD benchmark in Std. and Conv. divided subarray designs with optimized LBWL and LBBL for each subarray size

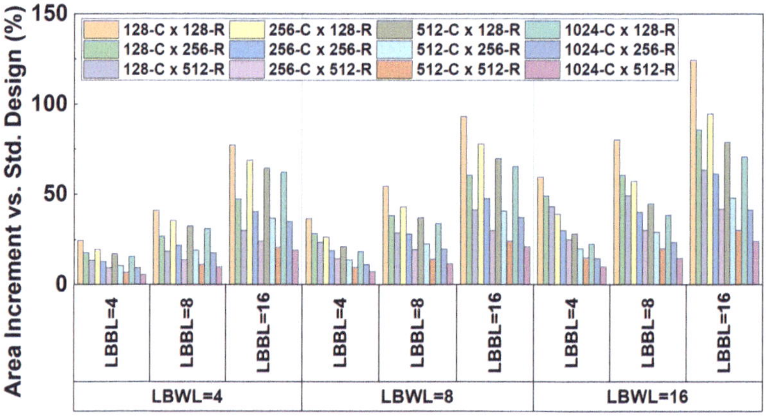

Fig. 8.16 Subarray area penalty of the Conv. divided subarray design in different LBWL and LBBL compared to the Std. counterpart

subarrays, as the baseline RC (#0) for both LBL and LWL in each LB is already sufficiently low.

For 512-C×128-R subarray (Fig. 8.11c), the optimized subarray design options include either: (1) Std. subarray design combined with prerequisite RC optimization (best case = #6, #8, #11, #12); or (2) Conv. divided subarray design featuring LBWL = 4 and LBBL = 4, with or without RC optimization. Similar to the Std. 128-C×128-R and 256-C×128-R subarrays, reducing C_{LBL} has marginal effects on WM compared to R_{LBL}. However, the decreases in both R_{LWL} (#6) and C_{LWL} (#8) can bring respective WM improvements of 120 mV and 117 mV due to the increased demand for RC reduction of the longer total WL length. Thus, the best WM result in Std. case appears when reducing both R_{LWL} and C_{LWL} for #12, providing 137 mV WM improvement compared to #0. In this subarray size, the baseline RC (#0) in each LB is still too low to observe any significant impacts from RC optimization.

Consequently, only 14 mV of WM improvement is realized in #10 compared to #0 for Conv. divided subarray design with LBWL = 4 and LBBL = 4. Although neither the Std. nor the Conv. divided subarrays suffer from the WFR, the latter offers superior WM compared to the former, considering #0 for both scenarios. Hence, in this initial phase, the Conv. divided subarray design is preferable concerning the uncertain approach of RC optimization, which may necessitate further endeavours in the future. Furthermore, the selection of LBWL = 4 and LBBL = 4 is made to mitigate superfluous subarray area overhead (17% shown in Fig. 8.16), exhibiting WM differences of less than 22 mV compared to alternative LB sizes.

For 1024-C×128-R subarray (Fig. 8.11d), most Std. subarray designs fail with write operations despite RC optimization efforts. Only the Std. subarrays featuring optimized RC #6, #8, #11, and #12 marginally exceed the WFR threshold. The presence of WFR is also observed in Conv. divided subarrays with LBWL values below 4, particularly when assuming #0. To overcome this limitation, it is imperative to pursue RC optimization strategies akin to #7, #8, #11, and #12 when LBWL falls below 4. Conversely, when LBWL exceeds 4, WFR ceases to be a concern for Conv. divided subarrays. Moreover, Conv. divided subarrays exhibit minimal variations in WM across different LBBL values, primarily due to the sufficiently low RC for LBL within 128-R subarray configuration. Notably, the WM of Conv. divided subarrays with LBWL = 16 and baseline RC (#0) surpasses that of Std. subarrays, even under the assumption of employing the best-optimized RC case (#12). Consequently, adopting the Conv. divided subarray design with LBWL = 16 and LBBL = 4 is recommended over the Std. counterpart to achieve superior WM by 786 mV (#0) at the cost of 23% subarray area overhead (Fig. 8.16). It is important to highlight that across all optimized scenarios within the 128-R subarray group, the worst-case WM (−30 to −150 mV) consistently remains above the specified WFR threshold.

For both 128-C×256-R and 256-C×256-R subarrays (Fig. 8.12a and b), the optimized configuration involves either the (1) Std. subarray design with targeted RC (#1, #2, #4, #9, #10) or the (2) Conv. divided subarray design featuring LBWL = 4 and LBBL = 8, with or without RC optimization. In line with observations for the 128-C×128-R and 256-C×128-R subarrays (Fig. 8.11a and b), diminishing R_{LBL} yields more pronounced enhancements than R_{LWL} and C_{LWL}, owing to the shorter overall WL length in the Std. subarray design with 128-C and 256-C. Notably, the reduction in C_{LBL} begins to positively impact WM upon transitioning from 128-R to 256-R subarray groups, while the influence of C_{LBL} on WM remains subordinate to that of R_{LBL}. Thus, the most optimized RC case is #12, demonstrating WM enhancements of 161 mV and 154 mV in the Std. 128-C×256-R and 256-C×256-R subarray designs, respectively, compared to #0. Although no WFR is evident in either Std. or divided subarrays, the latter may be preferable when assuming #0 for both cases, showcasing a superior WM of 24 mV for both subarray sizes. However, it is noteworthy that the WM of Conv. divided subarrays may deteriorate compared to their Std. counterparts with the best-optimized RC values for both scenarios. This suggests that the Std. subarray design would be advantageous if potential RC optimization methods are identified in the future.

As the total LBL length increases from 128-R to 256-R, the WM exhibits more significant enhancements with increasing LBBL at fixed LBWL. Furthermore, the results indicate that reducing RC in LBL can lead to greater WM enhancements than LWL when N_{col} = 128/256. Consequently, the configuration of LBWL = 4 and LBBL = 8 is selected for both Conv. divided 128-C×256-R and 256-C×256-R subarrays to enhance WM by 40 mV and 52 mV, respectively, at the expense of 27% and 22% subarray area penalty compared to Std. counterparts (Fig. 8.16) under baseline RC (#0).

For 512-C×256-R and 1024-C×256-R subarrays (Fig. 8.12c and d), the effectiveness of reducing R_{LWL} and C_{LWL} becomes comparable to that of R_{LBL} and more pronounced than C_{LBL} as the total WL length increases to 512-C. Upon transitioning to 1024-C, the relative enhancements in WM become notably dominated by the RC of LWL rather than LBL. Thus, the targeted RC cases are #10 for 512-C and #12 for 1024-C in Std. subarrays, yielding WM improvements of 159 mV and 572 mV, respectively. Despite observing significant WM enhancements with optimized RC in Std. subarrays, the WM of Conv. divided subarrays featuring LBWL = 4 for 512-C and 16 for 1024-C, LBBL = 8, and #0, consistently outperforms their Std. counterparts under any RC assumptions. Furthermore, Std. subarrays still encounter write failures at both the baseline RC (#0) and numerous optimized RC values. Consequently, the Conv. divided technique emerges as the optimized design choice for these subarray sizes. The selection of LBWL = 4/16 and LBBL = 8 for Conv. divided subarrays of 512-C/1024-C×256-R (#0) enhances WM by 820 mV/637 mV, incurring 19%/23% subarray area overhead compared to Std. counterparts (Fig. 8.16). All optimized configurations at 256-R ensure functional write operation, as the worst-case WM values (−60∼−250 mV) exceed the WFR constraint.

For 128-C×512-R and 256-C×512-R subarrays (Fig. 8.13a and b), the optimized approach entails either adopting the (1) Std. subarray design with targeted RC adjustments (#2, #9, #10) or the (2) Conv. divided subarray design with LBWL = 4 and LBBL = 16, with or without RC optimization. It has been previously highlighted that reducing RC for LBL represents the only effective method to enhance WM for shorter total WL lengths within the 128-C and 256-C subarray groups. Furthermore, to surpass the WFR boundary, the targeted RC should achieve a minimum of 1/4 R_{LBL} (#2) or 1/2 for both R_{LBL} and C_{LBL} (#9 and #10). optimized RC adjustment (#2) in the Std. subarray yields WM improvements of 441 mV and 461 mV in the cases of 128-C and 256-C, respectively. Similarly, the augmentation of LBBL offers greater significance in enhancing WM, owing to the expanded total BL length to 512-R. Thus, LBBL should be set to a minimum of 16 to prevent WFR under any RC assumption. Conversely, increasing LBWL does not contribute to WM enhancement due to sufficiently low RC for LWL. Consequently, the recommended LB sizes for the Conv. divided subarray design are LBWL = 4 and LBBL = 16 to reach WM improvements of 305 mV (#0) and 316 mV (#0) at the cost of subarray area overhead of 30% and 24% (Fig. 8.16) for 128-C and 256-C, respectively, compared to Std. counterparts.

For the 512-C×512-R and 1024-C×512-R subarrays (Fig. 8.13c and d), the optimized configuration involves employing Conv. divided subarray design with LBWL = 4 and 16, respectively, alongside LBBL = 16. This optimization underscores the necessity of RC optimization in the Std. subarray design for the 512-R subarray group, as all Std. designs under the baseline RC (#0) exhibit WFR issues. Specifically, to overcome this WFR constraint and enhance WM by 421 mV in the Std. 512-C×512-R subarray design, the best-optimized RC case of #10 is required. However, RC optimization appears pointless for Std. 1024-C×512-R subarray due to significantly inferior baseline RC (#0) characteristics for both LWL and LBL in such a large subarray. Even with the best-optimized RC assumption, WFR persists in the Std. 1024-C×512-R subarray, thus favouring the Conv. divided subarray design for these subarray sizes. In the case of 512-C subarrays, augmenting LBL proves more efficacious than LWL in improving WM. Consequently, LBWL = 4 and LBBL = 16 are selected for Conv. divided 512-C×512-R subarrays, resulting in a 393 mV WM enhancement at the expense of a 21% subarray area penalty compared to their Std. counterparts (Fig. 8.16). Conversely, when transitioning to 1024-C, increasing LWL becomes crucial and comparable to LBL to mitigate WFR. Hence, the optimized LB size for Conv. divided 1024-C×512-R subarrays is LBWL = 16 and LBBL = 16, incurring a 24% subarray area penalty compared to Std. counterparts (Fig. 8.16). The write functionalities of all the optimized cases at 512-R are guaranteed, as the worst-case WMs (−150∼ − 300 mV) fulfill the successful write criteria.

Figure 8.14 illustrates the WM comparison between the Std. and Conv. divided subarray designs with a baseline RC (#0) across various configurations of LBBL and LBWL within each subarray size. It is crucial to highlight that the conservative assumption sets the boundary of WFR at −300 mV (indicated by the red dashed line), as discussed in Sect. 8.3. A WM value below this red dashed line indicates the presence of WFR in the SRAM subarray. The optimized results for LB size are presented in Table 8.4.

To summarize the results of RC impacts, reducing the RC of LBL seems to be more helpful than LWL in less N_{col} (128-C, 256-C, and 512-C), and vice versa in larger N_{col} (1024-C) at the same N_{row}. However, the WM improvements by RC optimization become insignificant as N_{col} increases to 512-C and 1024-C. Moreover, the WMs of the divided subarray design are not obviously distinct from each other since the RC has already been reduced effectively. As these optimized RC targets should be proven achievable in other work, only baseline RC (#0) is compared between Std. and Conv. divided subarray designs for an overview of read delay (RD), WM, and write delay (WD) in Fig. 8.15 (27 °C and TT corner). Through the application of Conv. divided technique, improvements of 338 mV in WM and 37%/68% in RD are achieved for the 512-C/1024-C×512-R subarray configuration. Additionally, the implemented design facilitates the recovery from write failures. It exhibits that the divided technique is necessary to enable a large-size subarray while maintaining or improving the WM and speed at the cost of area overhead and extra parasitic routing.

8.4.2 Area Penalty at Subarray Level

Figure 8.16 illustrates the percentage increase in subarray area when transitioning from Std. to Conv. divided subarray designs, considering varying LBWL, LBBL, and subarray sizes. Significantly, augmenting LBBL at a constant LBWL has resulted in a substantial subarray area increase, in contrast to the scenario where LBWL is increased while LBBL is fixed. Consequently, a larger subarray design favours a greater N_{col} with an appropriate LBWL. Additionally, the subarray area overhead diminishes as the subarray size increases while maintaining the same LB size, attributable to heightened array area efficiency. Hence, it is more effective to apply the divided technique to a bigger subarray with larger N_{col} than N_{row}. This subarray area penalty may serve as a bottleneck of macro-level PPA enhancements, which will be discussed in the upcoming chapter. Hence, a novel divided approach is introduced to ameliorate the subarray area penalty associated with large-size subarray design and improve macro-level EDAP. This aspect will be further discussed in the upcoming chapter.

Takeaways

- **Conservative Write Failure Boundary:**
  - **WM Criteria:** Conservatively set above −300 mV to potentially cover advanced nodes.
- **Standard Subarray Design:**
  - **Challenges:** Subarrays of sizes 128-C×512-R, 256-C×512-R, 512-C×256/512-R, and 1024-C×128/256/512-R encounter write failure issues.
  - **Motivated Design Strategy:** Divided technique is necessary to mitigate write failure risk.
- **Conventional Divided Subarray Design:**
  - **Benefit:** Enables large subarray sizes while maintaining or improving WM.
  - **Drawback:** Introduces area overhead and extra parasitic routing, potentially degrading macro-level PPA results (Chap. 9).
  - **Impact of RC:**
    - Reducing the RC of LBL is more beneficial than LWL in smaller N_{col} configurations (128-C, 256-C, 512-C).
    - In larger N_{col} configurations (1024-C), reducing the RC of LWL is more effective.

(continued)

- **Impact of LB Size on Area:**
  - **Effect of increasing LBBL at constant LBWL:** Significant increase in subarray area.
  - **Effect of increasing LBWL at constant LBBL:** Less pronounced increase in subarray area.
  - **Motivated Design Strategy:** Larger subarray design favours greater N_{col} with appropriate LBWL.

8.5 Potential Design Directions and Their Major Challenges for Enabling Larger Subarray Size Design

8.5.1 Non-Recommended Approach: Conv. Divided Technique

Figures 8.16 and 8.17b illustrate that the subarray area penalty is substantial due to the additional circuits required compared to the Std. design (Fig. 8.17a). This increased area will cause reverse macro-level PPA scaling despite improvements at the subarray level, primarily due to the added interconnect overhead. The next chapter (Chap. 9) provides further details, demonstrating that the Conv. divided technique is not recommended for enabling larger subarray sizes.

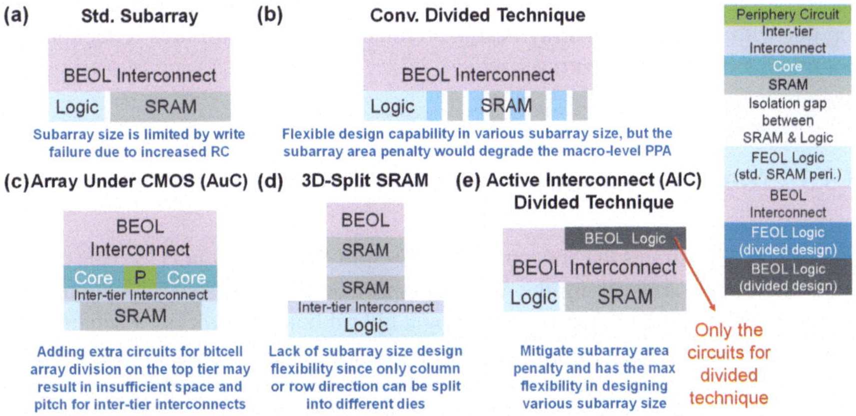

Fig. 8.17 Overview of subarrays with (**a**) Std. design, (**b**) Conv. divided technique, (**c**) Array under CMOS (AuC), (**d**) 3D-Split SRAM, and (**e**) Active interconnect (AIC) divided technique

8.5.2 Non-Recommended Approach: Array under CMOS (AuC)

Figure 8.17c illustrates the concept of Array under CMOS (AuC), which involves positioning part of the periphery (e.g., WL decoder, write driver, and sense amplifier) and core logic circuits on the top tier, while placing the SRAM bitcell array and the rest of periphery (e.g., column select, pre-charge, etc) along with signal lines on the bottom tier [19]. This design allows for independent optimization of the top and bottom tiers and can significantly reduce the subarray area. Notably, placing the WL decoder in the centre of the bitcell array can potentially reduce the RC delay of the WL by 50% [19]. To avoid overlapping the routing space with the WL decoder, the column selects, sense amplifiers, and write drivers should be positioned at the edge since they run perpendicular to the WL decoder. Consequently, the RC reduction is applicable only to the WL and not to the BL. The RC reduction in the WL can ideally double N_{col} compared to the Std. subarray design. However, beyond this factor of 2, the subarray size remains constrained by the increased RC of both WL and BL as the subarray is upsized. If additional circuits for divided techniques or distributed WL decoders are placed on the top tier to reduce the effective RC, there would be insufficient space for interconnect routing between different LBs and the pitch of the inter-tier interconnects. This would necessitate additional area extension to accommodate the routing space and pitch requirement. The core logic circuits will have to be positioned outside the SRAM footprint, resulting in increased area demand. Moreover, the cost would increase significantly, and the thermal impact on the bottom tier could become problematic due to thermal stability issues when fabricating the top tier with both peripheral and divided circuits. Therefore, this approach is only recommended for placing peripheral circuits and core logic on the top tier within the Std. subarray design architecture, aiming for area reduction but not for enabling larger subarray size.

8.5.3 Non-Recommended Approach: 3D-Split SRAM

Figure 8.17d illustrates the concept of the 3D-Split SRAM design [11]. This design splits the bitcell array along the WL (BL-split) or BL (WL-split) directions and places them on different dies, followed by a hybrid bonding process. Each column or row of different dies requires a vertical connection to connect the WL or BL, respectively. Consequently, the pitch of hybrid wafer bonding (HWB) is a crucial factor in determining the additional area overhead due to inter-tier interconnects. The pitch requirement is also constrained by the cell height and cell width, as defined by the technology design rules. If the HWB pitch and scaling are not compatible with bitcell area scaling, additional area overhead is necessary to avoid insufficient routing space between the dies. This design ideally reduces the effective

RC by 50% in either the WL or BL directions, potentially doubling the subarray N_{col} or N_{row} if the HWB pitch requirement is met. However, flexibility in varying subarray sizes is limited when more stacked layers of the bitcell array are needed for larger subarray sizes. This limitation arises due to increased routing complexity between different dies, higher costs, and additional RC overhead from HWB. Furthermore, only one direction can be optimized for benefits. Consequently, the 3D-Split SRAM is not recommended for enabling larger subarray sizes.

8.5.4 Non-Recommended Approach: Different Bitcell Topologies

Apart from utilizing a 6T bitcell design with varying PU:PG:PD ratios, exploring different bitcell topologies like 8T and 10T is a well-established research direction for enhancing margin and PP [1]. However, Chap. 7 emphasizes that increasing the bitcell and subarray area can negatively impact the overall macro-level PPA results due to significant interconnect overhead. Incorporating more transistors in SRAM bitcell design requires additional efforts at the macro-level analysis to prevent undesirable degradation. Consequently, we will continue to concentrate on 6T bitcell designs in resistance-dominated nodes.

8.5.5 Recommended Approach: Active Interconnect (AIC) Technique

Figure 8.17e depicts the active interconnect (AIC) divided subarray design concept [10]. By relocating necessary circuits for the divided technique from FEOL to the BEOL active layer, the subarray area penalty is reduced. This approach preserves potential WM and PP improvements at both subarray and macro levels while allowing maximum flexibility in designing various subarray sizes compared to other options such as AuC and 3D-Split SRAM. Moreover, our focus is on ultra-scaled nodes. The previously mentioned options might have been effective for 14 nm and, at best, for 7 nm nodes, considering the current HWB pitch. Older nodes provide more space and area for placing inter-tier interconnects. However, for nodes smaller than 1 nm, these options are clearly no longer viable. Therefore, the AIC divided subarray design is highly recommended among potential design directions. Further details are provided in Chap. 9.

Takeaways

- **Motivated Design Strategy:** Enable larger subarray sizes for macro-level PPA improvements without incurring significant subarray area penalties.
- **Non-Recommended Approach:** Subarray designs such as Std., Conv. Divided, AuC, 3D-Split, and 8T/10T bitcell present challenges like high WFR, significant subarray area penalties, insufficient inter-tier routing space and pitch, and limited flexibility in designing various subarray sizes.
- **Recommended Approach:** AIC divided subarray design shows promise by potentially improving WM and PP at both the subarray and macro levels, while mitigating subarray area overhead by relocating the divided circuits to the BEOL active layer.

8.6 Conclusions

To achieve macro-level EDAP with a larger subarray size, various subarray sizes are systematically evaluated while ensuring write ability and improved speed using the simulation circuit and methodologies presented in Chap. 2. This chapter investigates potential WFRs, including PG leakage, limited C_{boost}-generated write-assist voltage, max operating V_{GS}, and substrate current. The application of the WFR examination framework yields precise values for WM criteria, specifically −300 mV in HD A14 NSFET SRAM subarray. Subsequently, multiple subarray designs are optimized based on design type (Std. or Divided), LB size, RC of WL and BL, and GBL width, taking into account this write failure boundary. The subarray-level benchmarks within the Conv. divided subarray design demonstrate notable improvements in write ability, providing the potential for designing larger subarray sizes compared to the Std. counterpart. This analysis motivates further exploration into alternative divided subarray designs capable of addressing the area penalty issue, as elaborated in the upcoming chapter.

References

1. Chen, J., et al. (2022). A reliable 8T SRAM for high-speed searching and logic-in-memory operations. *IEEE Transactions on Very Large Scale Integration (VLSI) Systems, 30*(6), 769–780. https://doi.org/10.1109/TVLSI.2022.3164756
2. Franco, J., et al. (2022). Low thermal budget PBTI and NBTI reliability solutions for multi-Vth CMOS RMG stacks based on atomic oxygen and hydrogen treatments. In *2022 International Electron Devices Meeting (IEDM)* (pp. 30.4.1–30.4.4). https://doi.org/10.1109/IEDM45625.2022.10019385

3. Ha, D., et al. (2017). Highly manufacturable 7 nm FinFET technology featuring EUV lithography for low power and high performance applications. In *2017 Symposium on VLSI Technology* (pp. T68–T69). https://doi.org/10.23919/VLSIT.2017.7998202
4. Jang, D., et al. (2017). Device exploration of nanosheet transistors for sub-7-nm technology node. *IEEE Transactions on Electron Devices, 64*(6), 2707–2713. https://doi.org/10.1109/TED.2017.2695455
5. Karandikar, A., & Parhi, K. K. (1998). Low power SRAM design using hierarchical divided bit-line approach. In *Proceedings International Conference on Computer Design. VLSI in Computers and Processors (Cat. No.98CB36273)* (pp. 82–88). https://doi.org/10.1109/ICCD.1998.727027
6. Liu, H.-H., et al. (2023). CFET SRAM DTCO, interconnect guideline, and benchmark for CMOS scaling. *IEEE Transactions on Electron Devices, 70*(3), 883–890. https://doi.org/10.1109/TED.2023.3235701
7. Liu, H.-H., et al. (2023). CFET SRAM with double-sided interconnect design and DTCO benchmark. *IEEE Transactions on Electron Devices* 1–8. https://doi.org/10.1109/TED.2023.3305322
8. Liu, H.-H., et al. (2023). DTCO of sequential and monolithic CFET SRAM. In R.-H. Kim & N. V. Lafferty (Eds.), *DTCO and computational patterning II* (Vol. 12495). International society for optics and photonics. SPIE. https://doi.org/10.1117/12.2657524
9. Liu, H.-H., et al. (2023). Extended methodology to determine SRAM write margin in resistance-dominated technology node. *IEEE Transactions on Electron Devices, 69*(6), 3113–3117. https://doi.org/10.1109/TED.2022.3165738
10. Liu, H.-H., et al. (2024). Future design direction for SRAM data array: Hierarchical subarray with active interconnect. *IEEE Transactions on Circuits and Systems I: Regular Papers* 1–12. https://doi.org/10.1109/TCSI.2024.3410518
11. Mathur, R., et al. (2021). 3D-split SRAM: Enabling generational gains in advanced CMOS. In *2021 IEEE Custom Integrated Circuits Conference (CICC)* (pp. 1–2). https://doi.org/10.1109/CICC51472.2021.9431528
12. Matsukawa, T., et al. (2012). Suppressing VT and GM variability of FinFETs using amorphous metal gates for 14 nm and beyond. In *2012 International Electron Devices Meeting* (pp. 8.2.1–8.2.4). https://doi.org/10.1109/IEDM.2012.6479002
13. Mukhopadhyay, S., et al. (2008). Capacitive coupling based transient negative bit-line voltage (Tran-NBL) scheme for improving write-ability of SRAM design in nanometer technologies. In *2008 IEEE International Symposium on Circuits and Systems (ISCAS)* (pp. 384–387). https://doi.org/10.1109/ISCAS.2008.4541435
14. Mukhopadhyay, S., et al. (2011). SRAM write-ability improvement with transient negative bit-line voltage. *IEEE Transactions on Very Large Scale Integration (VLSI) Systems, 19*(1), 24–32. https://doi.org/10.1109/TVLSI.2009.2029114
15. Ortiz-Conde, A., et al. (2002). A review of recent MOSFET threshold voltage extraction methods. *Microelectronics Reliability, 42*(4), 583–596. ISSN: 0026-2714. https://doi.org/10.1016/S0026-2714(02)00027-6
16. Pei, Z., et al. (2023). Emerging interconnect exploration for SRAM application using non-conventional H-tree and center-pin access. In *2023 24th International Symposium on Quality Electronic Design (ISQED)* (pp. 1–1). https://doi.org/10.1109/ISQED57927.2023.10129316
17. Pei, Z., et al. (2023). Graphene-based interconnect exploration for large SRAM caches for ultrascaled technology nodes. *IEEE Transactions on Electron Devices, 70*(1), 230–238. https://doi.org/10.1109/TED.2022.3225512
18. Rahman, A., et al. (2018). Reliability studies of a 10 nm high-performance and low-power CMOS technology featuring 3rd generation FinFET and 5th generation HK/MG. In *2018 IEEE International Reliability Physics Symposium (IRPS)* (pp. 6F.4-1-6F.4–6). https://doi.org/10.1109/IRPS.2018.8353648
19. Salahuddin, S. M., et al. (2020). Thermal stress-aware CMOS–SRAM partitioning in sequential 3-D technology. *IEEE Transactions on Electron Devices, 67*(11), 4631–4635. https://doi.org/10.1109/TED.2020.3023923

20. Sinangil, M. E., Mair, H., & Chandrakasan, A. P. (2011). A 28 nm high-density 6T SRAM with optimized peripheral-assist circuits for operation down to 0.6 V. In *2011 IEEE International Solid-State Circuits Conference* (pp. 260–262). https://doi.org/10.1109/ISSCC.2011.5746310
21. Yoshimoto, M., et al. (1983). A divided word-line structure in the static RAM and its application to a 64K full CMOS RAM. *IEEE Journal of Solid-State Circuits, 18*(5), 479–485. https://doi.org/10.1109/JSSC.1983.1051981

Chapter 9
SRAM Hierarchical Subarray Design with Active Interconnect Technique

In sub 10-nm nodes, the growing dominance of interconnects in chips poses challenges in designing large-size static random-access memory (SRAM) subarrays [18, 21, 22]. The primary concern involves the write failure problem arising from the increased resistance and capacitance (RC) for bitline (BL) and wordline (WL) [18] within the standard (Std.) subarray design, as discussed in the previous chapter. Chapter 8 tackles this issue by incorporating conventional (Conv.) divided WL (DWL) [34] and divided BL (DBL) [12] techniques into SRAM subarray design based on nanosheet (NS) in 14-Å-compatible technology (A14). The examination encompasses the analysis of the source of write failure, the establishment of write failure criteria derived from distinct sources, and the optimization of RC and local block (LB) size within the subarray design. This approach facilitates the subarray design of various subarray sizes, ensuring successful write operations while enhancing subarray-level performance and power (PP), as outlined in Fig. 8.15 of Chap. 8. However, Chap. 9 demonstrates that the additional logic gates come with subarray area penalty that may degrade the overall performance, power, and area (PPA) at the macro level due to increased inter-subarray interconnect overhead.

To overcome this limitation, the active interconnect (AIC) design is proposed with the features of fabricating another or multiple active regions at the BEOL layers in this chapter [19]. It is important to highlight that the AIC concept aligns with the principles of three-dimensional (3D) sequential stacking architectures, as demonstrated in prior works [2, 7, 23, 27, 28, 30, 32, 33]. By moving these extra logic gates from FEOL to BEOL in the AIC divided subarray design, the subarray area penalty is significantly mitigated without compromising PP compared to the Std. and Conv. divided counterparts. To achieve this concept, A14 carbon nanotube (CNT) [9][1] gate-all-around transistor (GAAFET) is explored as potential BEOL-compatible device. Notably, the model calibration and background knowledge of CNFET were supported by Carlo Gilardi and Subhasish Mitra from Stanford

[1] Acknowledgment is given to Carlo Gilardi and Subhasish Mitra for the CNFET model.

H.-H. Liu, F. Catthoor, *Circuit-Technology Co-Optimization of SRAM Design in Advanced CMOS Nodes*, https://doi.org/10.1007/978-3-031-76109-6_9

University. The assumption of successful manufacturing integration may currently be overly optimistic due to the unknown challenges of fabricating CNTs in a realistic and ultra-scale manner. However, the AIC concept remains valid as the process matures or if another promising candidate for BEOL-compatible devices is found. In this research, a comprehensive design-technology co-optimization (DTCO) analysis is conducted to verify the value and potential benefits of up to 65% macro-level energy-delay-area-product improvement by AIC divided subarray design compared to the Std. counterpart. It is crucial to note that the 2.9-fold improvement is achieved through the optimal inter-subarray design, specifically the centre pin with E-tree (CP-E) technology detailed in Sect. 7.3 of Chap. 7. When accounting for the combined impact of CP-E and AIC advancements, the total enhancement reaches 18.9 times when compared to the side pin with H-tree (SP-H) and Std. subarray design.

In this chapter, multiple high-density (HD) A14 NSFET SRAM subarray options are thoroughly optimized across three design types (Std., Conv. divided, and AIC divided techniques) with distinct subarray sizes and numbers of local blocks (LBs). Furthermore, to validate the value of AIC divided subarray design and motivate more concrete process development, corresponding DTCO PPA analysis is carried out at both subarray and macro levels. Section 9.1 outlines the concern of macro-level PPA degradation resulting from the subarray area penalty introduced by Conv. divided technique compared to Std. subarray design when upsizing SRAM subarray, as discussed in Chap. 8. In Sect. 9.2, a novel AIC divided subarray design is proposed as a solution to mitigate the subarray area penalty and potentially enhance macro-level PPA. Section 9.3 details the layouts of subarrays and bitcells for Std., Conv. divided, and AIC divided subarray designs, accompanied by an assessment of estimated parasitic RC overhead. Moving forward, Sect. 9.4 optimizes the AIC divided subarray design by employing CNFET as the top-tier devices to facilitate LB selection signals. Section 9.5 succinctly presents a summary of subarray- and macro-level PPA benchmarks for all three subarray designs: Std., Conv. divided, and AIC divided subarray designs. Furthermore, this section includes a comparative analysis involving optimistic A14 Si NS, realistic A14 CNT, and conservative N22 FDSOI device models within the AIC divided subarray design framework. Section 9.6 concludes that the AIC divided subarray design stands out as a promising candidate for future SRAM data array design, showcasing the potential to achieve a noteworthy 65% macro-level EDAP improvement through the utilization of A14 CNT devices at the BEOL layer.

9.1 Previous Challenges in Conventional (Conv.) Divided Subarray Design

As depicted in Fig. 8.1 of Chap. 8, a notable 56% reduction in macro-level energy-delay-area product (EDAP) is obtained during the transition from a small- to a

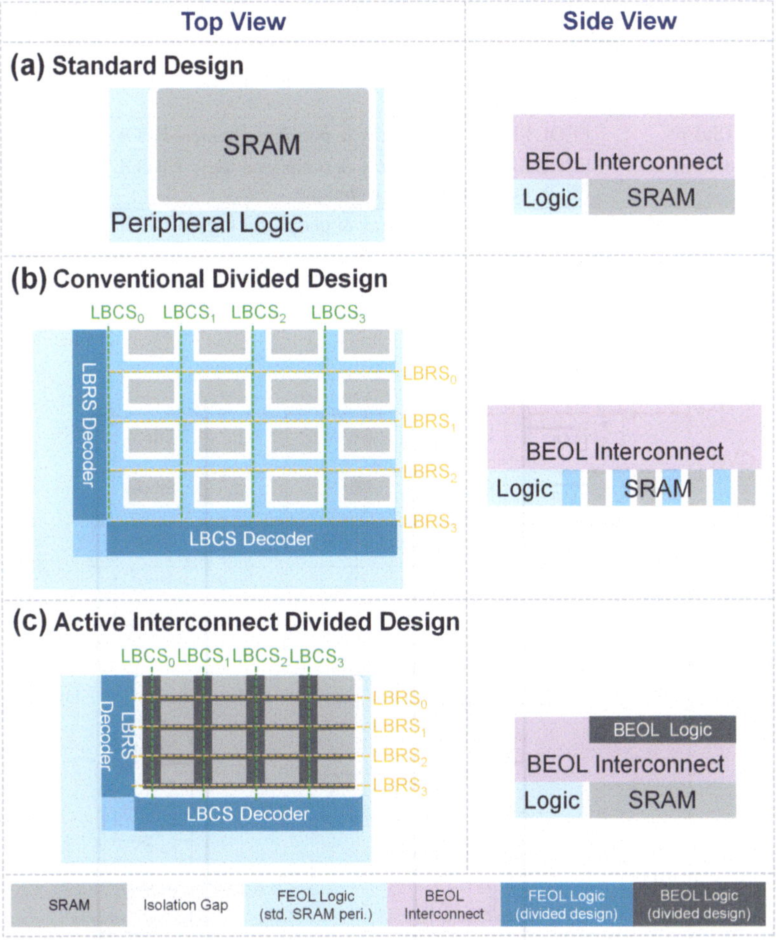

Fig. 9.1 Schematics of top and side views of SRAM subarray designs. (**a**) Standard (Std.) subarray design: undivided SRAM bitcell array. (**b**) Conventional (Conv.) divided subarray design: SRAM bitcell array is divided into several local blocks by extra FEOL logic gates. (**c**) Active interconnect (AIC) divided subarray design: SRAM bitcell array is divided into several local blocks by additional BEOL logic gates

large-size SRAM subarray, irrespective of potential write failure risk (WFR) [18]. However, the implementation of a large-size Std. subarray design (Fig. 9.1a and Table 9.1) presents challenges attributable to substantial parasitic RC overhead on both BL and WL. To conduct a comprehensive DTCO analysis involving various subarray options, multiple sizes of subarrays are achieved based on the Conv. DWL [34] and DBL [12] techniques (Fig. 9.1b, Table 9.1, and Fig. 9.2). For simplicity, the terms "standard (Std.)" and "divided" designs are used to refer to the SRAM subarray with and without DWL and DBL techniques, respectively, in this

Table 9.1 Design types and descriptions for SRAM subarray. (LBCS: local block column select; LBRS: local block row select)

Subarray design Tech	Descriptions
Standard (Std.)	FEOL SRAM bitcell array & peripheral logic + BEOL interconnect
Conventional (Conv.) Divided	FEOL SRAM bitcell array & peripheral logic + BEOL interconnect + FEOL logic for divided technique
Active Interconnect (AIC) Divided	FEOL SRAM bitcell array & peripheral logic + BEOL interconnect + FEOL LBRS and LBCS decoders + BEOL logic for divided technique w/o LBRS and LBCS decoders

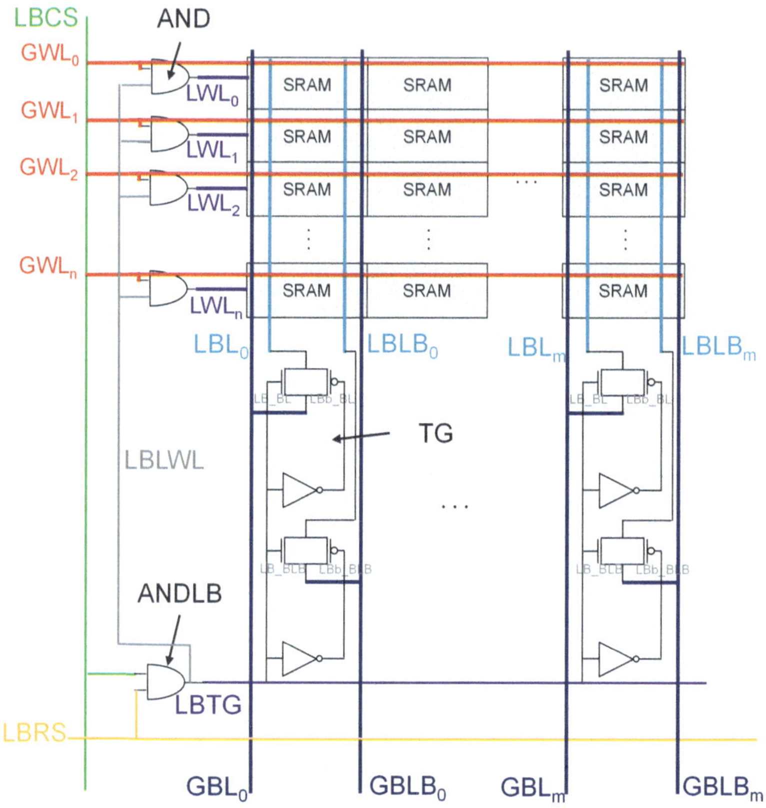

Fig. 9.2 Circuit diagram of a single LB in divided subarray design for (WL, BL) = (0~n, 0~m). (LB: local block; TG: transmission gate; AND: AND gate for LWL; ANDLB: AND gate for LB select; LBRS: local block row select; LBCS: local block column select; GBL/GBLB: global BL/BLB; LBL/LBLB: local BL/BLB; GWL: global WL; LWL: local WL; LBLWL: LWL select in LB)

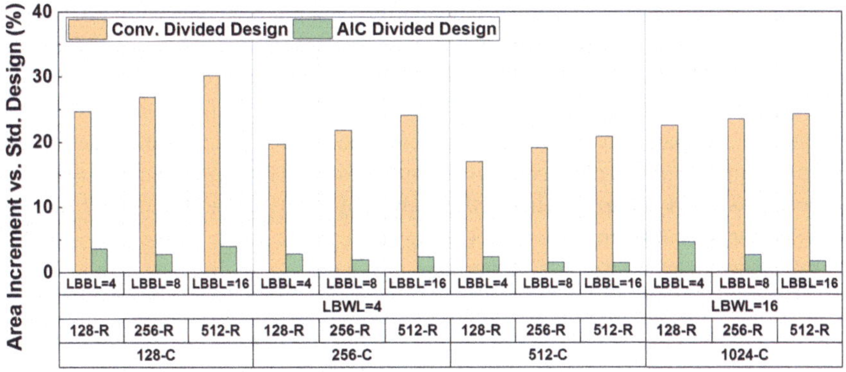

Fig. 9.3 Subarray area comparison of Conv. and AIC divided subarray designs with the optimized LB in different subarray sizes

work. By implementing selecting and gating circuits to partition the SRAM bitcell array into several LBs, the effective RC for BL and WL is decreased. As a result, the Conv. divided subarray design offers significant subarray-level improvements in PP and reduces the risk of write failures. However, this improvement comes at the cost of a noticeable subarray area overhead of up to 30% compared to the Std. subarray design (Fig. 9.3), which may negatively impact the macro-level PPA.

9.2 Proposed Solution – Active Interconnect (AIC) Divided Subarray Design

To address this subarray area penalty and maintain similar subarray-level PP enhancements, the AIC is proposed for implementation within the divided subarray design (Fig. 9.1c and Table 9.1) [19]. The AIC approach shares similarities with the concept of 3D sequential non-bonding stacking architecture, which involves a blanket active layer transfer onto prefabricated bottom devices [2, 7, 23, 27, 28, 30, 32, 33]. Note that it is sometimes referred to as monolithic 3D (M3D) integration [23] where multiple layers are fabricated sequentially, which is the opposite of parallel 3D technology based on wafer bonding [7]. 3D stacking technology has emerged as a promising alternative to extend Moore's law scaling [7], offering increased device density per chip area and reduced interconnect length. Among 3D techniques, through-silicon-via-based 3D (TSV-3D) stacking is a notable example of parallel 3D technology. However, it faces challenges such as low alignment precision and limited TSV dimension scaling, resulting in an increase in overall chip volume [7]. Consequently, 3D sequential (or M3D) technology has gained popularity due to its elimination of alignment issues and the incorporation of fine-grained monolithic inter-via (MIV) connections [7].

To prevent bottom-level device degradation, maintaining low thermal budget of the top-level device and layer transfer processing is crucial, typically below 500 °C [2, 23, 27, 28, 30, 32, 33]. The term "BEOL-compatible devices" specifically denotes the top-level device in this context. The key distinction of the AIC technology, compared to existing 3D sequential technology, lies in the insertion of single or multiple layers of active devices within the BEOL, which remains a part of the bottom SRAM subarray. These BEOL devices contribute to the overall SRAM subarray functionality instead of operating as separate logic or memory blocks. Therefore, we use the term AIC rather than 3D sequential to emphasize this unique conceptual framework.

The proposed AIC technology tackles the subarray area penalty issue in the Conv. divided subarray design by integrating BEOL-compatible devices. Several 1- or 2-dimensional (1- or 2D) options, such as CNT (cylinders) [6, 9, 14, 35], MoS_2 [37], and WS_2 [25]-based devices, could be potential candidates. Concerns regarding 2D-based FET [20] arise from higher contact resistance and lower mobility compared to Si FET, which may potentially lead to PP degradation in AIC divided subarray design. Among these devices, CNT FETs (CNFETs) have been projected to achieve large benefits compared to Si FETs [9] due to their superior electrostatic control and large injection velocity [6, 9, 14, 27, 35]. Moreover, CNFETs have already been integrated in the BEOL within multiple commercial silicon fabrication facilities at mature nodes (e.g., 90 nm node) [1, 29], leveraging low-temperature fabrication processes (i.e., <400 °C). For these reasons, this work considers gate-all-around CNFETs with HD CNTs [9] as BEOL devices. By incorporating the feature of creating an active region within the SRAM subarray's BEOL layer, the extra logic gates can be relocated from FEOL to BEOL. Hence, the AIC divided subarray design is predicted to reduce the subarray area overhead from up to 30% to less than 5% (Fig. 9.3) without any degradation in subarray-level PP compared to the Conv. counterparts.

Takeaways

- **Challenges in Conventional Divided Subarray Designs:** Significant subarray area increases due to additional logics, potentially degrading macro-level PPA results with increased interconnect overheads.
- **Motivated Design Strategy:** Find an alternative subarray design to mitigate area penalty while retaining WM and PP improvements of divided design.
- **Proposed Solution – Active Interconnect (AIC) Technique:** To mitigate the subarray area penalty, relocate additional logics from the FEOL to BEOL active layer using BEOL-compatible devices.

9.3 SRAM Subarray Design Overview

This study presents the HD 6T SRAM bitcell array and periphery designs based on A14 NS [11] and A14 CNT [9] technology for devices at FEOL and BEOL, respectively. A quick overview of the three different SRAM subarray designs (Fig. 9.1 and Table 9.1): Std., Conv. divided, and AIC divided subarray designs, are described in this section. Figures 9.4 and 9.5 illustrates the schematic views of the single LB layout in Conv. and AIC divided subarray designs, respectively, according to the circuit diagram shown in Fig. 9.2. Their physical parameters for each signal line and the corresponding estimated RC per cell (without device capacitance) extracted by the QuickCap tool [31] are listed in Tables 9.2 and 9.3.

9.3.1 Standard (Std.) Subarray Design

This subarray design consists of peripheral logic, including address/timing control, precharge circuits, row decoder, column multiplexer, sense amplifiers, and write

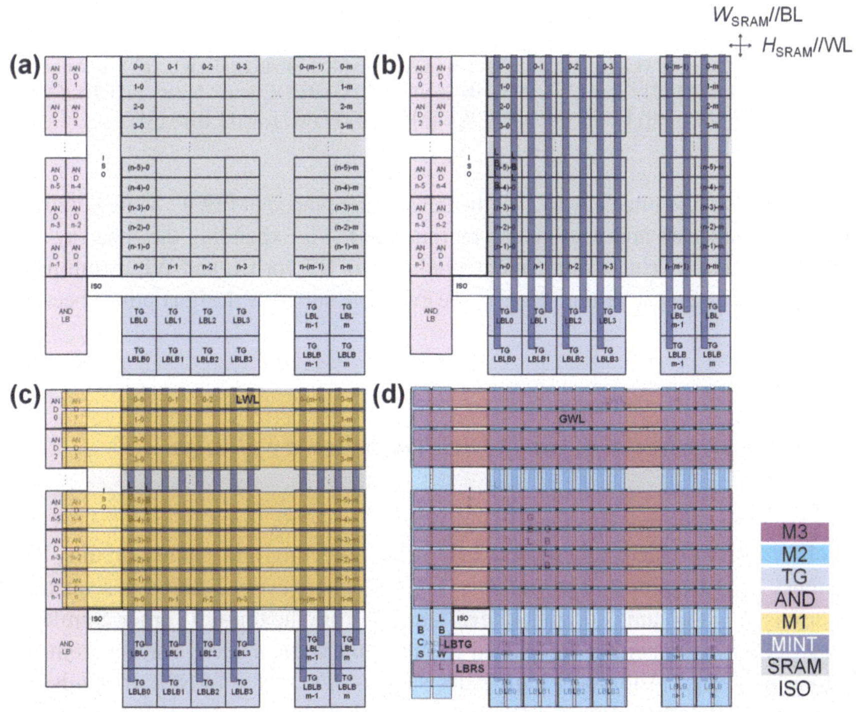

Fig. 9.4 Schematic views of a single LB layout in Conv. divided subarray design. (**a**) SRAM bitcell array, isolation gap between SRAM bitcell array and logic (ISO), AND and TG (FEOL active layer) located around LB. (**b**) LBL/LBLB (MINT) (**c**) LWL (M1). (**d**) GBL/GBLB (M2), GWL (M3), LBCS (M2), LBLWL (M2), LBTG (M3), and LBRS (M3)

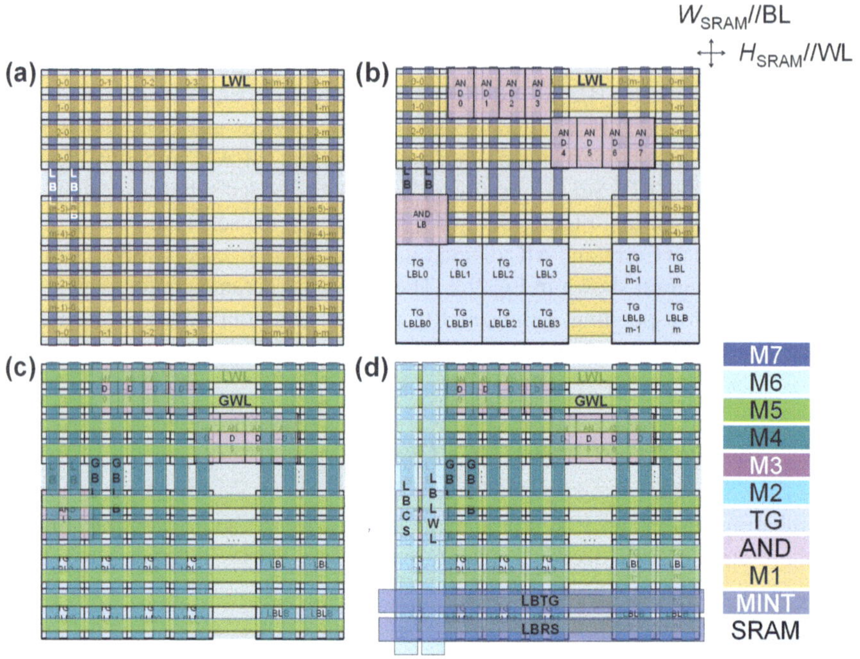

Fig. 9.5 Schematic views of a single LB layout in AIC divided subarray design. (**a**) SRAM bitcell array, LBL/LBLB (MINT), and LWL (M1). (**b**) AND and TG (BEOL active layer). (**c**) GBL/GBLB (M4) and GWL (M5). (**d**) LBCS (M6), LBLWL (M6), LBTG (M7), and LBRS (M7)

drivers, and SRAM bitcell array. All the devices are placed at FEOL. More details of the SRAM subarray architecture can be found in [18]. Expanding the subarray size in the Std. subarray causes dramatic degradation in RC for BL and WL, resulting in slower speed and write failure. Hence, the available subarray design options become limited.

9.3.2 Conventional (Conv.) Divided Subarray Design

To reduce the effective RC for BL and WL for avoiding write failure, the SRAM bitcell array is divided into several LBs using selecting and gating circuits. The reference layout example is presented in Fig. 9.4, where every 2 ANDs in sequences are arranged at the same horizontal level starting from the top-left corner. Similarly, every 2 TGs in sequences are placed at the same vertical level starting from the bottom-left corner. Additional logic gates such as AND gates (AND), transmission gates (TG), LB column select (LBCS) decoder, and LB row select (LBRS) decoder are added beyond the Std. subarray design (Fig. 9.2). Moreover, the isolation gaps (ISO) are required to be inserted in between the logic and SRAM bitcell

Table 9.2 Physical parameters and estimated RC per cell (without device capacitance) for each signal line (ruthenium is assumed) in Conv. divided subarray design. (T_{wire}: wire thickness; W_{wire}: wire width; L_{wire}: wire length)

	Metal Layer	T_{wire}(nm)	W_{wire}(nm)	L_{wire}(nm)	R (Ω)	C (aF)
BL	MINT	27	27	84	16	56.6
WL	M1	49	70	150	5.55	177
GBL	M2	27	63	84	6.54	37.3
GWL	M3	49	70	150	5.55	51.3
LBCS	M2	27	81	168	10.08	25.5
LBRS	M3	49	70	150	5.55	26.8
LBLWL	M2	27	81	168	10.08	25.5
LBTG	M3	49	70	150	5.55	26.8

Table 9.3 Physical parameters and estimated RC per cell (without device capacitance) for each signal line (ruthenium is assumed) and 3D vias in AIC divided subarray design. (T_{wire}: wire thickness; W_{wire}: wire width; L_{wire}: wire length)

	Metal Layer	T_{wire}(nm)	W_{wire}(nm)	L_{wire}(nm)	R (Ω)	C (aF)
BL	MINT	27	27	84	16	56.6
WL	M1	49	42	150	9.45	164
GBL	M4	28	47	84	3.68	22
GWL	M5	28	58	150	5.05	30
LBCS	M6	80	80	168	1.06	26.1
LBRS	M7	80	80	150	0.95	23.1
LBLWL	M6	80	80	168	1.06	26.1
LBTG	M7	80	80	150	0.95	23.1
$Via_{LBL\text{-}TG}$	V0–V2	16	16	144	88	20.8
$Via_{TG\text{-}GBL}$	V1–V3	16	16	152	84	19.6
$Via_{LWL\text{-}AND}$	V1–V2	16	16	75	46	22.7
$Via_{AND\text{-}GWL}$	V1–V4	16	16	208	101	26.8

array regions. Therefore, the Conv. divided subarray design leads to subarray area increment of up to 30% among the optimized subarrays compared to the Std. counterpart (Fig. 9.3). Although the subarray-level PP improves, the overall PP at macro level may be degraded due to the consequent larger inter-subarray interconnect and logics overhead [21, 22].

To achieve the division into several local WL (LWL) and BL (LBL), the ANDs and TGs are connected to global WL (GWL) and BL (GBL), respectively. AND is used to generate the LWL signal with the input of GWL and LBLWL, and TG is the switch between GBL/GBLB and LBL/LBLB. Each LB is activated by selecting the corresponding LBCS and LBRS signals. The number of LBs in WL (LBWL) and BL (LBBL) determines the number of LB in the WL and BL directions, respectively. The number of columns ($N_{col\text{-}LB}$) and rows ($N_{row\text{-}LB}$) in each LB is calculated by dividing the total number of column (N_{col}) and row Nrow) in subarray by LBWL and LBBL, respectively. In this initial work, full swing on GBL and LBL is assumed, but there is potential to extend it to full-swing GBL and low-swing LBL for lower-energy designs in the future.

9.3.3 Active Interconnect (AIC) Divided Subarray Design

To overcome the subarray area bottleneck observed in the Conv. divided subarray design, the AIC divided technique is introduced for SRAM subarray design. The AIC concept entails the introduction of an additional active layer at BEOL and the integration of BEOL-compatible and low-temperature growth (250∼450 °C) CNFET [7, 27, 28, 30]. Consequently, it brings significant area savings to less than 5% overhead (Fig. 9.3) at the FEOL by relocating a portion of the devices to the BEOL layer. It's important to emphasize that CNFET is exclusively utilized for the additional logic gates in the divided subarray design for specific reasons. Firstly, it aids in managing fabrication costs [14, 35], thereby preventing an undue escalation in comparison to the Std. subarray design. Secondly, while several process and design techniques (i.e., imperfection-immune paradigm [26]) have been introduced to mitigate CNFET variability, an assessment of the impact of variability in A14 is still lacking. For example, targets of chiral purification of CNTs (demonstrated in [8, 10, 11, 26] and necessary to achieve low variability in leakage and threshold voltage) are still unclear in A14. As a result, this study predominantly concentrates on integrating CNFET into specific circuit components while excluding SRAM bitcell array and other peripheral elements. The approach aims to successfully incorporate CNFET in SRAM AIC divided subarray designs, avoiding excessively high fabrication costs and operating under less strict variability constraints. Future research should explore hybrid design strategies integrating both Si and CNT technologies in the SRAM bitcell array and its peripheral design. This involves a more in-depth investigation, specifically focusing on mitigating CNFETs variability in A14. Alternatively, relocating supplementary logics from the frontside BEOL to the backside BEOL presents a viable design strategy. However, this approach may necessitate substrate thinning and pose challenges with mask alignment. Therefore, our study focuses exclusively on the AIC concept.

In this AIC divided approach, all the ANDs and TGs used for the divided techniques are moved to the BEOL active layer and placed on top of the SRAM bitcell array as depicted in the reference layout example (Fig. 9.5). The arrangement consists of placing every 4 ANDs in sequences at the same horizontal level starting from the top-left corner and every 2 TGs in sequences at the same vertical level starting from the bottom-left corner. Note that the LBCS and LBRS decoders are retained at the FEOL due to inadequate space available in the BEOL active layer (Fig. 9.1c and Table 9.1). Figure 9.6 and Table 9.3 show the schematic views and physical parameters of 3D vias assumed in AIC divided subarray design. The layers from bottom to top include FEOL active region (SRAM bitcell array and peripheral logics), 1st BEOL (VINT to V2), BEOL active region (AND and TG), and 2nd

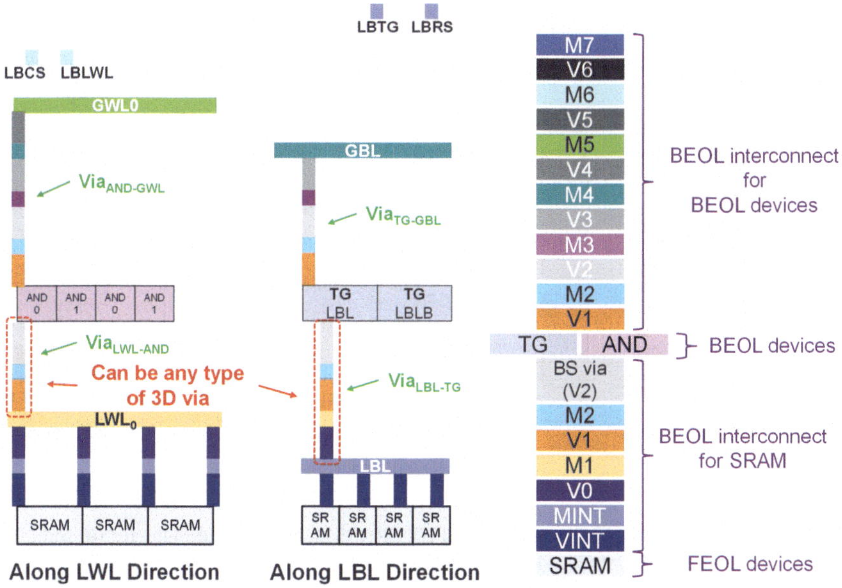

Fig. 9.6 Schematic views of 3D vias along LWL and LBL directions in AIC divided subarray design

BEOL (V1 to M7). Notably, the 2nd BEOL initiates from V1 layer due to the pitch requirement of customized subarray layout designs shown in Fig. 9.5. It does not require the densest pitch metal layer, such as MINT and M1, to complete the routing design, allowing it to save costs by eliminating the manufacturing of lower metal layers. The term "3D via" represents any type of vias (e.g., stacked via, buried power rail, nanoscale through-silicon via, etc.) that vertically connects BEOL devices to the 1st BEOL layer without additional subarray area cost. For a simple RC approximation of 3D vias, stacked via is assumed for the connection between LBL to TG, TG to GBL, LWL to AND, and AND to GWL without any shift in between (Fig. 9.6). Accurate RC values should be extracted by a full-custom subarray layout design in the future, which is not the primary focus of this study.

9.3.4 SRAM Bitcell Layout Designs

Figure 9.7 displays the three layout designs of HD A14 6T NSFET SRAM bitcell for Std. [17], Conv. divided, and AIC divided subarray designs, each with distinct signal routing requirements. In all designs, the ground voltage (V_{SS}) is assumed to be placed at the backside using buried power rail (BPR) technique [15, 17, 24]. The internal nodes (Q and QB), supply voltage (V_{DD}), and local signals (BL/LBL,

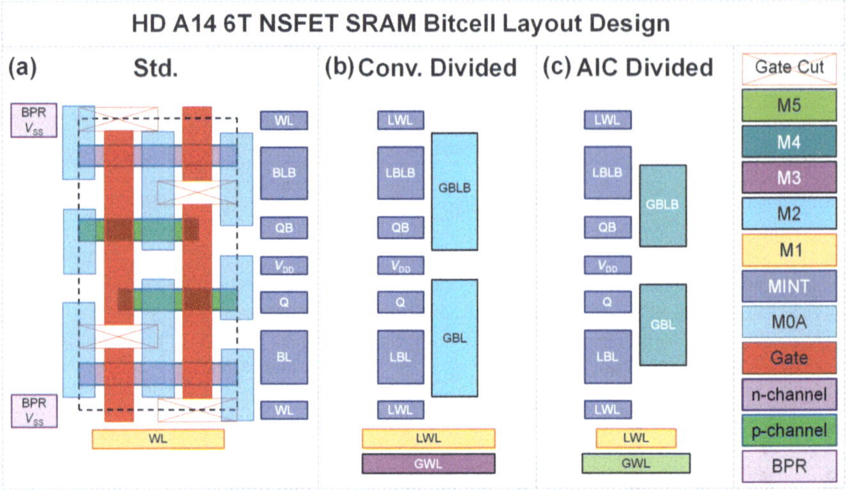

Fig. 9.7 Schematic views of HD A14 6T NSFET SRAM bitcell layout designs for (**a**) Std., (**b**) Conv. divided, and (**c**) AIC divided subarray designs

BLB/LBLB, and WL/LWL) are located at the MINT and M1 layers. The main difference between the Conv. and AIC divided subarray designs is the width of LWL, GBL, and GBLB due to distinct pitch limitations (Tables 9.2 and 9.3). In Conv. divided subarray design (Fig. 9.7b), the global signals (GWL and GBL) are connected from the AND and TG located outside the SRAM bitcell regions (Fig. 9.5). This arrangement leads to M1, M2, and M3 being exclusively occupied by LWL, GBL/GBLB, and GWL, respectively. However, in AIC divided subarray design (Fig. 9.7c), the AND and TG are placed right above the SRAM bitcell regions (Fig. 9.7). Therefore, the metal widths in AIC divided subarray design become narrow since extra spaces are required for the vertical connections from GBL/GBLB and GWL to the bitcell compared to the Conv. counterpart.

9.4 CNFET Optimization in AIC Divided Subarray Design

A14 CNFET has demonstrated the ability to deliver the on current that is three times as large as that of A14 Si NSFET at $V_{DD} = 0.45$ V and off current (I_{off}) = 2 nA [9]. This higher current capability of CNFET may help offset the additional parasitic RC introduced by the AIC divided subarray design. Therefore, CNFETs are introduced in AIC divided design for further PPA analysis. In this chapter, the model proposed in [9] was utilized, including both nanosheet (NS) and forksheet (FS) based layouts (NS and FS CNFET). The simulations were conducted utilizing Cadence Virtuoso [4] and Cadence Spectre [3].

9.4.1 *Model Calibration*

In this section, the simulations are performed at nominal $V_{DD} = 0.7$ V, temperature = 27 °C, and typical-typical (TT) corner. Except for the SRAM bitcell, where the I_{off} is assumed to be 0.02 nA, the assumption of $I_{off} = 0.2$ nA for all the peripheral logics remains identical in both Si and CNFETs models. The optimization of CNFETs is conducted considering their distinct applied voltage, on-current, subarray sizes, and application in comparison to the previous work [9]. To achieve the optimized read delay (RD) and write margin (WM) while maintaining identical I_{off} between CNFET and Si NSFET models, the number of stacked nanosheets (*NNS*) of CNFET (NNS_{CNT}) and CNT diameter (d_{CNT}) were systematically optimized. Table 9.4 lists the optimized results of physical parameters for the devices in AIC divided subarray design. After completing this model optimization, the subarray- and macro-level DTCOs are analysed in the following section. It is important to note that the LB sizes used here have been optimized for each size of subarray as discussed in Chap. 8.

9.4.1.1 Impact of NNS_{CNT}

Figures 9.8 and 9.9 show the impact of the NNS_{CNT} on RD and WM in the AIC divided subarray design with various subarray sizes. These simulations are taken under fixed NS width (W_{NS}) = 12 nm, $d_{CNT} = 0.8$ nm, CNT pitch (p_{CNT}) = 2 nm, and *NNS* of Si NSFET (NNS_{Si}) = 4. The read and write metrics are determined by the balance between the increased strength of the transistors and the larger device capacitance as *NNS* increases [35]. As a result, the optimized *NNS* is found to be 3 and 2 for NS and FS CNFET, respectively, as values tend to saturate beyond these *NNS* (Table 9.4).

Table 9.4 Optimized physical parameters for A14 Si NSFET, A14 CNFET, and N22 FDSOI devices in AIC divided subarray design at $V_{DD} = 0.7$ V. (NS: nanosheet; FS: forksheet; CPP: contacted poly pitch; L_g: gate length; W_{ch}: channel width; *NNS*: number of stacked nanosheets; d_{CNT}: CNT diameter; p_{CNT}: CNT pitch; I_{on}: on current, I_{off}: off current)

Parameters	A14 Si NSFET		A14 CNFET		N22 FDSOI
	SRAM	Logic	NSFET	FSFET	Planar
CPP (nm)	42				104
L_g (nm)	15	14	14	14	20
W_{ch} (nm)	11	12	12	21	170
NNS	4	4	3	2	–
d_{CNT} (nm)	–	–	0.85	0.85	–
p_{CNT} (nm)	–	–	2	2	–
V_{DD} (V)	0.7				0.8
I_{on} (μA/FET)	69	91	164	199	113
I_{off} (nA/FET)	0.02	0.2	0.2	0.2	0.2
Model Ref.	[11]		[9]		[5]

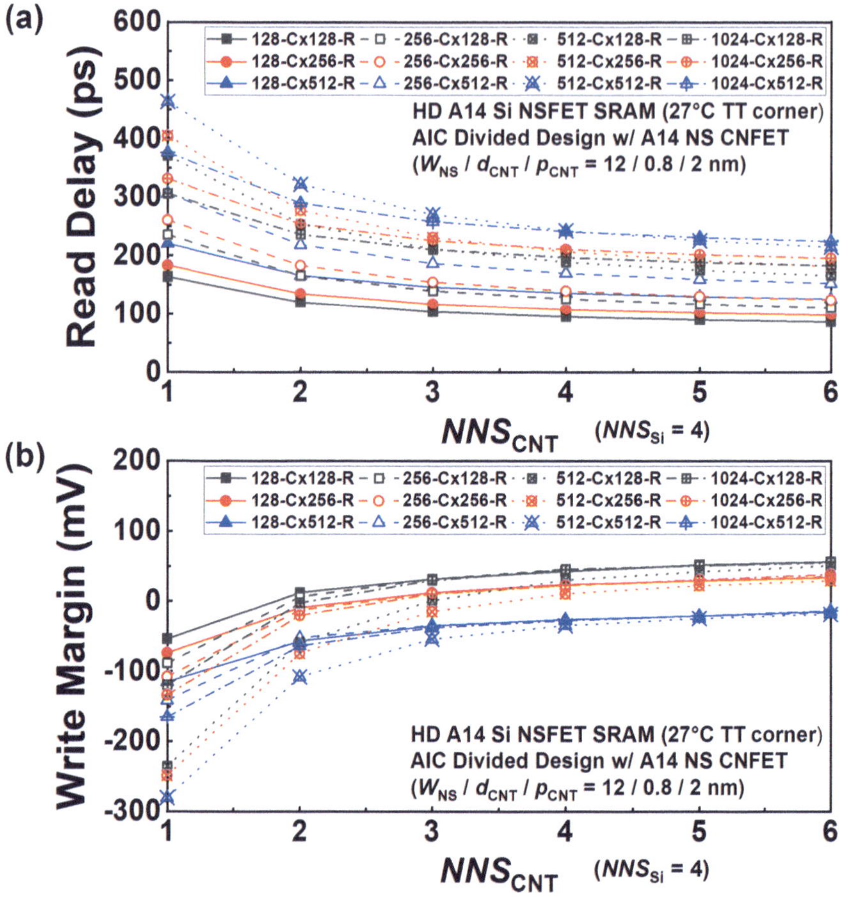

Fig. 9.8 (**a**) RD and (**b**) WM with varying NNS_{CNT} of A14 NS CNFET ($W_{NS}/d_{CNT}/p_{CNT}$ = 12/0.8/2 nm) for HD A14 Si NSFET (W_{NS} = 12 nm; NNS_{Si} = 4) AIC divided subarray design

9.4.1.2 Impact of d_{CNT}

Figures 9.10 and 9.11 demonstrate the impact of d_{CNT} on RD and WM in the AIC divided subarray design with various subarray sizes. These simulations are performed with fixed values of W_{NS} = 12 nm, p_{CNT} = 2 nm, and NNS_{Si} = 4. The sweeping range for d_{CNT} is from 0.7 to 0.89 nm, ensuring the same leakage assumption (0.2 nA) [13]. As d_{CNT} increases, both the transistor's strength and device capacitance increase. However, saturation or degradation occurs in RD and WM when d_{CNT} exceeds 0.85 nm due to the increased device capacitance. Thus, an optimized value of d_{CNT} at 0.85 nm is considered suitable for both NS and FS CNFETs (Table 9.4).

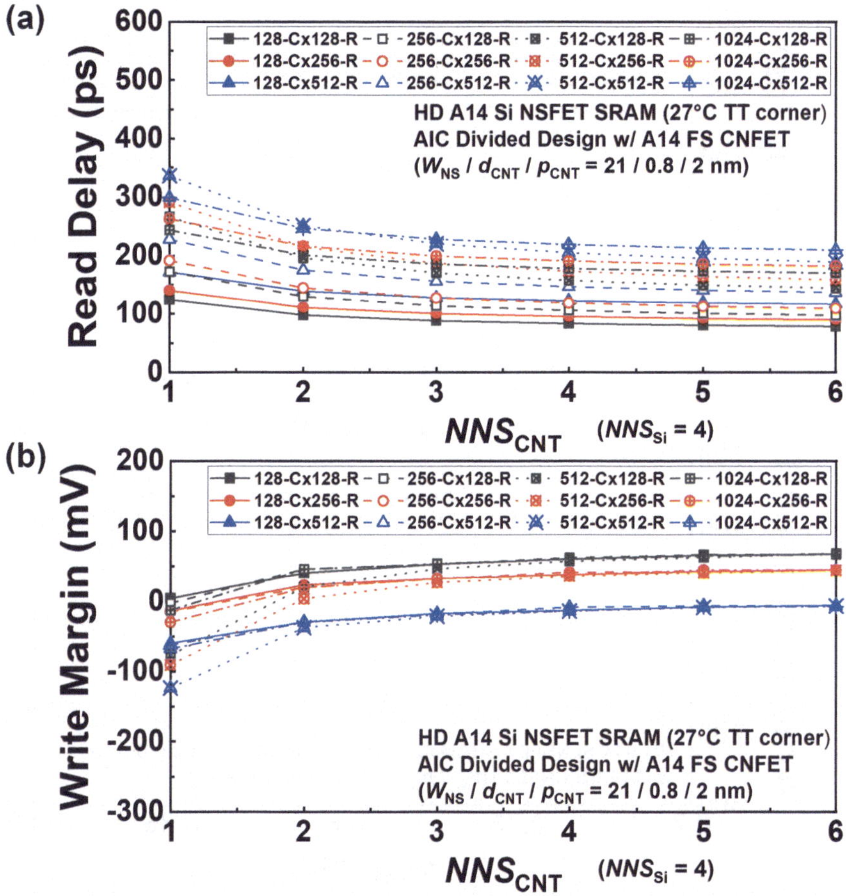

Fig. 9.9 (**a**) RD and (**b**) WM with varying NNS_{CNT} of A14 FS CNFET ($W_{NS}/d_{CNT}/p_{CNT}$ = 21/0.8/2 nm) for HD A14 Si NSFET (W_{NS} = 12 nm; NNS_{Si} = 4) AIC divided subarray design

9.4.2 Variability

Several key factors such as the variation in d_{CNT}, chirality (yield of metallic or semiconducting CNTs, density, bandgap), and doping, play a vital role in shaping the characteristics of CNFET [35]. These sources of variation add to the traditional variability encountered in silicon CMOS, which may affect the overall PP results. While existing techniques for mitigating these variations were well-demonstrated [6, 10, 26, 36], the process for CNT deposition and variation management stands out. However, this discussion goes beyond the primary focus of this work. It is imperative to extend the investigation of the variation's impact on CNFETs associated with the SRAM AIC divided subarray design in future studies.

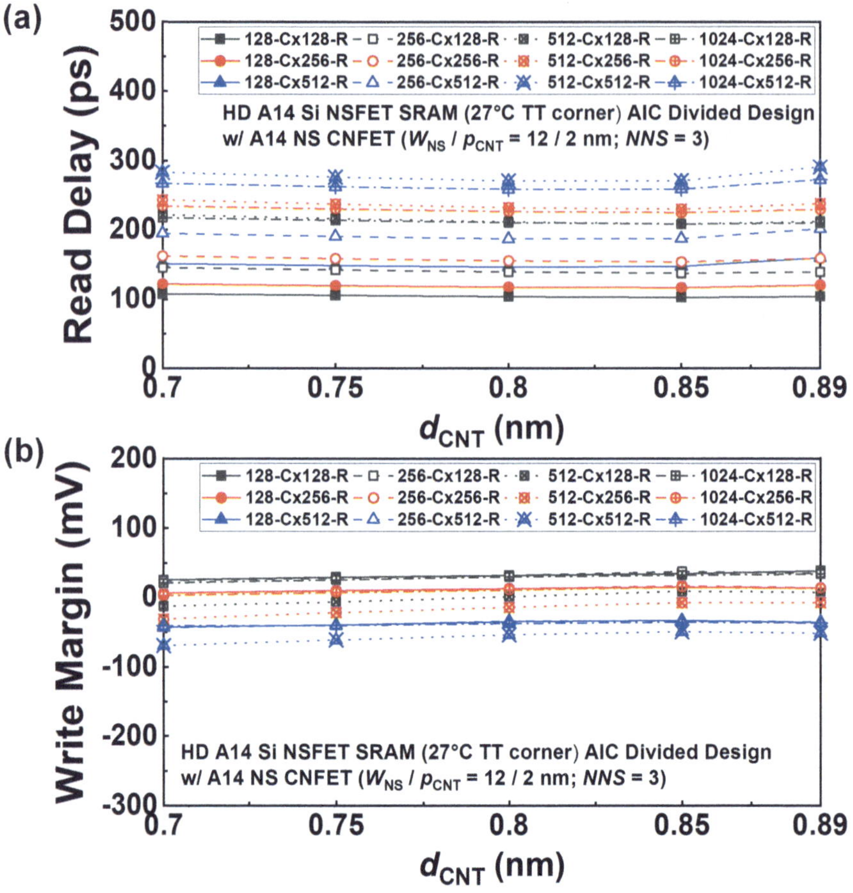

Fig. 9.10 (**a**) RD and (**b**) WM with varying d_{CNT} of A14 NS CNFET (W_{NS}/p_{CNT} = 12/2 nm; NNS_{CNT} = 3) for HD A14 Si NSFET (W_{NS} = 12 nm; NNS_{Si} = 4) AIC divided subarray design

9.5 SRAM Subarray- and Macro-level PPA Benchmark

The optimized design types, along with their corresponding LB sizes for each subarray in the Conv. divided subarray design, have undergone a thorough investigation of write-ability. The outcomes are succinctly presented in Table 8.4 within Chap. 8. This chapter consistently utilizes these optimized configurations for the upcoming subarray- and macro-level PPA analysis. Figures 9.12, 9.13, 9.14, and 9.15 present the comparison for various subarray row (-R) and column (-C) with optimized LB sizes between Std., Conv. divided, and AIC divided subarray designs at both subarray and macro levels. The SRAM subarray design featuring WFR is delineated within the red frame in this section.

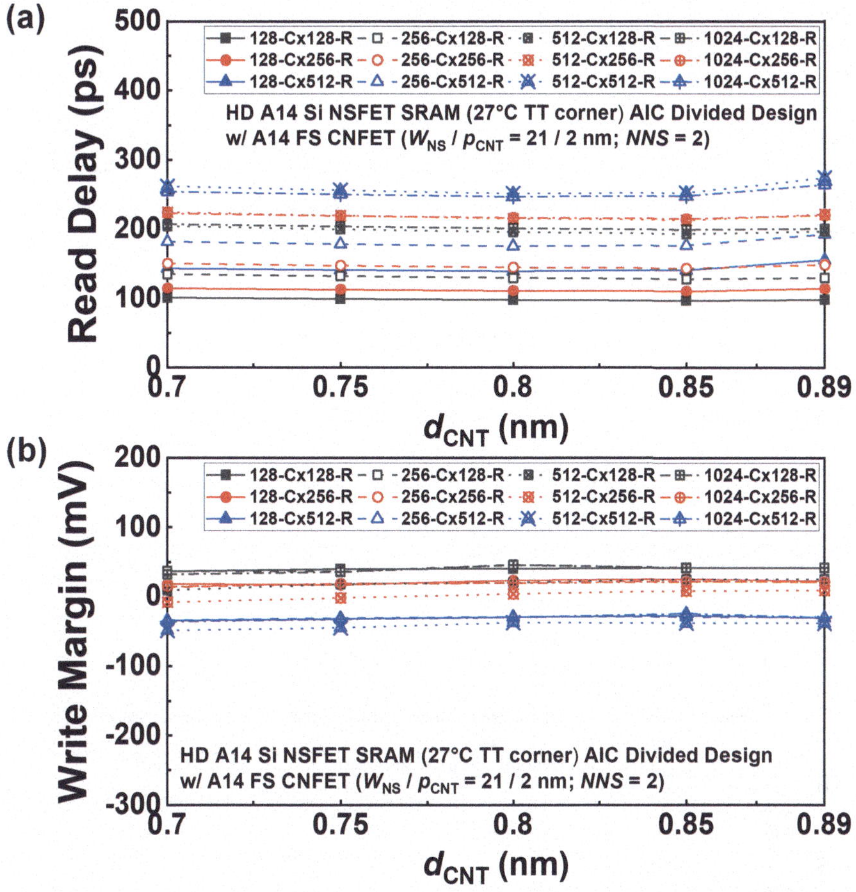

Fig. 9.11 (**a**) RD and (**b**) WM with varying d_{CNT} of A14 FS CNFET (W_{NS}/p_{CNT} = 21/2 nm; NNS_{CNT} = 2) for HD A14 Si NSFET (W_{NS} = 12 nm; NNS_{Si} = 4) AIC divided subarray design

9.5.1 Benchmark at Subarray Level

For the subarray sizes of 128-C/256-C×128-R (Fig. 9.12), the Std. subarray design demonstrates similar RD, WM, and write delay (WD) as the Conv. and AIC divided counterparts without compromising the subarray area. As a result, the Std. subarray design is considered the optimized design type for 128-C/256-C×128-R. Notably, the comparable margin and performance observed in smaller subarrays between Si and CNT models can be attributed to the compensatory effect of the delay overhead from additional logics' RC even though CNFET has relatively higher Ion than Si NSFET in A14. However, for the remaining N_{row} and N_{col} configurations, the Conv. and AIC divided subarray designs outperform the Std. counterpart. Hence, although the Std. subarray designs for 256-C×128/256-R and 512-C×128-R configurations

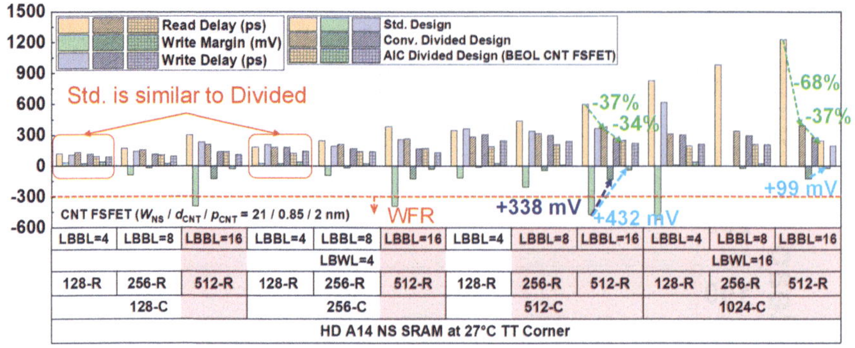

Fig. 9.12 SRAM subarray-level RD, WM, and WD benchmark in Std., Conv. divided, and AIC divided subarray designs with optimized LBWL and LBBL for each subarray size

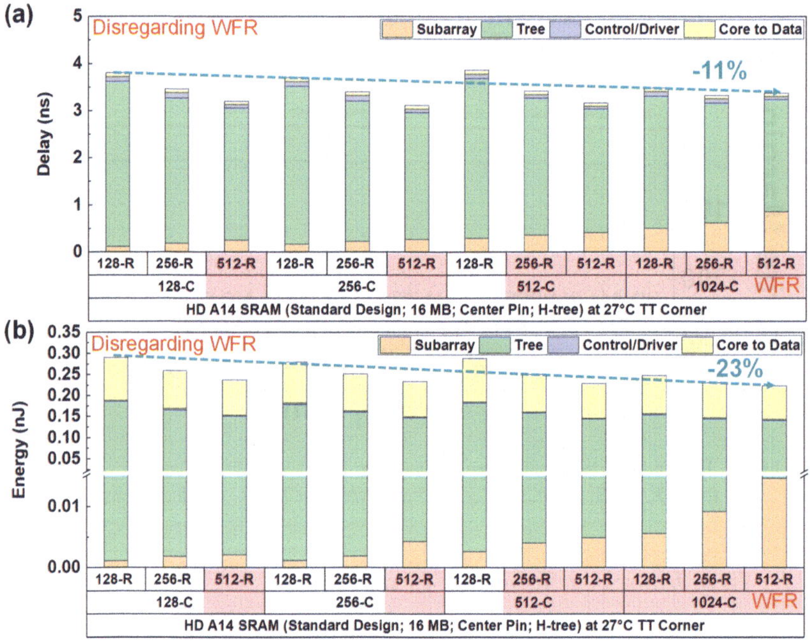

Fig. 9.13 SRAM data macro-level (**a**) delay and (**b**) energy breakdown in the Std. subarray designs (disregarding WFR) for each subarray size with data capacity of 16 MB, centre pin access, and H-tree technology

do not suffer from WFR, they are not regarded as optimized designs. Particularly, in the case of 512-R, the Std. subarray design encounters write failures necessitating the divided techniques to ensure successful write operations (Fig. 8.14 in Chap. 8). By utilizing Conv. divided technique into subarray design, the WM and RD improve by 338 mV and 37%/68%, respectively, for 512-C/1024-C×512-R. Among all the designs, the AIC divided subarray design stands out as it offers the best performance

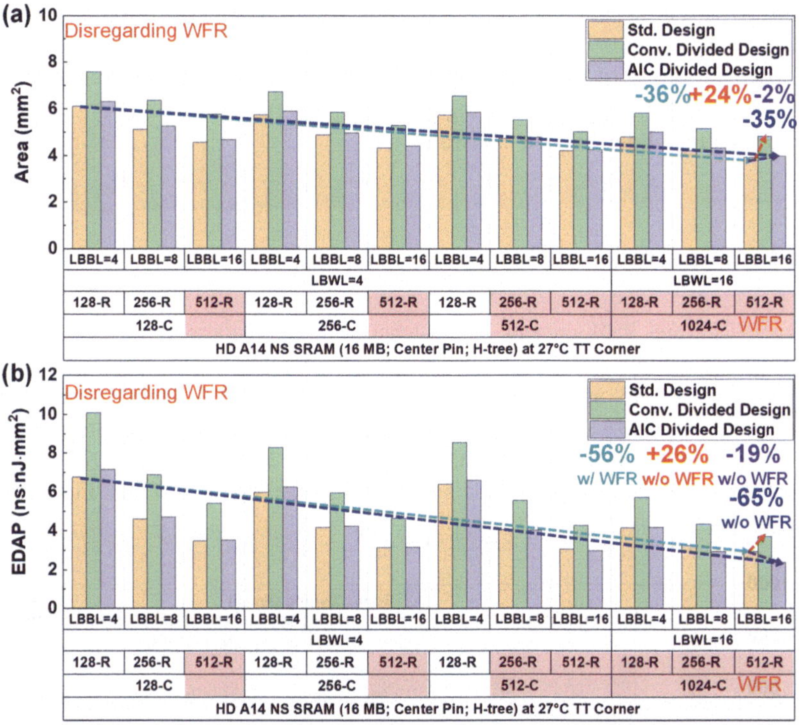

Fig. 9.14 SRAM data macro-level (a) area and (b) EDAP benchmark in the Std. (disregarding WFR), Conv. divided, and AIC divided subarray designs with optimized LBWL and LBBL for each subarray size with data capacity of 16 MB, centre pin access, and H-tree technology

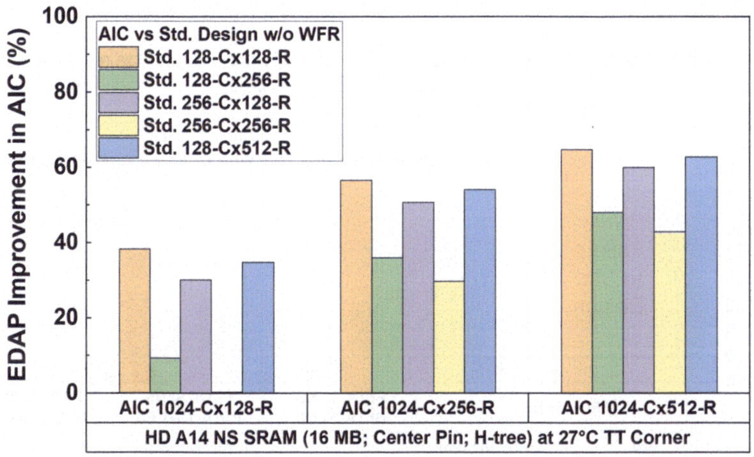

Fig. 9.15 EDAP improvements in AIC divided subarray design for a subarray size of 1024-C×128/256/512-R compared to Std. subarray design (regarding WFR) for subarray size of 128/256-C×128-R, 128/256-C×256-R, and 128-C×512-R with data capacity of 16 MB, centre pin access, and H-tree technology

and write margin with only a slight increase in area of less than 5% compared to the Std. counterpart (Fig. 9.3). AIC divided subarray design even provides additional enhancements in WM (99 mV) and RD (37%) compared to Conv. counterparts due to faster CNFET. This promising result prompts further PPA exploration at the macro level to validate its potential value.

9.5.2 Benchmark at Data Array Macro Level

Figures 9.13 and 9.14 present macro-level simulation data based on a 16 MB data capacity, centre pin access, and H-tree interconnect technology [21, 22]. As shown in Figs. 9.13 and 9.14, a notable 11%, 23%, 36%, and 56% of delay, energy, area, and EDAP gain, respectively, can already be achieved in Std. subarray design by simply transitioning from small (128-C×128-R) to large (1024-C×512-R) subarray size due to less H-tree contribution (disregarding WFR). Figure 9.16

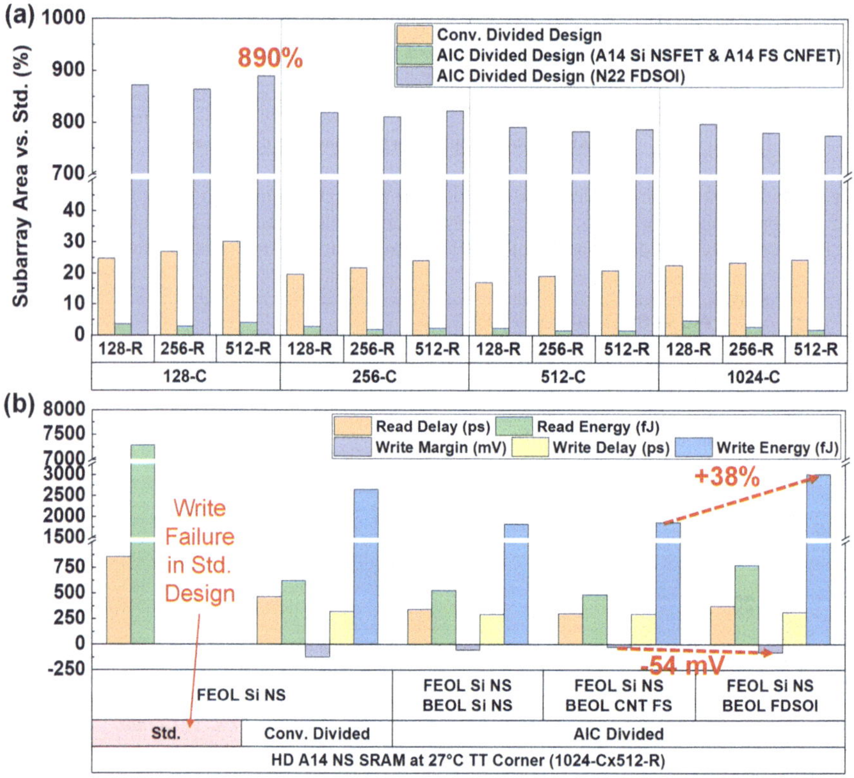

Fig. 9.16 SRAM subarray-level (**a**) area, (**b**) margin, performance, and power comparison in Std., Conv. divided, and AIC divided subarray designs assuming A14 Si, A14 CNT, and N22 FDSOI models for BEOL logics

clearly shows that the inter-subarray interconnect has a decisive influence on the overall macro-level PPA. To optimize this aspect, choosing a bigger-size subarray is the preferred approach to reduce the interconnect overhead. Further macro-level PPA improvement is still achievable even if they involve a larger subarray-level PPA overhead, which is relatively less impactful than inter-subarray interconnect. It is worth noting that increasing Nrow is more effective for PP improvement compared to Ncol since the longest wire is on horizontal direction [16, 21, 22].

While employing a bigger-size subarray with Std. subarray design may suffer from the WFR, this problem can be effectively addressed through the implementation of divided design. In the previous subsection, the Conv. divided subarray design substantially enhances subarray-level performance and WM compared to the Std. counterpart (Fig. 9.12). However, this enhancement comes at the cost of unignorable area penalty at the subarray level that could potentially lead to larger inter-subarray interconnects and logic gates overhead at macro level. Consequently, this 24% subarray area overhead introduced by Conv. divided subarray design causes 26% macro-level EDAP degradation compared to Std. counterpart (Fig. 9.14). To address these challenges, AIC divided subarray design has been proposed to mitigate the macro-level area overhead to less than 3% (Fig. 9.14a). AIC divided subarray design successfully provides up to 65% macro-level EDAP improvement compared to a realistic Std. counterpart that does not have WFR (Fig. 9.15). Figure 9.15) indicates the preferred selection between AIC divided subarrays, either with a subarray size of 1024-C×256-R or 1024-C×512-R. This decision stems from the negligible EDAP gain (0.2%) in a 1024-C×128-R subarray compared to the Std. 256-C×256-R subarray. Even assuming a non-failing write in Std. subarray design, AIC divided subarray design still achieves a notable 19% macro-level enhancement with a 1024-C×512-R subarray size (Fig. 9.14b). Notably, the increased macro area in the AIC divided subarray design associated with CNFET fabrication may cause a higher overall die cost when compared to the Std. counterpart and the traditional Si FET process. A more comprehensive cost analysis is advisable for future research, which falls outside the central scope of this work.

9.5.3 Comparison of Si, FDSOI, and CNT Models

In this section, the examination focuses on the comparison of utilizing A14 Si NSFET [11], 22-nm (N22) fully depleted silicon-on-insulator (FDSOI) [5], and A14 CNFET [9] models for realistic SRAM AIC divided subarray design to highlight the benefits of employing CNFET devices (Figs. 9.16 and 9.17). Over the past few years, several experimental studies on BEOL-compatible and low-temperature-fabricated CNFETs for 3D integration have been reported [7, 27, 28, 30], establishing them as realistic candidates for further 3D integrated-circuit (IC) development. Additionally, investigations into low-temperature-fabricated FDSOI as top-level transistors (below 500 °C) for 3D stacking technology have been conducted including electrical characteristics and reliability measurements [2, 32, 33]. However, it's

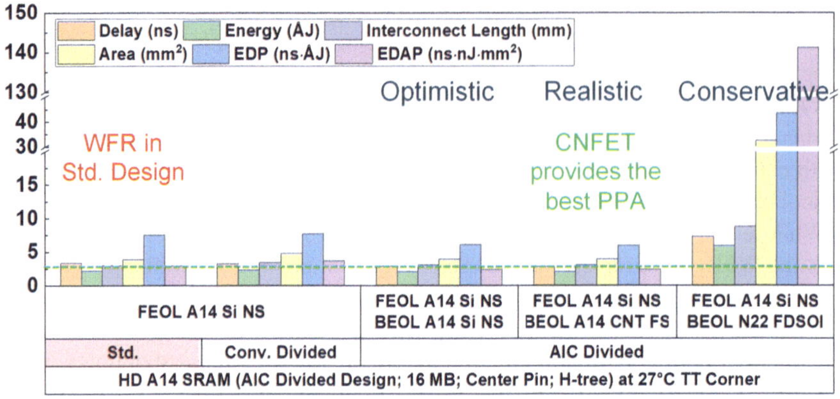

Fig. 9.17 SRAM data macro-level benchmark comparison in Std., Conv. divided, and AIC divided subarray designs assuming A14 Si, N22 FDSOI, and A14 CNT models for BEOL logics. (EDP: energy-delay product)

important to note that the FDSOI device is a planar-type transistor, with mass production limited to 22 nm by GlobalFoundries until now [5]. Apparently, the device dimensions of N22 FDSOI transistors could not align with A14 design rules in this study. Therefore, the FEOL-compatible A14 Si NSFET model (identical model as SRAM bitcell in this work) [11], regardless of potential degradation from the low-thermal-budget process, is also considered as an optimistic upper bound for BEOL logics in our realistic AIC technology comparison.

Figures 9.16 and 9.17 show the subarray- and data array macro-level PPA benchmark, respectively, across five distinct designs: Std., Conv. divided, AIC divided with A14 Si NSFET, AIC divided with A14 FS CNFET, and AIC divided with N22 FDSOI. The parameters for the N22 FDSOI device, including V_{DD}, gate length, width, contacted poly pitch, and metal pitch, are set at 0.8 V, 20 nm, 170 nm, 104 nm, and 80 nm, respectively, based on low V_T (LVT) transistors from GlobalFoundries' 22FDX^{TM} technology [5]. However, the implementation of the AIC divided subarray design with the N22 FDSOI model results in a considerable subarray area overhead, reaching up to 890% compared to the Std. counterpart (Fig. 9.16a). This amplified area incurs substantial macro-level PPA degradation due to longer inter-subarray interconnects (Fig. 9.17). Moreover, the N22 FDSOI model shows a 54 mV decrease in WM and a 38% increase in write energy at the subarray level (1024-C×512-R) in comparison to the A14 FS CNFET counterparts, attributed to higher V_{DD} and inferior device performance (Fig. 9.16b). Consequently, while the N22 FDSOI represents an existing solution for 3D integration, it proves unsuitable and serves as pessimistic lower bound for the AIC divided subarray design in the ultra-scaled A14 technology. This result indicates that BEOL-compatible logic still has strict requirements in terms of logic density.

Conversely, the AIC divided subarray design with A14 Si NSFET model demonstrates comparable (less than 2% difference) performance/power and identical area results to the A14 FS CNFET model at both subarray and macro levels. This validates the inherent advantages of employing CNFET as BEOL-compatible devices in the AIC divided subarray design, surpassing the optimistic yet unfeasible upper bound showcased by the A14 Si NSFET counterpart.

Takeaways

- Although conventional divided subarray design improves subarray-level WM and PP, it introduces significant macro-level EDAP degradation due to the area overhead of the divided circuits.
- AIC divided subarray design can improve macro-level EDAP by up to 65% by enlarging the subarray size to 1024-C×512-R, compared to a standard design of 128-C×128-R without write failure issues.
- Among all BEOL-compatible devices, the A14 CNFET is the most promising candidate for implementing AIC divided subarray design in deeply scaled technology nodes.

9.6 Conclusions

This chapter presents a comprehensive DTCO analysis, establishing the potential of the AIC divided subarray design integrated with CNFET as BEOL-compatible devices for future PPA improvements of SRAM data array design. In interconnect-dominated nodes, macro-level PPA is heavily influenced by the inter-subarray interconnect, a factor intricately linked to the subarray size. By scaling up the subarray size from 128-C×128-R to 1024-C×512-R, remarkable 11%, 23%, 36%, and 56% of delay, energy, area, and EDAP improvements, respectively, are achieved, albeit with the trade-off of a write failure risk in the Std. subarray design. To address this risk, the Conv. divided subarray design has been explored in Chap. 8, offering an improved WM of 338 mV. However, this enhancement comes at the cost of up to 30% additional subarray area overhead. To circumvent this subarray area penalty and maintain it below 3%, the AIC divided technique emerges as a promising solution for SRAM subarray design, as discussed in this chapter. The creation of an BEOL active layer above the SRAM bitcell array facilitates the relocation of additional logic gates utilized for divided techniques from FEOL to BEOL, thereby reducing FEOL area occupancy. The concept of AIC is enabled through the consideration of A14 CNFET due to its BEOL-compatible characteristics, scalability, and superior device performance compared to other candidates. AIC divided subarray design with 1024-C×512-R not only enhances the WM by 99 mV compared to the Conv. counterpart at the same subarray size but also yields up to a 65% improvement in macro-level EDAP compared to the Std. subarray design with 128-C×128-R.

To further contextualize these results, A14 Si NS and N22 FDSOI models are introduced as optimistic and realistic/pessimistic references, respectively, in comparison to A14 CNT models. AIC divided subarray with N22 FDSOI devices exhibits a significant macro area overhead of up to 890% compared to the Std. counterpart due to the substantial technology node gap. Additionally, it leads to a degradation of 54 mV in WM and a 38% decrease in subarray-level write energy compared to A14 FS CNT counterparts for a subarray size of 1024-C×512-R due to the higher nominal VDD and inferior device performance. Thus, N22 FDSOI is found to be unsuitable and serves as a pessimistic lower bound for the AIC divided subarray design within the ultra-scaled A14 technology, despite it is already an existing solution for 3D integration. On the contrary, AIC divided subarray design utilizing A14 Si NS model exhibits performance and power characteristics comparable to the A14 FS CNT counterpart, with a difference of less than 2%. This equivalence is observed in both subarray and macro levels while maintaining identical area results. Such validation underscores the inherent advantages of employing CNFET as BEOL-compatible devices in the AIC divided subarray design, surpassing the optimistic yet unfeasible upper bound showcased by the A14 Si NSFET counterpart. The provided comparison highlights the potential value of implementing AIC divided subarray design alongside A14 CNT devices in advancing the PPA of future SRAM cache design.

References

1. Bishop, M., et al. (2020). Fabrication of carbon nanotube field-effect transistors in commercial silicon manufacturing facilities. *Nature Electronics, 3*, 1–10. https://doi.org/10.1038/s41928-020-0419-7
2. Brunet, L., et al. (2016). First demonstration of a CMOS over CMOS 3D VLSI CoolCube™integration on 300 mm wafers. In *2016 IEEE Symposium on VLSI Technology* (pp. 1–2). https://doi.org/10.1109/VLSIT.2016.7573428
3. Cadence®. (2021). Spectre®Circuit Simulator https://www.cadence.com/en_US/home/tools/custom-ic-analog-rf-design/circuit-simulation.html
4. Cadence®(2021). Virtuoso®Layout Suite. https://www.cadence.com/en_US/home/tools/custom-ic-analog-rf-design/layout-design/virtuoso-layout-suite.html
5. Carter, R., et al. (2021). 22 nm FDSOI technology for emerging mobile, Internet-of- Things, and RF applications (pp. 2.2.1–2.2.4). https://doi.org/10.1109/IEDM.2016.7838029
6. Chen, R., et al. (2022). Carbon nanotube SRAM in 5-nm technology node design, optimization, and performance evaluation—part I: CNFET transistor optimization. *IEEE Transactions on Very Large Scale Integration (VLSI) Systems, 30*(4), 432–439. https://doi.org/10.1109/TVLSI.2022.3146125
7. Datta, S., et al. (2019). Back-end-of-line compatible transistors for monolithic 3-D integration. *IEEE Micro, 39*(6), 8–15. https://doi.org/10.1109/MM.2019.2942978
8. Fagan, J. (2019). Aqueous two-polymer phase extraction of single-wall carbon nanotubes using surfactants. *Nanoscale Advances*. https://doi.org/10.1039/C9NA00280D
9. Gilardi, C., et al. (2021). Extended scale length theory targeting low-dimensional FETs for carbon nanotube FET digital logic design-technology cooptimization. In *2021 IEEE International Electron Devices Meeting (IEDM)* (pp. 27.3.1–27.3.4). https://doi.org/10.1109/IEDM19574.2021.9720672

10. Hills, G., et al. (2015). Rapid co-optimization of processing and circuit design to overcome carbon nanotube variations. *IEEE Transactions on Computer-Aided Design of Integrated Circuits and Systems, 34*, 1. https://doi.org/10.1109/TCAD.2015.2415492
11. Jang, D., et al. (2017). Device exploration of nanosheet transistors for sub-7-nm technology node. *IEEE Transactions on Electron Devices, 64*(6), 2707–2713. https://doi.org/10.1109/TED.2017.2695455
12. Karandikar, A., & Parhi, K. K. (1998). Low power SRAM design using hierarchical divided bit-line approach. In *Proceedings International Conference on Computer Design. VLSI in Computers and Processors (Cat. No.98CB36273)* (pp. 82–88). https://doi.org/10.1109/ICCD.1998.727027
13. Lin, Q., et al. (2023). Band-to-band tunneling leakage current characterization and projection in carbon nanotube transistors. In *ACS Nano, 17*. https://doi.org/10.1021/acsnano.3c04346
14. Liu, F., et al. (2021). High-yield and low-cost separation of high-purity semiconducting single-walled carbon nanotubes with closed-loop recycling of raw materials and solvents. *Nano Research, 14*, 4281–4287. https://doi.org/10.1007/s12274-021-3671-x
15. Liu, H.-H., et al. (2023). CFET SRAM DTCO, interconnect guideline, and benchmark for CMOS scaling. *IEEE Transactions on Electron Devices, 70*(3), 883–890. https://doi.org/10.1109/TED.2023.3235701
16. Liu, H.-H., et al. (2023). CFET SRAM with double-sided interconnect design and DTCO benchmark. *IEEE Transactions on Electron Devices* 1–8. https://doi.org/10.1109/TED.2023.3305322
17. Liu, H.-H., et al. (2023). DTCO of sequential and monolithic CFET SRAM. In R.-H. Kim & N. V. Lafferty (Eds.), *DTCO and computational patterning II* (Vol. 12495). International society for optics and photonics. SPIE. https://doi.org/10.1117/12.2657524
18. Liu, H.-H., et al. (2023). Extended methodology to determine SRAM write margin in resistance-dominated technology node. *IEEE Transactions on Electron Devices, 69*(6), 3113–3117. https://doi.org/10.1109/TED.2022.3165738
19. Liu, H.-H., et al. (2024). Future design direction for SRAM data array: Hierarchical subarray with active interconnect. *IEEE Transactions on Circuits and Systems I: Regular Papers* 1–12. https://doi.org/10.1109/TCSI.2024.3410518
20. Mir, S. H., Yadav, V. K., & Singh, J. K. (2020). Recent advances in the carrier mobility of two-dimensional materials: A theoretical perspective. *ACS Omega, 5*(24), 14203–14211. https://doi.org/10.1021/acsomega.0c01676
21. Pei, Z., et al. (2023). Emerging interconnect exploration for SRAM application using non-conventional H-tree and center-pin access. In *2023 24th International Symposium on Quality Electronic Design (ISQED)* (pp. 1–1). https://doi.org/10.1109/ISQED57927.2023.10129316
22. Pei, Z., et al. (2023). Graphene-based interconnect exploration for large SRAM caches for ultrascaled technology nodes. *IEEE Transactions on Electron Devices, 70*(1), 230–238. https://doi.org/10.1109/TED.2022.3225512
23. Sabry Aly, M. M., et al. (2019). The N3XT approach to energy-efficient abundant-data computing. *Proceedings of the IEEE, 107*(1), 19–48. https://doi.org/10.1109/JPROC.2018.2882603
24. Salahuddin, S., et al. (2020). Buried power SRAM DTCO and system-level benchmarking in N3. In *2020 IEEE Symposium on VLSI Technology* (pp. 1–2). https://doi.org/10.1109/VLSITechnology18217.2020.9265076
25. Schram, T., et al. (2021). High yield and process uniformity for 300 mm integrated WS2 FETs. In *2021 Symposium on VLSI Technology* (pp. 1–2)
26. Shulaker, M. M., et al. (2015). Efficient metallic carbon nanotube removal for highly-scaled technologies. In *2015 IEEE International Electron Devices Meeting (IEDM)* (pp. 32.4.1–32.4.4). https://doi.org/10.1109/IEDM.2015.7409815
27. Shulaker, M. M., et al. (2014). Monolithic 3D integration of logic and memory: Carbon nanotube FETs, resistive RAM, and silicon FETs. In *2014 IEEE International Electron Devices Meeting* (pp. 27.4.1–27.4.4). https://doi.org/10.1109/IEDM.2014.7047120

28. Shulaker, M. M., et al. (2017). Three-dimensional integration of nanotechnologies for computing and data storage on a single chip. *Nature, 547*, 74–78. https://doi.org/10.1038/nature22994
29. Srimani, T., et al. (2023). Foundry monolithic 3D BEOL transistor + memory stack: Iso-performance and iso-footprint BEOL carbon nanotube FET+RRAM vs. FEOL silicon FET+RRAM. In *2023 IEEE Symposium on VLSI Technology and Circuits (VLSI Technology and Circuits)* (pp. 1–2). https://doi.org/10.23919/VLSITechnologyandCir57934.2023.10185414
30. Srimani, T., et al. (2020). Heterogeneous integration of BEOL logic and memory in a commercial foundry: Multi-tier complementary carbon nanotube logic and resistive RAM at a 130 nm node. In *2020 IEEE Symposium on VLSI Technology* (pp. 1–2). https://doi.org/10.1109/VLSITechnology18217.2020.9265083
31. Synopsys. (2020). QuickCap R-2020.09-SP4. https://www.synopsys.com/implementation-and-signoff/signoff/quickcap-nx.html
32. Vandoore, A., et al. (2021). 3D sequential CMOS top tier devices demonstration using a low temperature Smart Cut™Si layer transfer. In *2021 Silicon Nanoelectronics Workshop (SNW)* (pp. 1–2)
33. Vandooren, A., et al. (2022). Demonstration of 3D sequential FD-SOI on CMOS FinFET stacking featuring low temperature Si layer transfer and top tier device fabrication with tier interconnections. In *2022 IEEE Symposium on VLSI Technology and Circuits (VLSI Technology and Circuits)* (pp. 330–331). https://doi.org/10.1109/VLSITechnologyandCir46769.2022.9830400
34. Yoshimoto, M., et al. (1983). A divided word-line structure in the static RAM and its application to a 64K full CMOS RAM. *IEEE Journal of Solid-State Circuits, 18*(5), 479–485. https://doi.org/10.1109/JSSC.1983.1051981
35. Zahoor, F., et al. (2023). Carbon nanotube field effect transistors: An overview of device structure, modeling, fabrication and applications. *Physica Scripta, 98*(8), 082003. https://doi.org/10.1088/1402-4896/ace855
36. Zhang, J., et al. (2011). Overcoming carbon nanotube variations through co-optimized technology and circuit design. In *Electron Devices Meeting, 1988. IEDM '88. Technical Digest., International*. https://doi.org/10.1109/IEDM.2011.6131490
37. Zhu, J., et al. (2023). Low-thermal-budget synthesis of monolayer molybdenum disulfide for silicon back-end-of-line integration on a 200 mm platform. *Nature Nanotechnology, 18*, 456–463. https://doi.org/10.1038/s41565-023-01375-6

Part V
Summaries

Chapter 10
Conclusions, Contributions, and Future Works

This book focuses on identifying the next promising candidate for L1, L2, L3 cache designs in sub 1 nm node. The scope of the research encompasses four distinct topics, as outlined in Sect. 1.4 of Chap. 1. Part I presents a design-technology co-optimization (DTCO) framework for power, performance, and area (PPA) analysis. Part II discusses several SRAM bitcell designs tailored for complementary transistor (CFET) technology, accompanied by comprehensive scaling trends, challenges, and routing solutions. Part III offers recommendations for future design candidates based on subarray- and macro-level PPA evaluations. In recognition of the limited impact of bitcell/subarray scaling on SRAM macro, Part IV explores subarray size optimization through a novel active interconnect (AIC) divided design, aiming for more substantial macro-level PPA enhancements. The integration of two key innovations, namely SRAM bitcell scaling and AIC divided subarray design, significantly enhances macro-level PPA. This finding serves as a valuable reference for future research in cache design exploration. This chapter summarizes the conclusions (Sect. 10.1), contributions (Sect. 10.2), and future works (Sect. 10.3) of this book. The conclusions and contributions are outlined within the framework of Parts I, II, III, and IV.

10.1 Conclusions

This section is structured into Part I (Sect. 10.1.1), Part II (Sect. 10.1.2), Part III (Sect. 10.1.3), and Part IV (Sect. 10.1.4), summarizing the findings and innovations of this book.

H.-H. Liu, F. Catthoor, *Circuit-Technology Co-Optimization of SRAM Design in Advanced CMOS Nodes*, https://doi.org/10.1007/978-3-031-76109-6_10

10.1.1 Part I – SRAM Basic Principles and Simulation Methodologies

Chapter 2 provides an overview of the DTCO PPA simulation framework (Fig. 2.1), essential definitions of simulation metrics (Table 2.4), and simulation methodologies ranging from the bitcell to macro levels. These methodologies contain critical terms such as hold static noise margin (HSNM), read static noise margin (RSNM), write margin (WM), read delay (RD), read energy (RE), write delay (WD), and write energy (WE). Additionally, dedicated simulation methodologies, including write failure examination (Chap. 2) and WM extraction (Chap. 3), have been specifically investigated for SRAM bitcells in deeply scaled nodes. The discussion on potential sources and the testing framework for write failures is crucial given the absence of silicon data in advanced nodes that are not yet commercially available. The proposed write failure examination aims to characterize sources such as pass-gate (PG) leakage, limited write-assist voltage, max operating gate volage, and substrate current, which can inform write-ability-aware subarray design in Chap. 8.

Chapter 3 delves into the challenges and proposed solutions for the WM methodology with detailed attention. It highlights the inadequacies of existing WM methodologies in resistance-dominant technology nodes due to their oversight of the time-dependent impacts of parasitic bitline (BL) resistance (R), capacitance (C), and pass-gate (PG) transistors. The study introduces the utilization of BL ramp-down slope (S_{BL}) and ramp-down time (t_{BL}), where S_{BL} = -voltage/t_{BL} is assumed as -2 V/ns to represent SRAM macro designs operating at a 2 GHz frequency for write operations [1, 2]. The test bench comprises high-density (HD) and high-performance (HP) subarrays, with subarray sizes of 256-C×256-R and 128-C×256-R, respectively, in 14-Å-ccomptaible technology (A14). Inclusion of PG devices ("w/ R & PG") leads to 13% and 10% degraded WM results for HD and HP subarray designs, respectively, due to additional PG leakage paths compared to pure resistance design ("w/ R"). Furthermore, considering BL capacitance (C_{BL}) in the BL parasitic circuits ("w/ R & C") leads to a 10% and 4% degradation in WM for HD and HP subarray designs, respectively, when t_{BL} = 1 ns. However, the impact of C_{BL} on WM diminishes when t_{BL} exceeds 50 ns. The combined impact of R, C, and PG devices ("w/ R & PG & C") results in a 17% (HD) and 16% (HP) degradation in WM, along with a 30% yield loss (HP). These findings underscore the necessity of employing the proposed WM methodology in the SRAM PPA evaluation framework. The relevant research is documented in the publication [10].

10.1.2 *Part II – SRAM Bitcell Scaling Roadmap Towards CFETs*

10.1.2.1 SRAM Bitcell Layout Design Evolution

Chapter 4 addresses the scaling bottleneck observed in nanosheet (NS) and fork-sheet (FS)-based SRAM bitcells, emphasizing the incorporation of CFET into SRAM bitcell designs in 5-Å-compatible technology (A5) and 3-Å-compatible technology (A3). The chapter elucidates the major challenges and delineates the process flow for sequential (seq) and monolithic (mono) CFET SRAM bitcell layout designs. In the seq CET case, maintaining extra spacing between the active and M0A merge regions is imperative to provide sufficient process window during lithography. Conversely, the mono CFET case necessitates the development of additional process steps to eliminate parasitic devices located on top of the bottom PG devices. Constructing the cross-coupled inverters architecture entails extending each inverter's gate to merge with the drain of the other inverter at the spacer region, i.e. spacer merge [19]. Furthermore, the active trim cut process is indispensable to establish a non-uniform NS width between transistors for both HP and low-power (LP) bitcell configurations. The HD A5 CFET SRAM bitcell design showcases a notable advantage in bitcell area scaling, offering up to 55% and 40% improvement compared to A14 NSFET and A10 FSFET counterparts, respectively. To further advance the scaling of CFET SRAM bitcells towards A3, the dielectric isolation wall (DIW) is introduced as a replacement for gate-cut (GC). Moreover, the self-aligned M0A/gate merge is proposed to eliminate the additional spacing in the sequential case, serving as another booster for bitcell area. By incorporating DIW and self-aligned M0A/gate merge, up to 29% and 21% bitcell area scaling is projected in HD and HP/LP A3 CFET SRAM bitcell designs, respectively, compared to their A5 counterparts. Further insights into the research are available in publications [11, 12].

The utilization of seq and mono CFET has yielded significant bitcell area scaling benefits down to A3 due to the stacked architecture. However, this advancement comes with the presence of tall via connections from the frontside back-end-of-line (BEOL) to the PG devices situated on the bottom tiers, resulting in increased BL and WL capacitance (C_{BL} and C_{WL}) compared to non-stacked devices. Additionally, the wordline (WL) width becomes narrower owing to the insufficient backside routing space in A3, leading to a higher WL resistance (R_{WL}) compared to A5 counterparts. Chapter 5 addresses these issues by introducing the hybrid CFET into the SRAM bitcell design roadmap, showcasing double-sided interconnect routing characteristics without incurring any bitcell area penalties. This chapter discusses major challenges and highlights process features, such as common active region formation and separate top and bottom device fabrication. Moreover, the hybrid CFET architecture introduces increased process complexity and costs due to additional fabrications such as wafer flipping, bonding, substrate removal, and double MINT layer formations. A3 hybrid NS-on-NS CFET SRAM bitcell demonstrates notable reductions of 41% in R_{WL}, 25% in C_{BL}, and 35% in C_{WL} by widening the WL width and shortening the via height, compared to seq and mono counterparts. Table 5.4 presents an overview of the physical terms associated with each bitcell design option. Further details on this topic can be found in the related work published in [13].

10.1.2.2 SRAM Bitcell Routing Design Evolution

The introduction of the buried power rail (BPR) technique [20] marked a groundbreaking advancement in SRAM bitcell design starting from A14. It effectively resolved the space constraints within the frontside BEOL layer by moving the power rail delivery network (PDN) from the frontside to the backside, allowing wider BL and WL widths. However, the continued scaling of cell height from A10 to A14 SRAM bitcells causes an inevitable decrease in BL width to accommodate the limited space within the frontside BEOL layer. To address this challenge, particularly in the MINT layer where BL resides, the concept of spacer merge emerged in CFET SRAM bitcell [19]. This innovation relocated the formation of cross-coupled inverters from BEOL to MEOL, leading to relatively wider BL width in A5 CFET SRAM bitcells compared to A10. However, the transition from A5 to A3 introduces DIW as GC replacement and self-aligned M0A/gate merge, as reported in Chap. 4 [11]. This change is accompanied by more aggressive bitcell area scaling, leading to an increase in BL width due to limited space availability within the MINT layer. Meanwhile, WL length consistently decreased as cell height scaled, with width and thickness maintaining relative adequacy. An exception was observed in A3 seq and mono cases, where WL width experienced a reduction due to the ground voltage (V_{SS}) power rail sharing the same M1 layer (Chap. 4). Nevertheless, this narrowed WL issue has been tackled through the implementation of the hybrid CFET SRAM bitcell design, as elaborated in Chap. 5. To obtain a concise overview of the modifications in the dimensions of BL and WL, please refer to Fig. 5.17 in Chap. 5.

10.1.3 Part III – SRAM Subarray- and Macro-level DTCO PPA Analysis

Chapter 6 examines the subarray-level PPA outcomes and offers optimization guidelines for HD subarray size and BEOL RC [11, 13]. The parasitic RC scaling trend reveals an unavoidable increase in R_{BL} due to faster scaling of cell height compared to cell width. With the supply voltage (V_{DD}) scaled down to 0.7V, the worst-case RSNM approaches the thermal voltage (approximately ~30mV at 80°C) suggesting a halt in V_{DD} scaling in deeply scaled nodes. Examination of the optimized fin height (H_{fin}) and number of stacked nanosheets (*NNS*) for SRAM bitcells aligns with logic assumptions, as detailed in Table 5.4 (Chap. 5). HP bitcell configurations consistently outperform HD and LP counterparts in SNM values in most bitcell designs due to increased read current. However, for A14 NSFET, A5 seq Fin-on-Fin CFET, and A5 seq NS-on-NS CFET SRAM bitcells, the optimized choice shifts to LP bitcell configurations owing to their more robust read path. As discussed in Part I, increased write failures require extra attention in evaluating WM and associated side effects. A more negative WM suggests a stronger need for write-assist techniques. Thus, the aim of improving WM prompts the inclusion of a shorter BL but longer WL subarray design (HD-2, 576-C×128-R) in the test bench. Comparative analysis shows that the HD-2 subarray exhibits superior WM over its HD-1 (288-C×256-R), HP (144-C×256-R), and LP (144-C×256-R) counterparts. However, HD-2 exhibits higher delay and energy consumption than HD-1 due to longer total WL length. This trade-off suggests that sacrificing delay and energy may be necessary to ensure reliable write functionality in ultra-scaled SRAM technology. In pursuit of an alternative approach to improving WM, A3 hybrid FS-on-FS CFET SRAM bitcells with increased NS width (W_{NS}) and BL width (W_{BL}) are incorporated in the examination. In A3, preferred bitcell designs for PMOS PG-based SRAM subarrays include: A3 hybrid FS-on-FS CFET with W_{BL} of 21 nm for HD-1, A3 hybrid NS-on-NS CFET for both HD-2 and HP, and A3 hybrid FS-on-FS CFET with W_{BL} of 12 nm for LP. Relevant works are published in [11, 13].

The final suggestions among several design options are provided following the subarray-level PPA results, which are further elucidated in Chap. 7 through macro-level analysis. The impact of inter-subarray overhead begins to surpass that of the subarray due to increased interconnect resistance in advanced technology nodes [16, 18]. Consequently, various combinations of pin positions and routing types have been examined. Centre-pin access with E-tree (CP-E) technology emerges as the most advantageous option at the macro-level, outperforming side-pin access with H-tree (SP-E) technology by 53% in delay, 45% in energy, 37% in interconnect length, and 41% in area. Among the four subarray designs (HD-1, HD-2, HP, and LP), HD-1 stands out due to its compact bitcell design, yielding minimal inter-subarray interconnect and area overheads. However, the promising HD-1 option necessitates greater power and area for a write-assist technique, a consideration not addressed in this study. To achieve a balance between PPA metrics and write failure risk, the HD-2 subarray design emerges as the optimized candidate with the least write-assist requirement compared to HD-1, HP, and LP counterparts. The transition from A14 to A5 SRAM bitcell design enhances the energy-delay-area product (EDAP) by up to 53%~76% owing to significant bitcell/subarray area scaling. Beyond A5, A3 CFET SRAM bitcell design offers only an additional improvement of 9%~20% in EDAP at the expense of a more complex process and heightened cost concerns. From an economic standpoint, it is not recommended to continue utilizing hybrid CFET structures and pursuing scaling beyond A5 for future SRAM design directions. Instead of centring efforts on traditional bitcell-level scaling, the focus should be redirected towards reducing inter-subarray interconnects, as they dominate the overall PPA results at the macro level. This observation prompts a deeper investigation into alternative PPA enhancement techniques, such as the novel SRAM subarray design introduced in Chaps. 8 and 9.

10.1.4 Part IV – Beyond Traditional SRAM Bitcell-level Scaling – Future Design Direction for SRAM Data Array Towards A14

Part IV aims to explore an alternative node-to-node PPA booster, departing from the conventional approach to SRAM bitcell scaling. In Chap. 8, it is demonstrated that transitioning from a smaller to a larger standard (Std.) subarray design can potentially improve macro-level EDAP by 56% due to reduced inter-subarray overhead. Nonetheless, the larger subarray size presents challenges, particularly regarding write failure risks (WFR) due to increased R_{BL} and R_{WL}. To mitigate this issue, conventional (Conv.) divided WL and divided BL techniques [8, 22] have been introduced in A14 subarray designs. To comprehensively optimize subarray design while considering WFR, a write failure examination was conducted to determine the limited number of columns and rows (N_{col} and N_{row}) of a subarray. Results indicate that the divided design can expand the upper limit of subarray size due to the subarray partitioning approach. Furthermore, various RC test cases were applied to optimize the local block (LB) size for the divided subarray design. Table 8.4 presents the optimized design combinations across 12 different subarray sizes. Through the implementation of Conv. divided technique, enhancements of 338 mV in WM and 37% in RD were achieved for the 512-C×512-R subarray configuration compared to the Std. subarray design (neglect WFR in Std.). However, the subarray area penalty resulting from the Conv. divided circuits may potentially degrade macro-level PPA. This underscores the need for further exploration of alternative divided subarray designs capable of mitigating the subarray area penalty, as discussed in Chap. 9.

Chapter 9 illustrates a 24% increase in subarray area overhead induced by Conv. divided subarray design, leading to a 26% degradation in macro-level EDAP compared to the Std. counterpart (neglect WFR in Std.) with a subarray size of 1024-C×512-R. To mitigate this subarray area penalty while preserving similar subarray-level power and performance enhancements in Conv. divide technique, the AIC is proposed for integration within the divided subarray design. By relocating the additional logic gate tailored for the divide design from the front-end-of-line (FEOL) to the BEOL, the subarray area penalty is expected to decrease from 30% to less than 5% during the subarray layout design phase. Among BEOL-compatible and low-temperature growth devices, A14 carbon nanotube (CNT) gate-all-around transistors (GAAFETs) have been chosen for their superior electrostatic control and high injection velocity [7]. According to subarray-level PPA analysis, AIC divided subarray design emerges as the most promising, offering optimized performance and WM with a marginal increase in subarray area of less than 5%. Furthermore, the AIC divided subarray design (1024-C×512-R) achieves a significant 65% enhancement in macro-level EDAP compared to the Std. counterpart (128-C×128-R). AIC has been validated as another effective PPA booster through DTCO analysis from device to array level, facilitating comprehensive optimization of SRAM subarray design in deeply scaled nodes.

10.2 Contributions

This section highlights the main contributions to the state-of-the-art made in this book across Parts I (Sect. 10.2.1), Part II (Sect. 10.2.2), Part III (Sect. 10.2.3), and Part IV (Sect. 10.2.4).

10.2.1 Part I – SRAM Basic Principles and Simulation Methodologies

- Develop a DTCO PPA simulation framework covering bitcell, subarray, and macro levels for advanced SRAM technology across A14 to A3.
- Design a computer-based examination flow to identify write failure criteria, addressing all potential sources of failure and the impacts of write-assist techniques.
- Propose an advanced WM methodology based on the BL sweeping method, customized for resistance-dominated technologies. This approach facilitates the extraction of time dependencies from BL parasitic circuits while mitigating the risk of underestimating yield loss.

10.2.2 Part II – SRAM Bitcell Scaling Roadmap Towards CFETs

- Develop 12 novel SRAM bitcell integration schemes and layout designs optimized for A3 technology utilizing seq, mono, and hybrid CFET architectures with stacked Fin, NS, and FS channels (Table 5.3). Each design variant features HD, HP, and LP bitcell configurations.
- Introduce DIW as a replacement for GC to enhance bitcell area scaling in A3 with corresponding process flow descriptions.
- Suggest the integration of self-aligned gate and M0A merges for A3 seq CFET SRAM bitcells for further bitcell area scaling.
- Propose innovative interconnect routing solutions to tackle the challenge of limited routing space in both MEOL and BEOL of bitcells during the transition from A5 to A3 technology.

10.2.3 Part III – SRAM Subarray- and Macro-level DTCO PPA Analysis

- Evaluate margins and PPA metrics at the subarray level to determine the best design options among various bitcell types (1, 1, 4, and 12 options in A14, A10, A5, and A3, respectively) and subarray configurations (HD-1, HD-2, HP, LP).
- Discuss comprehensively the trade-offs of each design choice, particularly addressing the degraded WM results resulting from scaling the bitcell area.
- Assess the macro-level PPA to identify the optimized design choice, considering factors such as access pin placement, inter-subarray routing, subarray configuration, data capacity, and bitcell scaling.

10.2.4 Part IV – Beyond Traditional SRAM Bitcell-level Scaling – Future Design Direction for SRAM Data Array Towards A14

- Investigate the write failure criteria and correlate them with WM value for both Std. and divided subarray designs using A14 technology.
- Optimize subarray designs for 12 specified sizes outlined in Table 8.4.
- Propose a novel AIC divided subarray design aimed at reducing the area penalty associated with Conv. counterparts.
- Realize significant macro-level PP benefits compared to Std. designs by implementing larger-size subarray designs utilizing the AIC divided technique.

10.3 Future Works

This section provides an overview of potential future work stemming from this book.

10.3.1 Advanced DTCO Simulation Framework

The compact models utilized in this study for the initial PPA analysis framework of nodes beyond A14 are calibrated using a limited set of silicon data and TCAD simulations. Specifically, the incorporation of realistic device variability and reliability terms in the FSFET and CFET models is still evolving and has not yet reached full maturity. Moreover, the integration capability of CNFETs needs to be demonstrated in the future under realistic manufacturing conditions to avoid overly optimistic model performance. Therefore, future efforts in advancing DTCO framework require an expanded dataset derived from experimental sources, adhering to the dimensions scaled as per the design guidelines delineated in this book. Identifying the optimized methodology for predicting read stability and write ability metrics in deeply scaled nodes is also imperative, drawing upon well-established compact models and silicon data. Furthermore, the current parasitic extraction (PEX) methodology lacks adaptability in response to geometric variations for conducting Monte Carlo (MC) simulations. Hence, the development of a novel PEX approach capable of aligning with the variability is imperative to be added into the current framework. Moreover, the investigation into time-dependent variability and reliability effects on ultra-scaled SRAM warrants attention, particularly in conjunction with the development of an updated MC simulation framework. To strengthen the macro-level simulator, design parameters such as workload-dependencies, tag array, pipeline, and multiport should be considered for the next-generation DTCO methodologies.

10.3.2 Alternative SRAM Bitcell Design

An alternative scheme for scaling SRAM bitcells, incorporating new integration methods, transistor architectures, and materials, should be developed to achieve further PPA enhancements beyond A3. The potential research scope includes the optimization of backside power delivery network (BSPDN) [21] and utilization of BEOL-compatible transistors for SRAM bitcell design [4–6]. Corresponding silicon data and PEX results should also be developed to facilitate consistent PPA benchmarking with the options provided in this study. It is noteworthy that the consideration of backside signals is deferred to this work due to the anticipated higher capacitance resulting from taller vias, leading to inferior PP. Additionally, the less dense backside metal pitch compared to the frontside may not adequately support the design specifications for tight BL and WL pitches. However, future work should consider using backside signals, assuming the substrate removal process is implemented. This approach would mitigate issues related to tall vias and insufficient pitch.

10.3.3 Alternative SRAM Subarray and Macro Designs

Future research in SRAM design should extend beyond traditional bitcell-level scaling and embrace 3-dimensional (3D) approaches, such as 3D-IC SRAM macro [3], 3D-split SRAM [14], and AIC extensions. Partitioning the memory system into memory on logic, logic on memory, and multi-tier SRAM arrays holds promise for significantly reducing footprint within the same node. Realizing these technologies necessitates concurrent development of bonding techniques to support dense via pitch. Furthermore, exploring alternative BEOL-compatible transistors, relocating peripheral components to the BEOL active layer, and extending the simulation environment beyond A14 can be pursued as extensions of AIC divided subarray designs. However, placing the write driver and sense amplifier at the BEOL active layer is not advisable due to analog circuit instability and potential routing issues (connection to BL metal layer needs to pass by the upper WL metal layer). Conversely, placing the row decoder at the BEOL active layer could be feasible, as the required WL connection is at the uppermost metal layer, minimizing potential routing challenges. Besides allocating the backside solely for power distribution, researchers may explore the feasibility of relocating other essential functionalities, such as the clock signal (BSCLK), to the backside. This strategic placement can effectively free up the frontside BEOL space.

10.3.4 Application-oriented Circuit Design

In this book, the peripheral circuitry within the subarray (such as sense amplifier, write driver, WL decoder, precharge circuit, etc.) is primarily designed for HD applications. In contrast, the HP and LP cases are limited to transistor sizing adjustments and subarray size reorganization. The differences in macro-level PPA results presented in Chap. 7 are marginal and do not correlate with various cache levels. Therefore, it is recommended to redesign the peripheral circuits with a specific focus on density, speed, and power optimization, respectively. For instance, while dual rail V_{DD} is common in LP SRAM design, its performance degradation at lower V_{DD} levels makes it less suitable for HP applications [9]. Further exploration into comprehensive circuit-level designs with trade-off insights should be pursued in future work.

10.3.5 Collaborative DTCO and STCO Works

In pursuit of a thorough DTCO and system-technology co-optimization (STCO) analysis within the CACTI^{++} timebox, collaborative efforts have been undertaken with Zhenlin Pei from the University of Texas at Arlington (UTA) and Mahta Mayahinia from Karlsruher Institut für Technologie (KIT). Within the domain of "interconnect-centric technology", research endeavours prioritize minimizing RC and interconnect overhead for both local and global signals. Emerging solutions, such as Graphene-based interconnects, E-tree, and centre-pin access [16–18], exhibit significant potential in this regard. Additionally, the adoption of the AIC concept for inter-subarray routing, situated atop the subarrays rather than at the same level, is proposed to mitigate interconnect-induced area overhead [18]. Investigations into electromigration are advocated, particularly concerning the scaling of metal line dimensions and the exploration of alternative metal materials [15]. Considerations for the design of non-uniformity in SRAM macros should take the frequency of data access into account to determine the optimized proximity to input/output port locations. Moreover, it is recommended to explore actual dynamic workloads with non-uniform data and instantiation scheduling for specific applications as part of future endeavours.

References

1. Chang, J., et al. (2017). 12.1 A 7 nm 256 Mb SRAM in high-k metal-gate FinFET technology with write-assist circuitry for low-VMIN applications. In *2017 IEEE International Solid-State Circuits Conference (ISSCC)* (pp. 206–207). https://doi.org/10.1109/ISSCC.2017.7870333
2. Chang, J., et al. (2013). A 20 nm 112 Mb SRAM in High- metal-gate with assist circuitry for low-leakage and low-VMIN applications. In *2013 IEEE International Solid-State Circuits Conference Digest of Technical Papers* (pp. 316–317). https://doi.org/10.1109/ISSCC.2013.6487750
3. Chen, R., Weckx, P., Salahuddin, S. M., Kim, S.-W., Sisto, G., Van der Plas, G., Stucchi, M., Baert, R., Debacker, P., Na, M. H., Ryckaert, J., Milojevic, D., & Beyne, E. (2020). 3d-optimized SRAM macro design and application to memory-on-logic 3d-IC at advanced nodes. In *2020 IEEE International Electron Devices Meeting (IEDM)* (15.2.1–15.2.4).
4. Chen, R., et al. (2022). Carbon nanotube SRAM in 5-nm technology node design, optimization, and performance evaluation—part I: CNFET transistor optimization. *IEEE Transactions on Very Large Scale Integration (VLSI) Systems, 30*(4), 432–439. https://doi.org/10.1109/TVLSI.2022.3146125
5. Chen, R., Chen, L., Liang, J., Cheng, Y., Elloumi, S., Lee, J., Xu, K., Georgiev, V. P., Ni, K., Debacker, P., Asenov, A., & Todri-Sanial, A. (2022). Carbon nanotube SRAM in 5-nm technology node design, optimization, and performance evaluation—part II: CNT interconnect optimization. *IEEE Transactions on Very Large Scale Integration (VLSI) Systems, 30*(4), 440–448.
6. Das, S., Sebastian, A., Pop, E., McClellan, C.J., Franklin, A.D., Grasser, T., Knobloch, T., Illarionov, Y., Penumatcha, A.V., Appenzeller, J., Chen, Z., Zhu, W., Asselberghs, I., Li, L., Avci, U., Bhat, N., Anthopoulos, T., & Singh, R. (2021). Transistors based on two-dimensional materials for future integrated circuits. *Nature Electronics, 4*, 786–799.
7. Gilardi, C., Chehab, B., Sisto, G., Schuddinck, P., Ahmed, Z., Zografos, O., Lin, Q., Hellings, G., Ryckaert, J., Wong, H. S., & Mitra, S. (2021). Extended scale length theory targeting low-dimensional FETs for carbon nanotube FET digital logic design-technology co-optimization. In *2021 IEEE International Electron Devices Meeting (IEDM)* (pp. 27.3.1–27.3.4).
8. Karandikar, A., & Parhi, K. K. (1998). Low power SRAM design using hierarchical divided bit-line approach. In *Proceedings International Conference on Computer Design. VLSI in Computers and Processors (Cat. No.98CB36273)* (pp. 82–88). https://doi.org/10.1109/ICCD.1998.727027
9. Kim, T. H., et al. (2020). An embedded level-shifting dual-rail SRAM for high-speed and low-power cache. *IEEE Access, 8*, 187126–187139. https://doi.org/10.1109/ACCESS.2020.3030099
10. Liu, H.-H., et al. (2023). Extended methodology to determine SRAM write margin in resistance-dominated technology node. *IEEE Transactions on Electron Devices, 69*(6), 3113–3117. https://doi.org/10.1109/TED.2022.3165738
11. Liu, H.-H., et al. (2023). CFET SRAM DTCO, interconnect guideline, and benchmark for CMOS scaling. *IEEE Transactions on Electron Devices, 70*(3), 883–890. https://doi.org/10.1109/TED.2023.3235701
12. Liu, H.-H., et al. (2023). DTCO of sequential and monolithic CFET SRAM. In R.-H. Kim & N. V. Lafferty (Eds.), *DTCO and computational patterning II* (Vol. 12495). International society for optics and photonics. SPIE. https://doi.org/10.1117/12.2657524

13. Liu, H.-H., et al. (2023). CFET SRAM with double-sided interconnect design and DTCO benchmark. *IEEE Transactions on Electron Devices* 1–8. https://doi.org/10.1109/TED.2023.3305322
14. Mathur, R., et al. (2021). 3D-split SRAM: Enabling generational gains in advanced CMOS. In *2021 IEEE Custom Integrated Circuits Conference (CICC)* (pp. 1–2). https://doi.org/10.1109/CICC51472.2021.9431528
15. Mayahinia, M., Liu, H. H., Mishra, S., Tokei, Z., Catthoor, F., & Tahoori, M. (2023). Electromigration-aware design technology co-optimization for SRAM in advanced technology nodes. In *2023 Design, Automation & Test in Europe Conference & Exhibition (DATE)* (pp. 1–6).
16. Pei, Z., et al. (2023). Graphene-based interconnect exploration for large SRAM caches for ultrascaled technology nodes. *IEEE Transactions on Electron Devices, 70*(1), 230–238. https://doi.org/10.1109/TED.2022.3225512
17. Pei, Z., et al. (2023). Technology/memory co-design and co-optimization using e-tree interconnect. In *Proceedings of the Great Lakes Symposium on VLSI 2023*. GLSVLSI '23. Knoxville, TN, USA (pp. 159–162). Association for Computing Machinery. ISBN: 9798400701252. https://doi.org/10.1145/3583781.3590311
18. Pei, Z., et al. (2023). Emerging interconnect exploration for SRAM application using non-conventional H-tree and center-pin access. In *2023 24th International Symposium on Quality Electronic Design (ISQED)* (pp. 1–1). https://doi.org/10.1109/ISQED57927.2023.10129316
19. Ryckaert, J., et al. (2018). The complementary FET (CFET) for CMOS scaling beyond N3. In *2018 IEEE Symposium on VLSI Technology* (pp. 141–142). https://doi.org/10.1109/VLSIT.2018.8510618
20. Salahuddin, S., et al. (2020). Buried power SRAM DTCO and system-level benchmarking in N3. In *2020 IEEE Symposium on VLSI Technology* (pp. 1–2). https://doi.org/10.1109/VLSITechnology18217.2020.9265076
21. Yang, S., Schuddinck, P., Garcia-Bardon, M., Xiang, Y., Veloso, A., Chan, B. T., Mirabelli, G., Hiblot, G., Hellings, G., & Ryckaert, J. (2023). PPA and scaling potential of backside power options in N2 and A14 nanosheet technology. In *2023 IEEE Symposium on VLSI Technology and Circuits (VLSI Technology and Circuits)* (pp. 1–2).
22. Yoshimoto, M., et al. (1983). A divided word-line structure in the static RAM and its application to a 64K full CMOS RAM. *IEEE Journal of Solid-State Circuits, 18*(5), 479–485. https://doi.org/10.1109/JSSC.1983.1051981

Appendix A
Appendix

Section A.1 presents the classic SRAM memory subarray design utilized in this study for evaluating power, performance, and area (PPA) metrics. This design encompasses the SRAM bitcell array, address control, timing control, precharge/predischarge, row decoder, column multiplexer (MUX), sense amplifier, and write driver. Table A.1 presents the abbreviations for both input and output signals, as well as the terms utilized in the circuits. The design of address and timing control relies on either NMOS or PMOS pass-gate (N-PG or P-PG) SRAM bitcell configurations. Notably, the circuit schematics referenced herein are adaptable to various subarray sizes through appropriate transistor sizing.

Sections A.2 and A.3 provide a comprehensive summary of the top views for A5 and A3 sequential (seq) and monolithic (mono) 6-transistor (6T) CFET SRAM bitcell layout designs. Section A.4 illustrates the top views for A3 hybrid CFET SRAM bitcell layout designs. The SRAM Bitcell Layout Designs, as detailed in Table A.2, include high-density (HD), high-performance (HP), and low-power (LP) options. These layouts show the integration of stacked Fin, NS, and FS channels in CFET SRAMs, allowing for a thorough examination of individual designs indicating their BL widths (W_{BL}), WL widths (W_{WL}), cell widths (W_{cell}), and cell heights (H_{cell}) (Chap. 5). Notably, stacked FS CFET SRAM offers additional area scaling by up to 25% at the cost of larger R_{BL} (Fig. 6.1 in Chap. 6) compared to the A3 fin or NS counterparts due to the removal of gate extensions. Due to the bitcell area scaling of up to 44% compared to A5 counterparts, stacked FS CFET SRAM bitcells are classified under A3 node.

H.-H. Liu, F. Catthoor, *Circuit-Technology Co-Optimization of SRAM Design in Advanced CMOS Nodes*, https://doi.org/10.1007/978-3-031-76109-6

Table A.1 Abbreviation of signals used in subarray design in Sect. A.1.

Abbreviation	Definition
-R/N_{row}	Number of row
-C/N_{col}	Number of column
DFF	D-flip-flop
CE	Chip enable
WE	Write enable
CS	Column select
CLK	Clock
WDATA	Input write data
DATA	Output read data
LBCS	Local block row select
LBRS	Local block column select

Table A.2 Summary of HD, HP, LP CFET SRAM bitcell layout designs in A5 and A3.

Chapter	Technology	Device	Channel	Section
4	A5	Seq CFET	Fin	A.2
			NS	
		Mono CFET	Fin	
			NS	
	A3	Seq CFET	Fin	A.3
			NS	
			FS	
		Mono CFET	Fin	
			NS	
			FS	
5	A3	Hybrid CFET	NS	A.4
			FS	

A.1 Classic SRAM Memory Design – Memory Architecture and Block Diagram

A.1.1 Memory Architecture

Figure A.1 illustrates the memory architecture employed in this book, encompassing surrounding logic cores, data array, banks, mats, and subarrays. The configurations of interconnect routing types and access pin positions are delineated in Sect. 2.1.4 of Chap. 2. The subarray comprises a 6T bitcell array, which can be either N-PG or P-PG, alongside address control, timing control, precharge (for N-PG) or predischarge (for P-PG) circuits, row decoder, column multiplexer, sense amplifiers, and write drivers. The forthcoming sections will present the subarray circuit schematics.

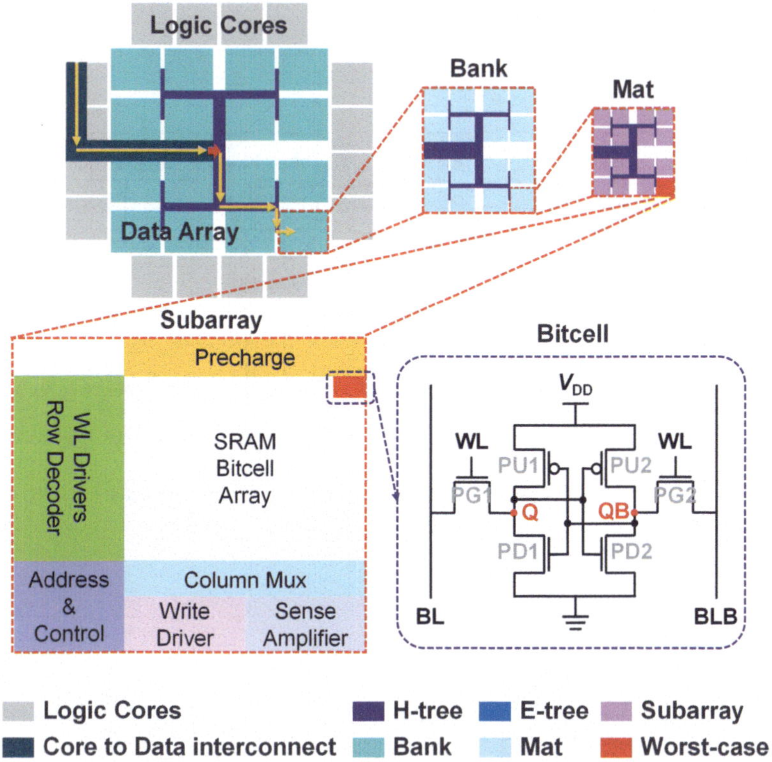

Fig. A.1 Schematic of SRAM data array, bank, mat, subarray, and bitcell

A.1.2 Block Diagram

Figures A.2 and A.3 depict the block diagrams of N-PG and P-PG-based standard (Std.) subarray designs, respectively, as employed in Chaps. 6, 8, and 9. Figure A.4 illustrates the block diagram of the conventional (Conv.) and active interconnect (AIC) N-PG-based divided subarray design, utilized in Chap. 8 and 9. The CLK signal interfaces with timing control, address control, row decoder, and write driver DFF to ensure timing synchronization. The distinctions between N-PG- and P-PG-based subarrays include: (1) an additional inverter design positioned between the row decoder and bitcell array; (2) precharge or predischarge circuitry; (3) incorporation of a sense amplifier for differential voltage sensing between BL with supply voltage (V_{DD}) or ground voltage (V_{SS}) as reference voltages; and (4) utilization of a write driver with different input write data signals. Additional decoders are necessary for the selection of LBCS and LBRS signals in the divided subarray design.

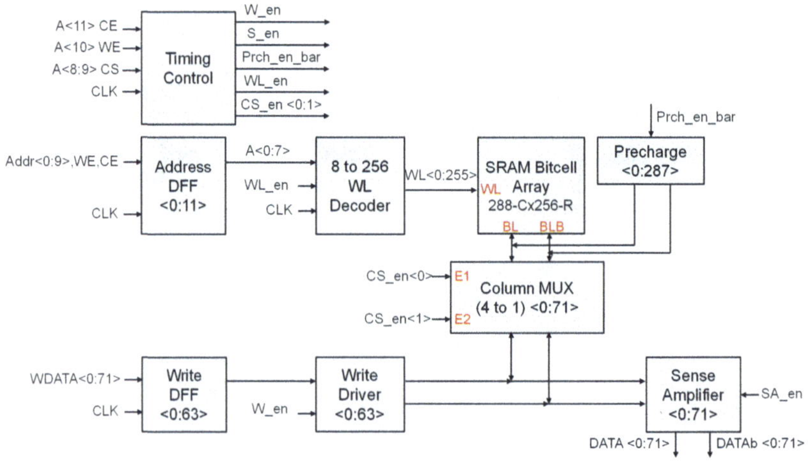

Fig. A.2 Schematic of N-PG-based SRAM standard subarray design with subarray size of 288-C×256-R

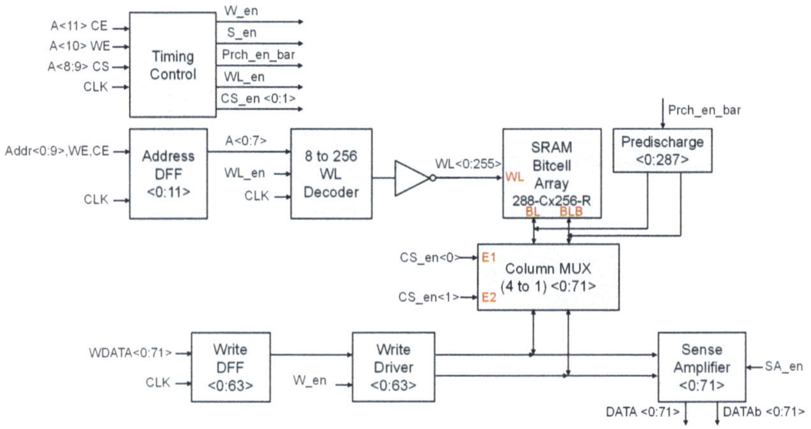

Fig. A.3 Schematic of P-PG-based SRAM standard subarray design with subarray size of 288-C×256-R

A.1.3 SRAM Bitcell

Figure A.5 depicts the schematics of N-PG- and P-PG-based 6T SRAM bitcells. N-PG bitcells encompass FinFET, NSFET, FSFET, and hybrid CFET configurations, whereas P-PG bitcells can be formed by seq, mono, and hybrid CFET configurations. Additional elaboration on these architectures is provided in Chaps. 4 and 5.

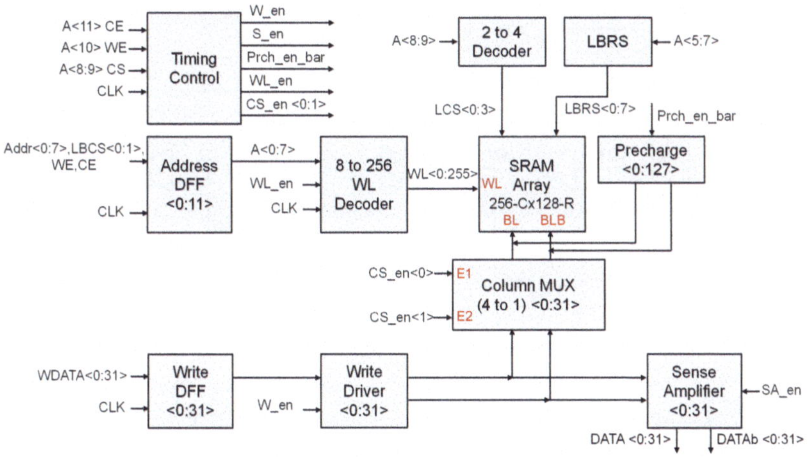

Fig. A.4 Schematic of N-PG-based SRAM divided subarray design with subarray size of 256-C×128-R

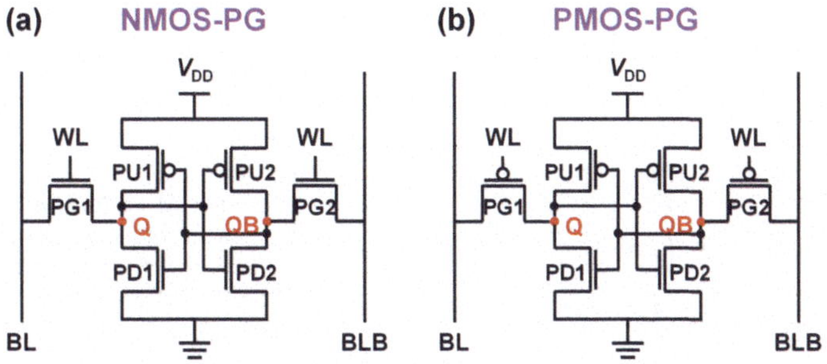

6T SRAM	Non-stacked FET (Fin, NS, FS)	Seq & Mono CFET	Hybrid CFET
N-PG	v		v
P-PG		v	v

Fig. A.5 Schematic of (**a**) N-PG and (**b**) P-PG-based SRAM Bitcell Layout Designs

A.1.4 Address Control

Figure A.6 illustrates the schematic view of a positive edge-triggered D Flip-Flop (DFF) utilized for address control. Each input signal including address, write enable (WE), chip enable (CE), and LBCS, is required to traverse this DFF prior to entering the subarray system to be synchronized with CLK. The positive-edge DFF facilitates the transfer of data from input D to output Q upon the rising edge of the CLK signal, transitioning from '0' to '1'. During CLK's low phase, the input data propagate from D to Q1, locking in the master latch (Q1 = D), while the slave latch preserves the

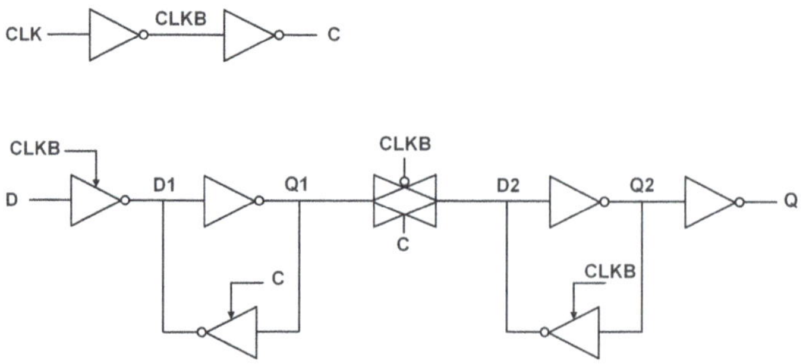

Fig. A.6 Schematic of positive edge triggering D-flip-flop (DFF)

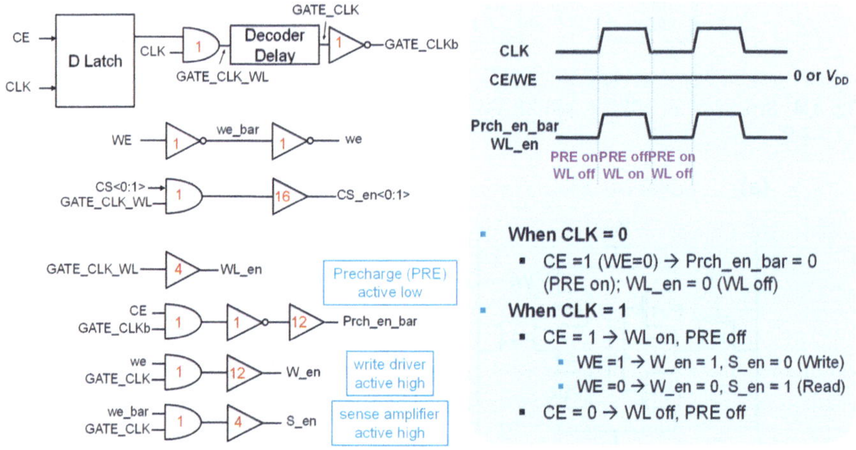

Fig. A.7 Schematic of timing control for N-PG-based SRAM subarray design

preceding state of D2 (Q2 = D2). Upon CLK's high phase, input data can't enter the circuit, prompting the transfer of data from Q1 to D2, ultimately reaching the output node Q.

A.1.5 Timing Control

Figures A.7 and A.8 depict the schematic view for timing control in N-PG- and P-PG-based SRAM subarray designs, respectively. The input signals encompass CE, WE, and column select (CS), while the output signals include write enable (W_en), sense enable (S_en), precharge enable bar (Prch_en_bar) or predischarge enable (Prch_en), WL enable (WL_en), and column select enable (CS_en). When CLK is low and CE is high, only precharge/predischarge (PRE) is enabled. When both CLK and CE are high, WL is activated while PRE is deactivated, with read and

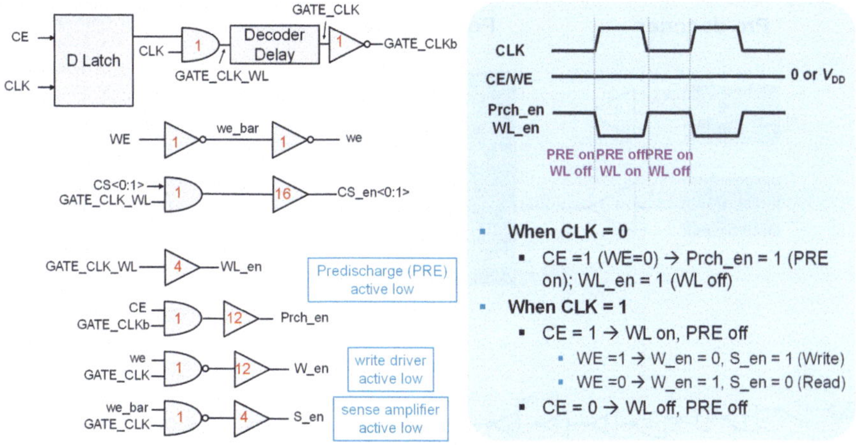

Fig. A.8 Schematic of timing control for P-PG-based SRAM subarray design

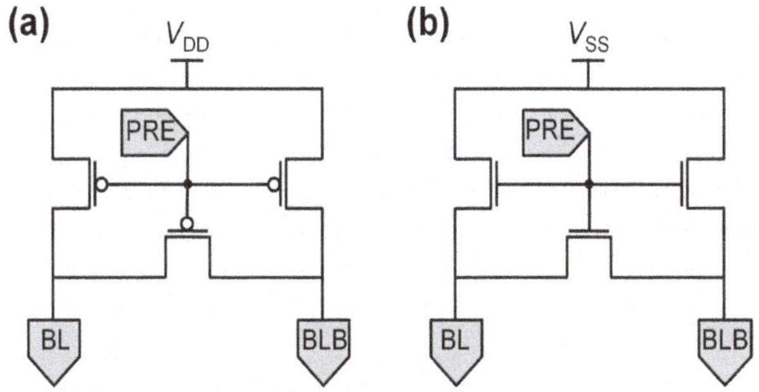

Fig. A.9 Schematic of (**a**) precharge and (**b**) predischarge circuit for N-PG- and P-PG-based SRAM subarray designs, respectively

write operations determined by WE being 0 and 1, respectively. When CLK is high and CE is low, both WL and PRE are deactivated. Notably, the decoder delay circuit is incorporated to synchronize the activation timing of the sense amplifier, write driver, and MUX circuits after the arrival of WL signals at the bitcell array.

A.1.6 Precharge and Predischarge

Figure A.9 depicts the precharge circuit (active low) utilized for N-PG-based subarrays and the predischarge circuit (active high) employed for P-PG-based subarrays, both featuring equalizers.

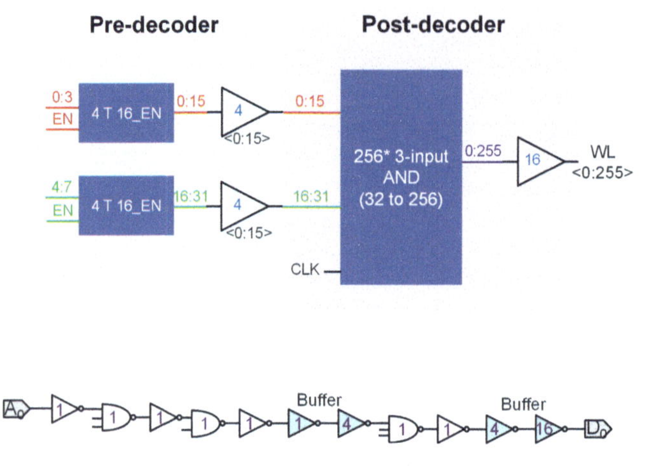

Data	Input
D_0	$\overline{A_3}\,\overline{A_2}\,\overline{A_1}\,\overline{A_0}$
D_1	$\overline{A_3}\,\overline{A_2}\,\overline{A_1}\,A_0$
D_2	$\overline{A_3}\,\overline{A_2}\,A_1\overline{A_0}$
D_3	$\overline{A_3}\,\overline{A_2}\,A_1\,A_0$
D_4	$\overline{A_3}A_2\overline{A_1}\,\overline{A_0}$
D_5	$\overline{A_3}A_2\overline{A_1}\,A_0$
D_6	$\overline{A_3}A_2A_1\overline{A_0}$
D_7	$\overline{A_3}A_2\,A_1\,A_0$
D_8	$A_3\overline{A_2}\,\overline{A_1}\,\overline{A_0}$
D_9	$A_3\overline{A_2}\,\overline{A_1}\,A_0$
D_{10}	$A_3\overline{A_2}\,A_1\overline{A_0}$
D_{11}	$A_3\overline{A_2}\,A_1\,A_0$
D_{12}	$A_3A_2\,\overline{A_1}\,\overline{A_0}$
D_{13}	$A_3A_2\,\overline{A_1}\,A_0$
D_{14}	$A_3A_2\,A_1\overline{A_0}$
D_{15}	$A_3A_2\,A_1\,A_0$

Fig. A.10 Schematic of row decoder for both N-PG and P-PG-based SRAM subarray designs

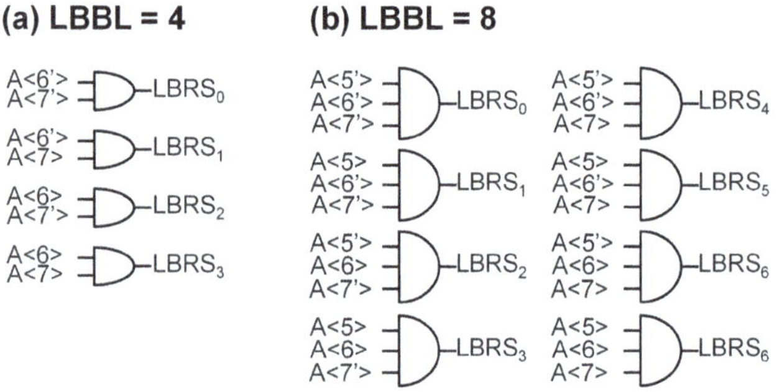

Fig. A.11 Schematic of local block row select (LBRS) decoder in SRAM divided subarray design with (**a**) LBBL = 4 and (**b**) LBBL = 8

A.1.7 Row Decoder

Figure A.10 illustrates the schematic view of WL row decoder, comprising a pre-decoder and a post-decoder. The pre-decoder consists of two 4-to-16 decoders and two buffers, while the post-decoder comprises 256 3-input AND gates and a buffer, all synchronized with the CLK signal. Figure A.11 illustrates the LBRS decoder with LBBL values of 4 and 8.

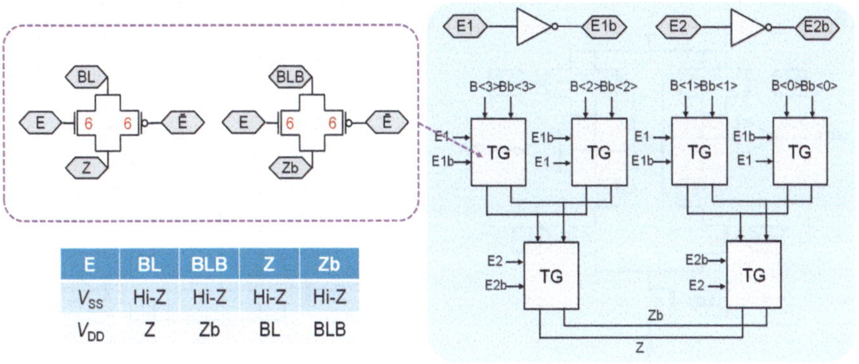

Fig. A.12 Schematic of 4 to 1 column multiplexer (MUX) composed by active-high transmission gate (TG) designs

A.1.8 Column Multiplexer

Figure A.12 depicts the schematic view of a column MUX, specifically a 4-input-to-1-output configuration, which comprises active-high transmission gate (TG) designs. When the CS_en is in a high state, the input data on BL and BLB lines are transmitted to the outputs Z and Zb, respectively. When the CS_en is in a low state, these routes are blocked. The quantities of sense amplifiers and write drivers are dictated by the input and output ratio of the column MUX.

A.1.9 Sense Amplifier and Output Latch

Figure A.13 illustrates the schematic views of current-controlled-latched sense amplifiers and output latches for both N-PG- and P-PG-based subarray designs. In the N-PG-based subarray, the sensing process unfolds as follows: (1) Precharge signals BL and BLB enable NMOS; (2) With SAE low, the precharge PMOS activates while the footer NMOS deactivates; (3) With SAE high, the precharge PMOS deactivates while both header PMOS and footer NMOS activate, forming a cross-coupled inverter to provide positive feedback and amplify the BL and BLB differential signals; (4) Discharging BL or BLB disables the path, halting current flow automatically to prevent redundant static power consumption; (5) Signals are output via an SR latch. In the P-PG-based subarray, the sensing process follows a similar pattern: (1) Predischarge signals BL and BLB enable PMOS; (2) With SAE low, the predischarge NMOS activates while the header PMOS deactivates; (3) With SAE high, the predischarge NMOS deactivates while both footer NMOS and header PMOS activate, forming a cross-coupled inverter to provide positive feedback and amplify the BL and BLB differential signals; (4) Charging BL or

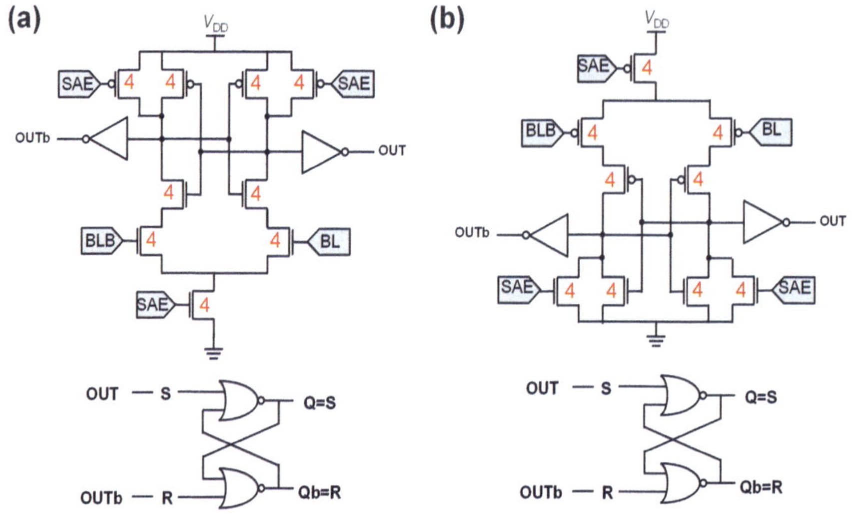

Fig. A.13 Schematic of current-controlled-latched sense amplifier and output latch for (**a**) N-PG and (**b**) P-PG-based SRAM subarray designs

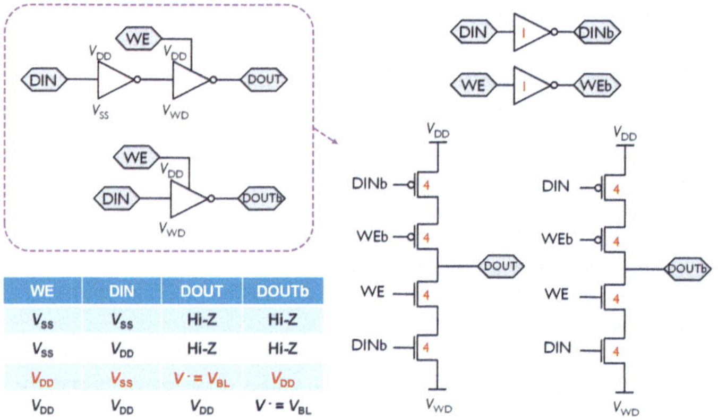

WE	DIN	DOUT	DOUTb
V_{SS}	V_{SS}	Hi-Z	Hi-Z
V_{SS}	V_{DD}	Hi-Z	Hi-Z
V_{DD}	V_{SS}	$V^{-} = V_{BL}$	V_{DD}
V_{DD}	V_{DD}	V_{DD}	$V^{-} = V_{BL}$

Fig. A.14 Schematic of write driver with active-high inverting tri-state buffer for N-PG-based SRAM subarray designs

BLB disables the path, halting current flow automatically to prevent redundant static power consumption; (5) Signals are output via an SR latch.

A.1.10 Write Driver

Figure A.14 depicts the schematic view of the write driver employing an active-high inverting tri-state buffer for N-PG-based subarray design. In the low state of

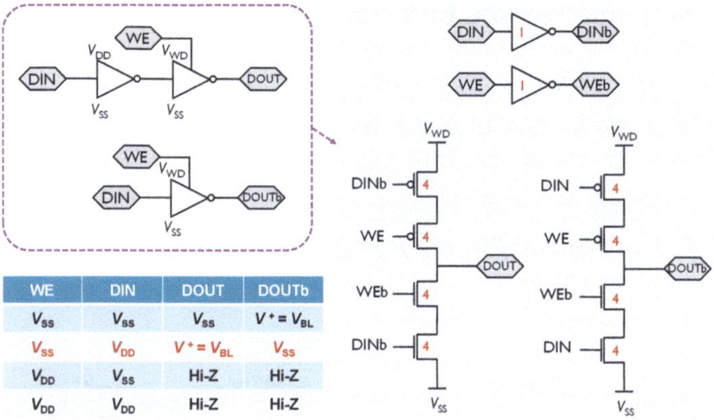

WE	DIN	DOUT	DOUTb
V_{SS}	V_{SS}	V_{SS}	$V^+ = V_{BL}$
V_{SS}	V_{DD}	$V^+ = V_{BL}$	V_{SS}
V_{DD}	V_{SS}	Hi-Z	Hi-Z
V_{DD}	V_{DD}	Hi-Z	Hi-Z

Fig. A.15 Schematic of write driver with active-low inverting tri-state buffer for P-PG-based SRAM subarray designs

WE, no operation occurs on the write path. When WE is high and DIN is low, the BL connects to the negative BL pulse, while BLB remains at V_{DD}. The opposite scenario unfolds when both WE and DIN are high. Figure A.15 illustrates the schematic view of the write driver with an active-low inverting tri-state buffer for P-PG-based subarray design. In the high state of WE, no activity takes place on the write path. When WE is low and DIN is high, the BL links to the positive BL pulse, while BLB remains at V_{SS}. The reverse situation arises when both WE and DIN are low.

A.2 A5 Seq and Mono CFET SRAM Bitcell Layout Design

This section presents A5 seq and mono 6T CFET SRAM bitcell layout designs, as shown in Figs. A.16, A.17, A.18, and A.19.

A.2.1 A5 Seq Fin-on-Fin CFET SRAM Bitcells

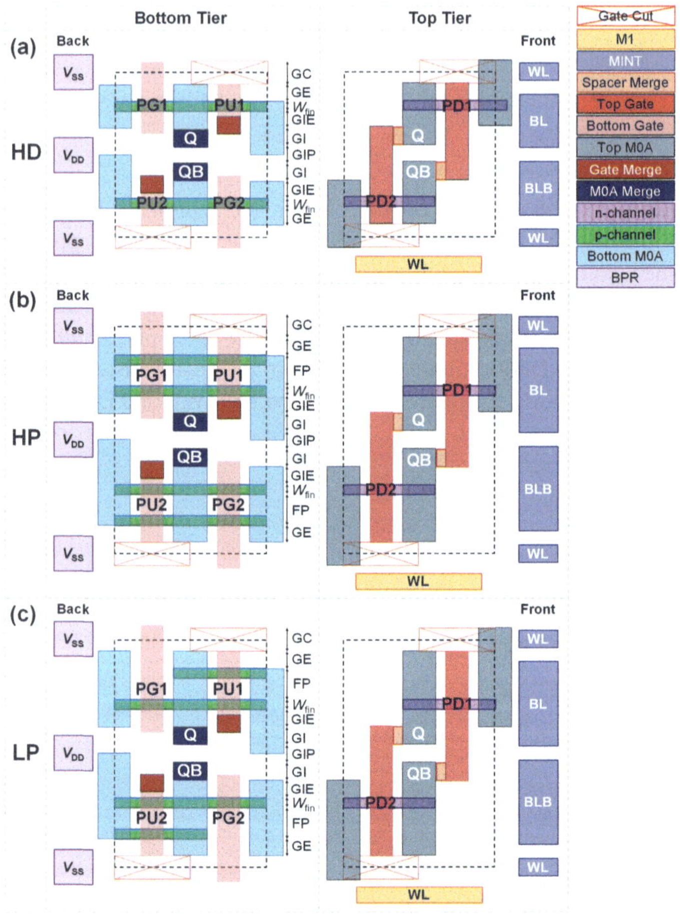

Fig. A.16 Top views of bottom and top tiers for (**a**) HD, (**b**) HP, and (**c**) LP A5 seq Fin-on-Fin CFET SRAM bitcell layout design. (HD/HP/LP: W_{BL} = 27.5/43.5/43.5 nm; W_{WL} = 65/65/65 nm; W_{cell} = 78/78/78 nm; H_{cell} = 85/117/117 nm)

A.2.2 A5 Seq NS-on-NS CFET SRAM Bitcells

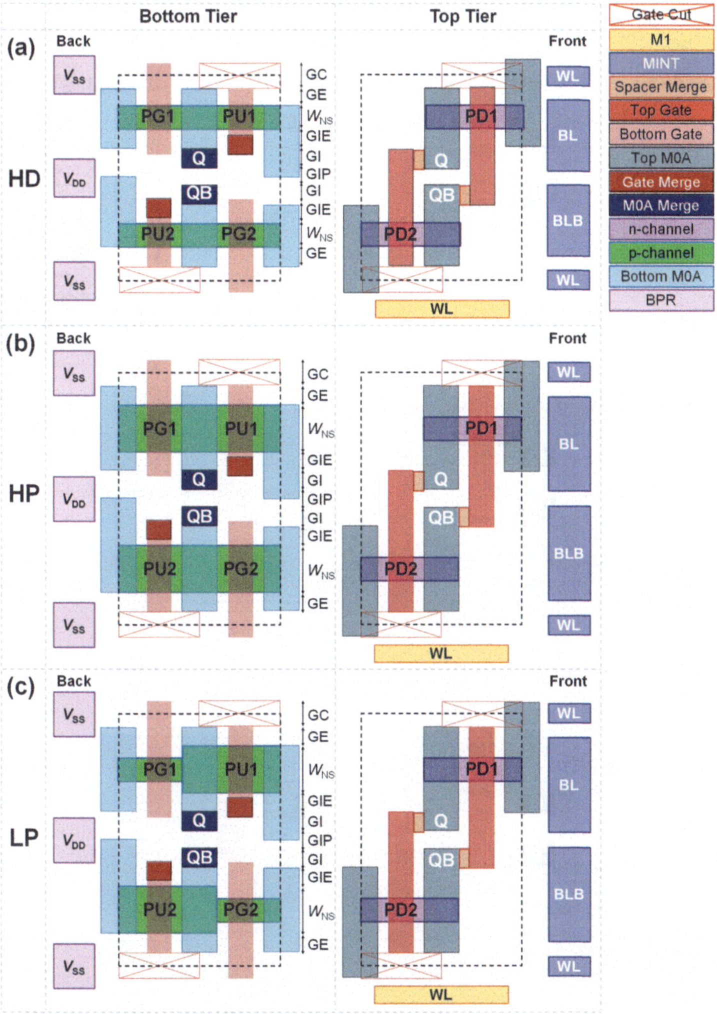

Fig. A.17 Top views of bottom and top tiers for (**a**) HD, (**b**) HP, and (**c**) LP A5 seq NS-on-NS CFET SRAM bitcell layout design. (HD/HP/LP: W_{BL} = 33.5/44.5/44.5 nm; W_{WL} = 65/65/65 nm; W_{cell} = 78/78/78 nm; H_{cell} = 97/119/119 nm)

A.2.3 A5 Mono Fin-on-Fin CFET SRAM Bitcells

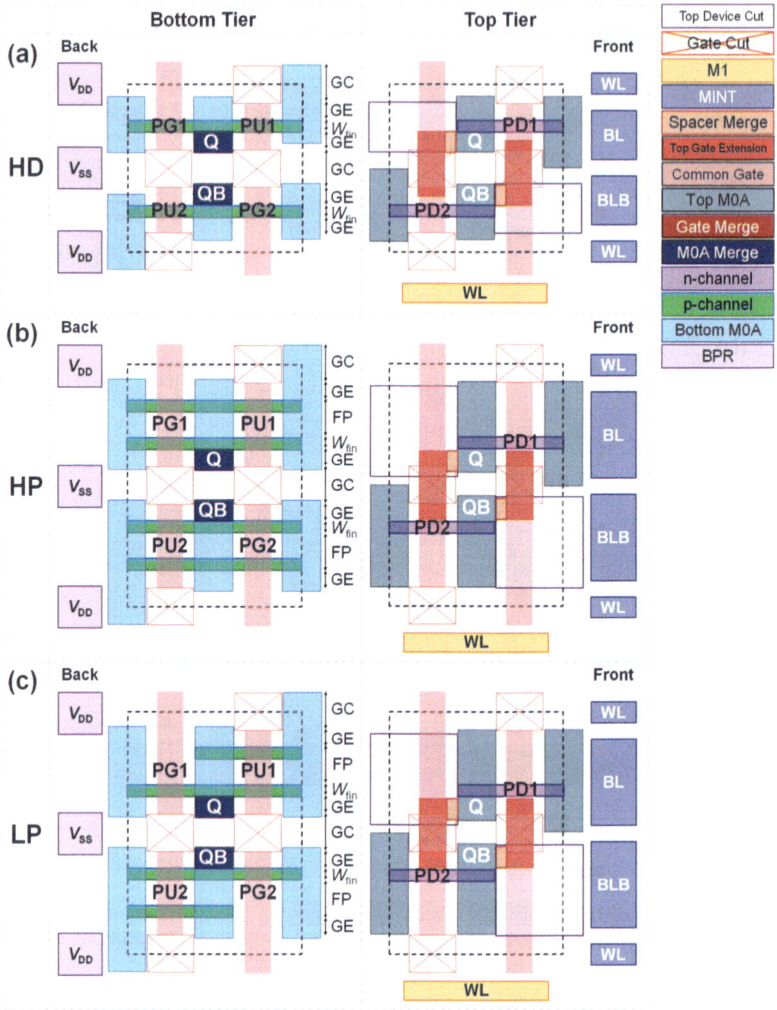

Fig. A.18 Top views of bottom and top tiers for (**a**) HD, (**b**) HP, and (**c**) LP A5 mono Fin-on-Fin CFET SRAM bitcell layout design. (HD/HP/LP: W_{BL} = 21/37/37 nm; W_{WL} = 65/65/65 nm; W_{cell} = 78/78/78 nm; H_{cell} = 72/104/104 nm)

A.2.4 A5 Mono NS-on-NS CFET SRAM Bitcells

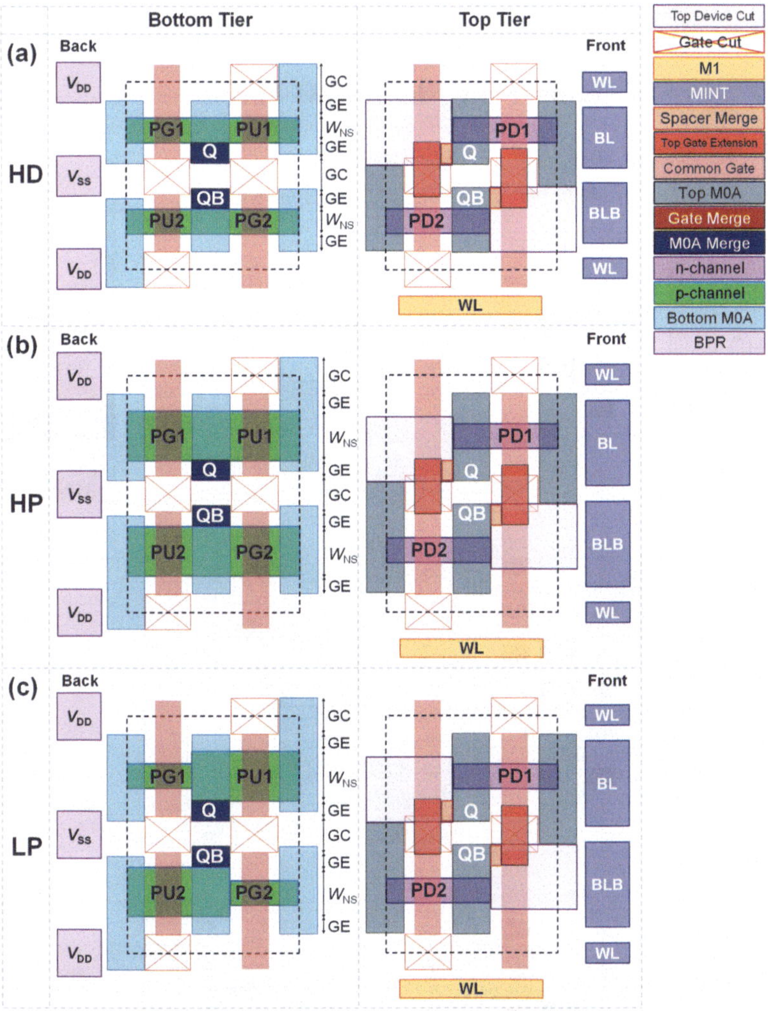

Fig. A.19 Top views of bottom and top tiers for (**a**) HD, (**b**) HP, and (**c**) LP A5 mono NS-on-NS CFET SRAM bitcell layout design. (HD/HP/LP: W_{BL} = 27/38/38 nm; W_{WL} = 65/65/65 nm; W_{cell} = 78/78/78 nm; H_{cell} = 84/106/106 nm)

A.3 A3 Seq and Mono CFET SRAM Bitcell Layout Design

This section presents A3 seq and mono 6T CFET SRAM bitcell layout designs, as shown in Figs. A.20, A.21, A.22, A.23, A.24, and A.25.

A.3.1 A3 Seq Fin-on-Fin CFET SRAM Bitcells

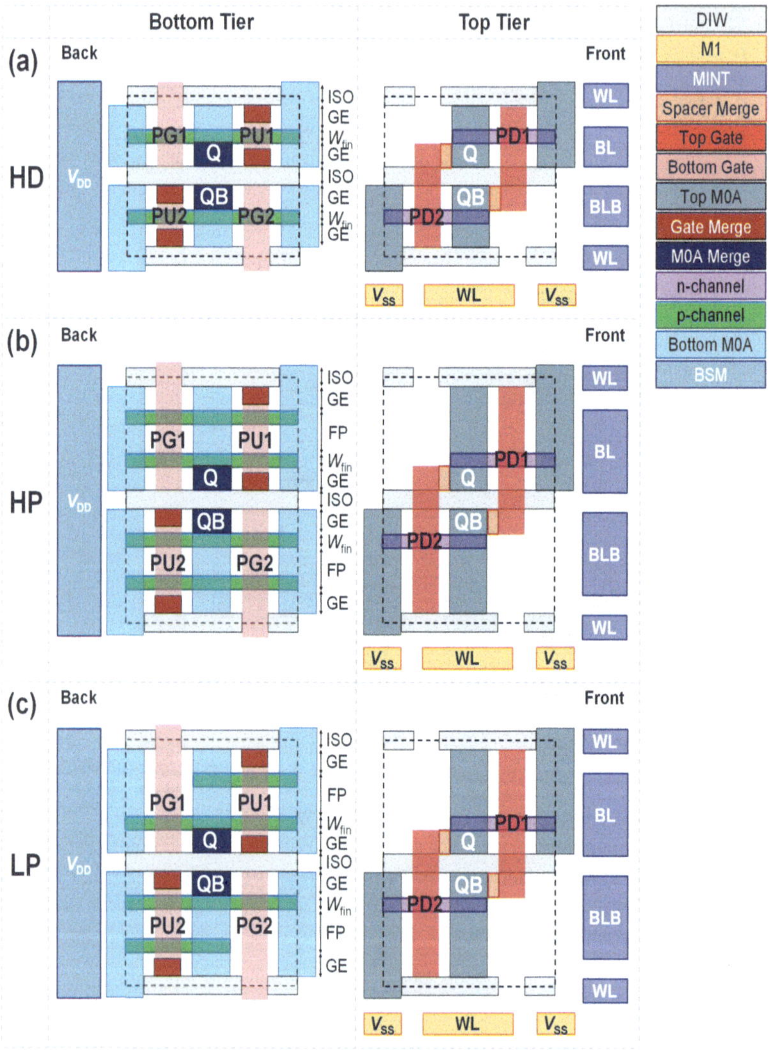

Fig. A.20 Top views of bottom and top tiers for (**a**) HD, (**b**) HP, and (**c**) LP A3 seq Fin-on-Fin CFET SRAM bitcell layout design. (HD/HP/LP: W_{BL} = 15/31/31 nm; W_{WL} = 41/41/41 nm; W_{cell} = 78/78/78 nm; H_{cell} = 60/92/92 nm)

A.3.2 A3 Seq NS-on-NS CFET SRAM Bitcells

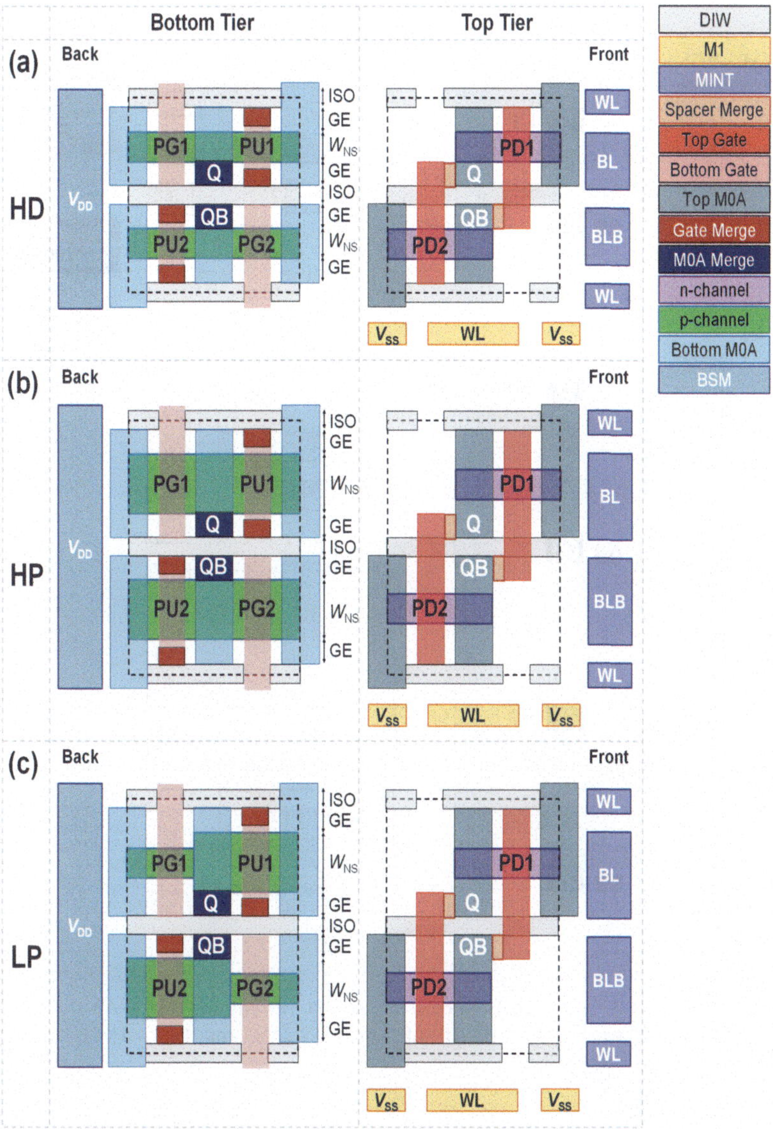

Fig. A.21 Top views of bottom and top tiers for (**a**) HD, (**b**) HP, and (**c**) LP A3 seq NS-on-NS CFET SRAM bitcell layout design. (HD/HP/LP: W_{BL} = 21/32/32 nm; W_{WL} = 41/41/41 nm; W_{cell} = 78/78/78 nm; H_{cell} = 72/94/94 nm)

A.3.3 A3 Seq FS-on-FS CFET SRAM Bitcells

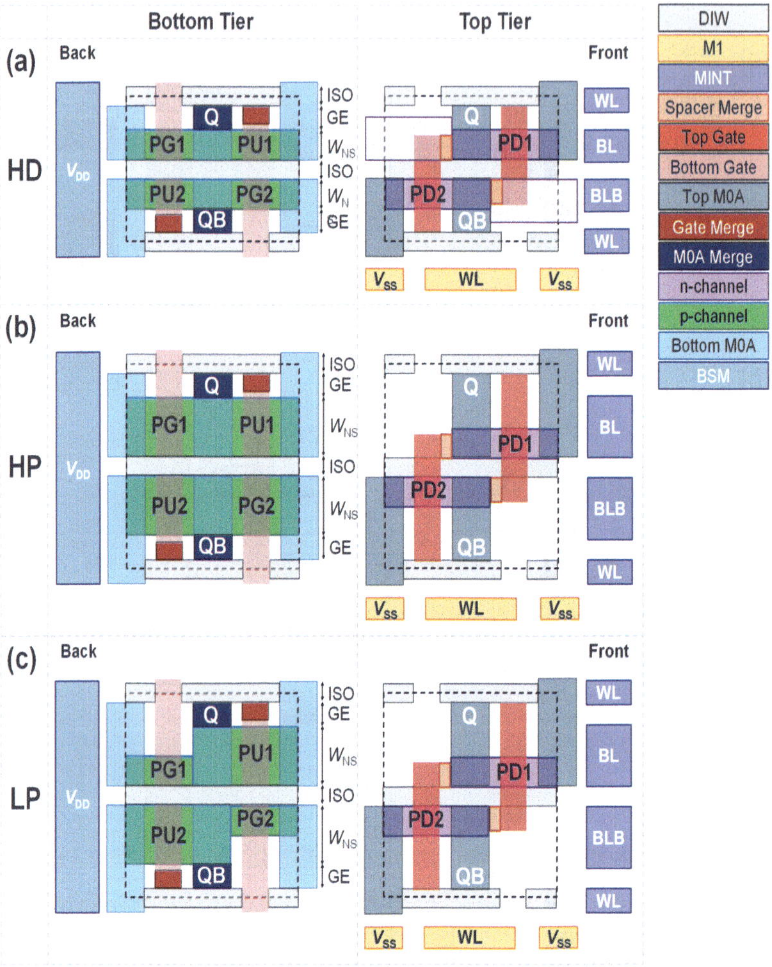

Fig. A.22 Top views of bottom and top tiers for (**a**) HD, (**b**) HP, and (**c**) LP A3 seq FS-on-FS CFET SRAM bitcell layout design. (HD/HP/LP: W_{BL} = 12/23/23 nm; W_{WL} = 41/41/41 nm; W_{cell} = 78/78/78 nm; H_{cell} = 54/76/76 nm)

A.3.4 A3 Mono Fin-on-Fin CFET SRAM Bitcells

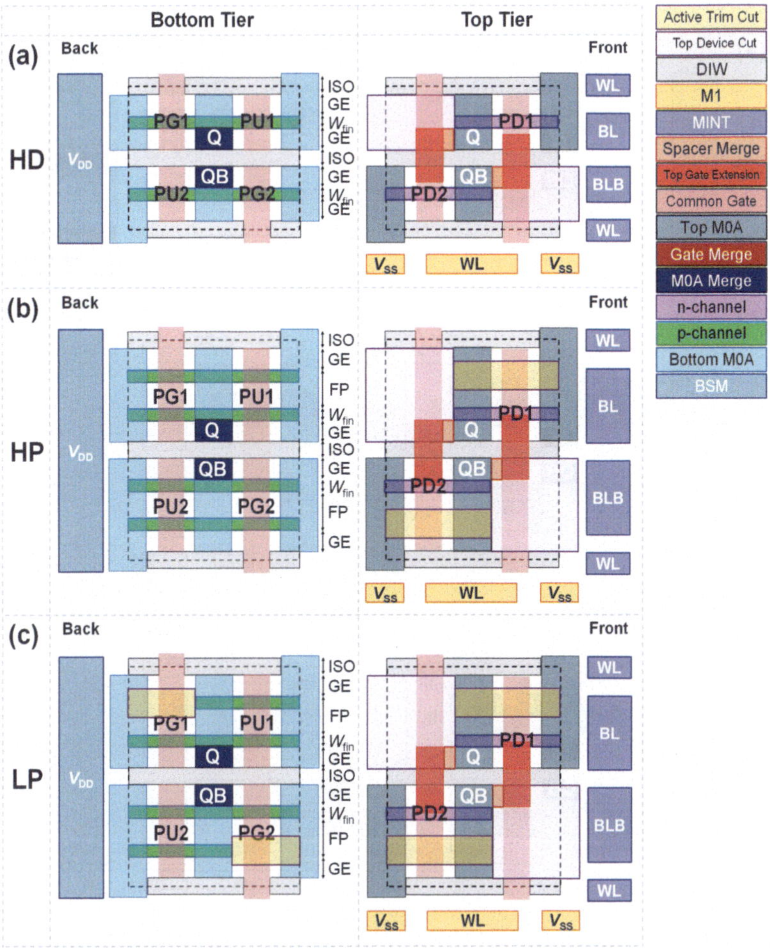

Fig. A.23 Top views of bottom and top tiers for (**a**) HD, (**b**) HP, and (**c**) LP A3 mono Fin-on-Fin CFET SRAM bitcell layout design. (HD/HP/LP: W_{BL} = 15/31/31 nm; W_{WL} = 41/41/41 nm; W_{cell} = 78/78/78 nm; H_{cell} = 60/92/92 nm)

A.3.5 A3 Mono NS-on-NS CFET SRAM Bitcells

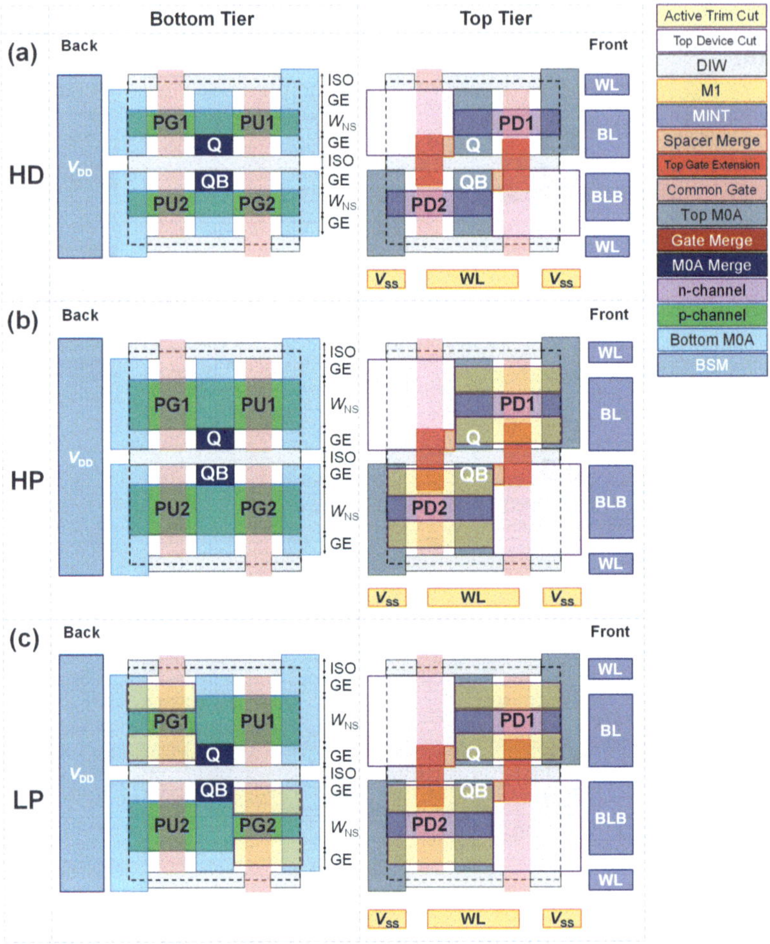

Fig. A.24 Top views of bottom and top tiers for (**a**) HD, (**b**) HP, and (**c**) LP A3 mono NS-on-NS CFET SRAM bitcell layout design. (HD/HP/LP: W_{BL} = 21/32/32 nm; W_{WL} = 41/41/41 nm; W_{cell} = 78/78/78 nm; H_{cell} = 72/94/94 nm)

A.3.6 A3 Mono FS-on-FS CFET SRAM Bitcells

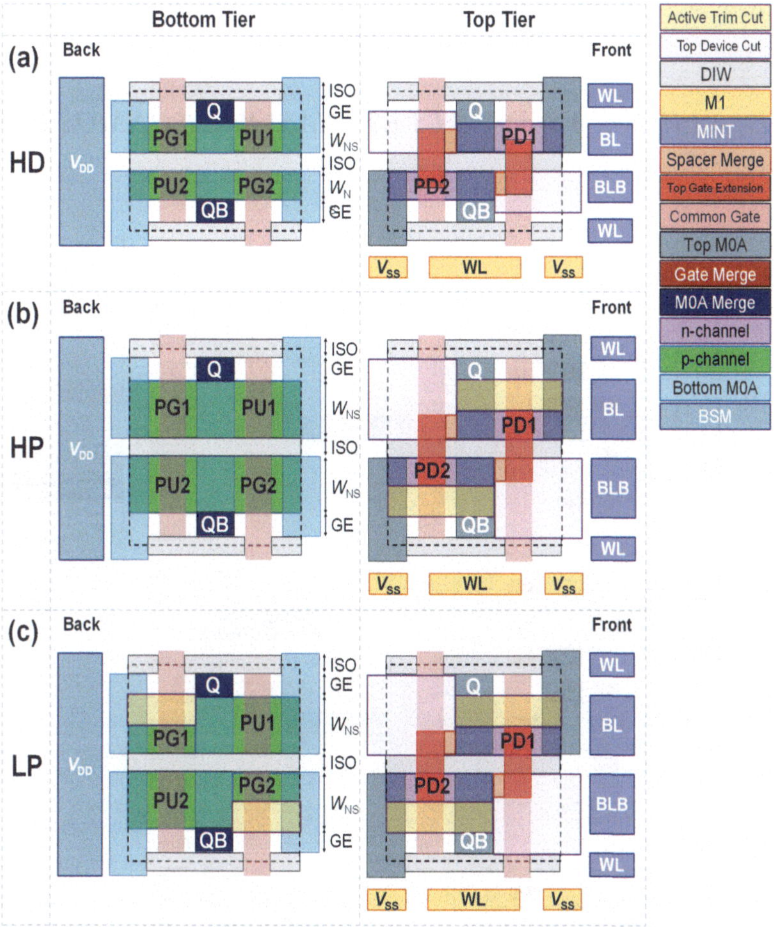

Fig. A.25 Top views of bottom and top tiers for (**a**) HD, (**b**) HP, and (**c**) LP A3 mono FS-on-FS CFET SRAM bitcell layout design. (HD/HP/LP: W_{BL} = 12/23/23 nm; W_{WL} = 41/41/41 nm; W_{cell} = 78/78/78 nm; H_{cell} = 54/76/76 nm)

A.4 A3 Hybrid CFET SRAM Bitcell Layout Design

This section presents A3 hybrid CFET SRAM bitcell layout designs, as shown in Figs. A.26, A.27, A.28, and A.29.

A.4.1 A3 Hybrid NS-on-NS CFET SRAM Bitcells (NMOS PG)

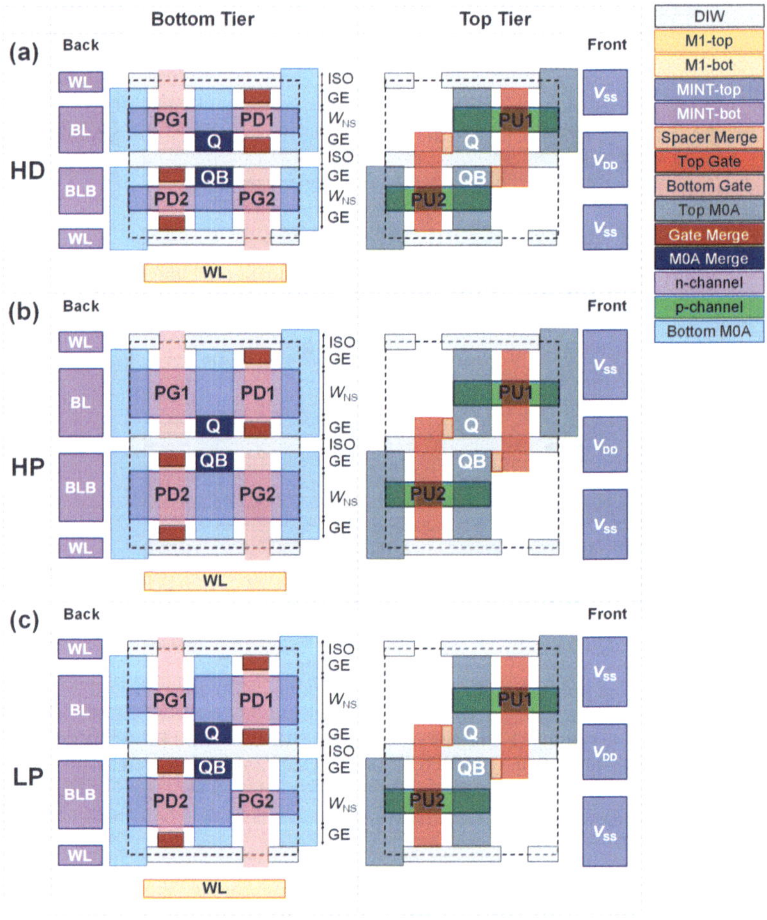

Fig. A.26 Top views of bottom and top tiers for (**a**) HD, (**b**) HP, and (**c**) LP A3 hybrid NS-on-NS CFET SRAM (w/ NMOS PG) bitcell layout design. (HD/HP/LP: W_{BL} = 21/31/31 nm; W_{WL} = 65/65/65 nm; W_{cell} = 78/78/78 nm; H_{cell} = 72/94/94 nm)

A.4.2 A3 Hybrid FS-on-FS CFET SRAM Bitcells (NMOS PG)

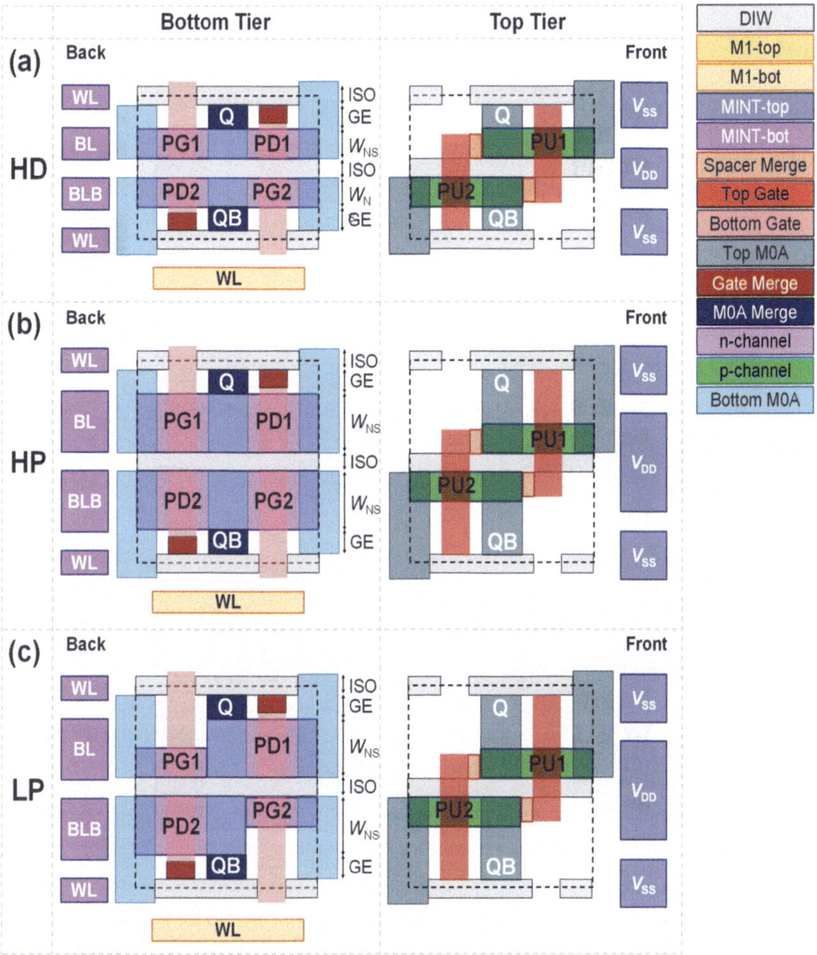

Fig. A.27 Top views of bottom and top tiers for (**a**) HD, (**b**) HP, and (**c**) LP A3 hybrid FS-on-FS CFET SRAM (w/ NMOS PG) bitcell layout design. (HD/HP/LP: W_{BL} = 12/23/23 nm; W_{WL} = 65/65/65 nm; W_{cell} = 78/78/78 nm; H_{cell} = 54/76/76 nm)

A.4.3 A3 Hybrid NS-on-NS CFET SRAM Bitcells (PMOS PG)

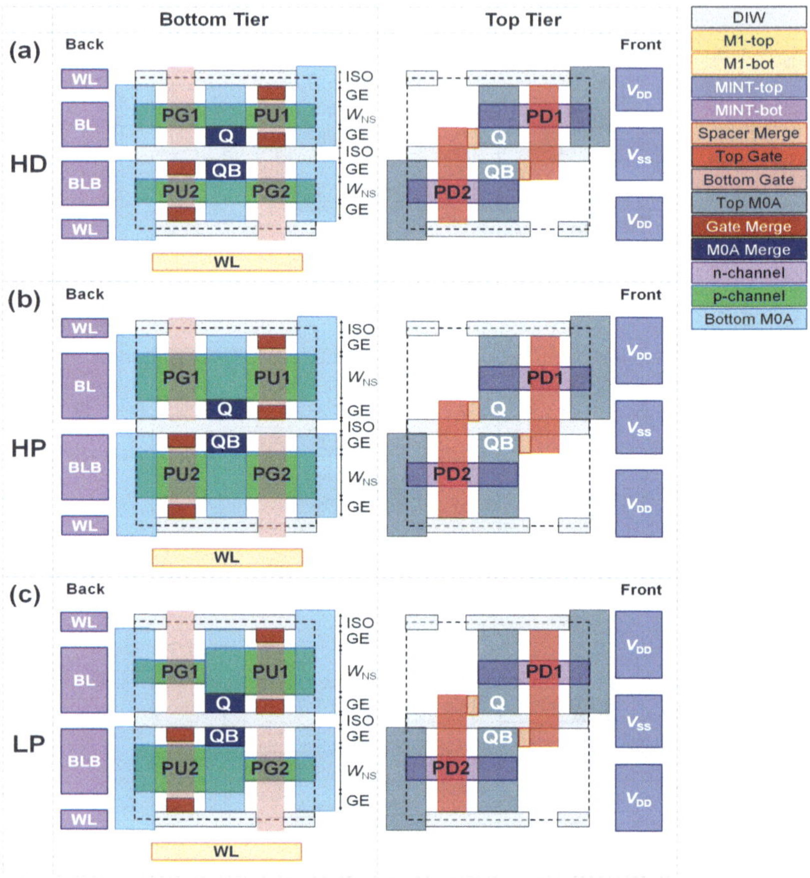

Fig. A.28 Top views of bottom and top tiers for (**a**) HD, (**b**) HP, and (**c**) LP A3 hybrid NS-on-NS CFET SRAM (w/ PMOS PG) bitcell layout design. (HD/HP/LP: W_{BL} = 21/31/31 nm; W_{WL} = 65/65/65 nm; W_{cell} = 78/78/78 nm; H_{cell} = 72/94/94 nm)

A.4.4 A3 Hybrid FS-on-FS CFET SRAM Bitcells (PMOS PG)

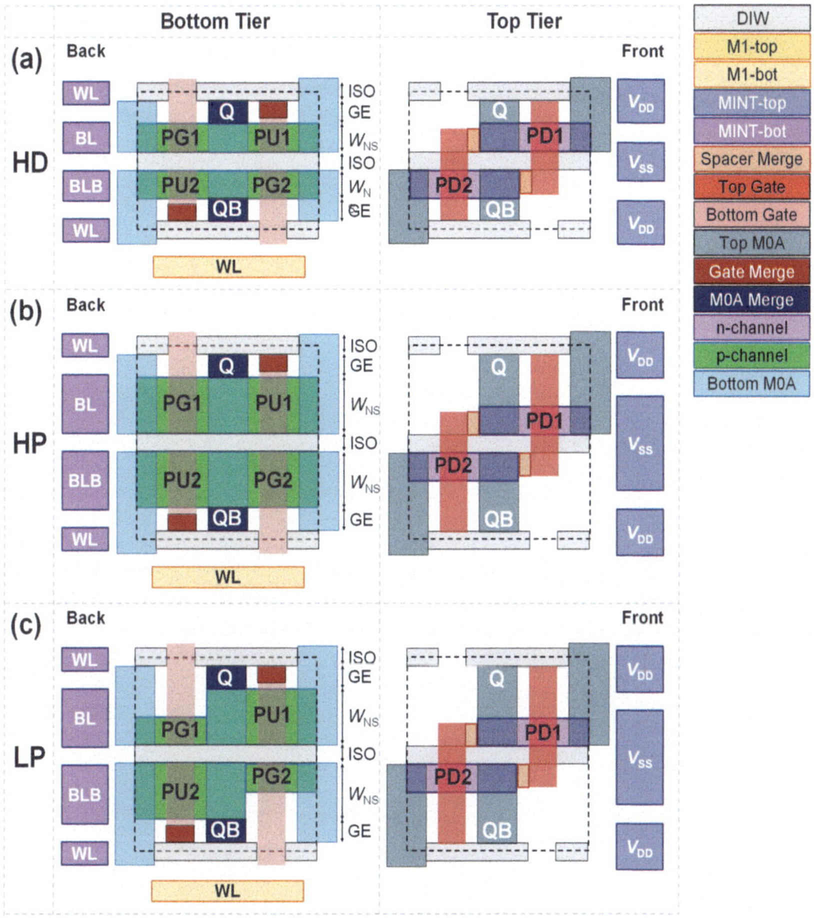

Fig. A.29 Top views of bottom and top tiers for (**a**) HD, (**b**) HP, and (**c**) LP A3 hybrid FS-on-FS CFET SRAM (w/ PMOS PG) bitcell layout design. (HD/HP/LP: W_{BL} = 12/23/23 nm; W_{WL} = 65/65/65 nm; W_{cell} = 78/78/78 nm; H_{cell} = 54/76/76 nm)

Index

H.-H. Liu, F. Catthoor, *Circuit-Technology Co-Optimization of SRAM Design in Advanced CMOS Nodes*, https://doi.org/10.1007/978-3-031-76109-6

SPRINGER NATURE

GPSR Compliance

The European Union's (EU) General Product Safety Regulation (GPSR) is a set of rules that requires consumer products to be safe and our obligations to ensure this.

If you have any concerns about our products, you can contact us on ProductSafety@springernature.com

In case Publisher is established outside the EU, the EU authorized representative is:

Springer Nature Customer Service Center GmbH
Europaplatz 3
69115 Heidelberg, Germany

The manufacturer's authorised representative in the EU is Springer Nature Customer Service Centre GmbH, Europaplatz 3, 69115 Heidelberg, Germany. If you have any concerns regarding our products, please contact ProductSafety@springernature.com

Printed and bound by CPI Group (UK) Ltd, Croydon, CR0 4YY
01/10/2026
02225859-0006